Thermoacoustics

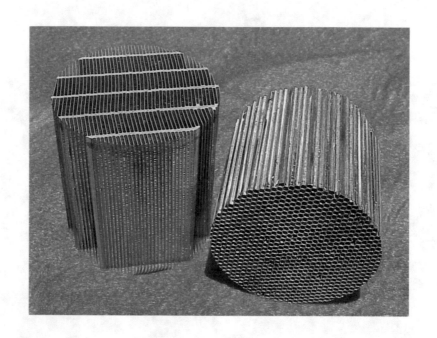

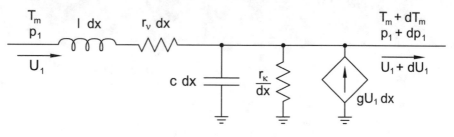

Gregory W. Swift

Thermoacoustics

A Unifying Perspective for Some Engines and Refrigerators

Second Edition

Gregory W. Swift
Los Alamos National Laboratory
Los Alamos
New Mexico, USA

Additional material to this book can be downloaded from https://www.springer.com/us/book/9783319669328.

or

http://www.lanl.gov/org/padste/adeps/materials-physics-applications/condensed-matter-magnet-science/thermoacoustics/animations.php

ISBN 978-3-319-88348-9   ISBN 978-3-319-66933-5   (eBook)
DOI 10.1007/978-3-319-66933-5

Printed on acid-free paper

This Springer imprint is published by Springer Nature
The registered company is Springer International Publishing AG
The registered company address is: Gewerbestrasse 11, 6330 Cham, Switzerland

The ASA Press

The ASA Press imprint represents a collaboration between the Acoustical Society of America and Springer dedicated to encouraging the publication of important new books in acoustics. Published titles are intended to reflect the full range of research in acoustics. ASA Press books can include all types of books published by Springer and may appear in any appropriate Springer book series.

Editorial Board

 **ASA Press**

The Acoustical Society of America

On 27 December 1928 a group of scientists and engineers met at Bell Telephone Laboratories in New York City to discuss organizing a society dedicated to the field of acoustics. Plans developed rapidly and the Acoustical Society of America (ASA) held its first meeting on 10–11 May 1929 with a charter membership of about 450. Today ASA has a worldwide membership of 7000.

The scope of this new society incorporated a broad range of technical areas that continues to be reflected in ASA's present-day endeavors. Today, ASA serves the interests of its members and the acoustics community in all branches of acoustics, both theoretical and applied. To achieve this goal, ASA has established technical committees charged with keeping abreast of the developments and needs of membership in specialized fields as well as identifying new ones as they develop.

The Technical Committees include acoustical oceanography, animal bioacoustics, architectural acoustics, biomedical acoustics, engineering acoustics, musical acoustics, noise, physical acoustics, psychological and physiological acoustics, signal processing in acoustics, speech communication, structural acoustics and vibration, and underwater acoustics. This diversity is one of the Society's unique and strongest assets since it so strongly fosters and encourages cross-disciplinary learning, collaboration, and interactions.

ASA publications and meetings incorporate the diversity of these Technical Committees. In particular, publications play a major role in the Society. *The Journal of the Acoustical Society of America* (JASA) includes contributed papers and patent reviews. *JASA Express Letters* (JASA-EL) and *Proceedings of Meetings on Acoustics* (POMA) are online, open-access publications, offering rapid publication. *Acoustics Today*, published quarterly, is a popular open-access magazine. Other key features of ASA's publishing program include books, reprints of classic acoustics texts, and videos.

ASA's biannual meetings offer opportunities for attendees to share information, with strong support throughout the career continuum, from students to retirees. Meetings incorporate many opportunities for professional and social interactions and attendees find the personal contacts a rewarding experience. These experiences result in building a robust network of fellow scientists and engineers, many of whom become lifelong friends and colleagues.

From the Society's inception, members recognized the importance of developing acoustical standards with a focus on terminology, measurement procedures, and criteria for determining the effects of noise and vibration. The ASA Standards Program serves as the Secretariat for four American National Standards Institute Committees and provides administrative support for several international standards committees.

Throughout its history to present day, ASA's strength resides in attracting the interest and commitment of scholars devoted to promoting the knowledge and practical applications of acoustics. The unselfish activity of these individuals in the development of the Society is largely responsible for ASA's growth and present stature.

Preface

I'm thrilled by the power density and efficiency recently achieved by thermoacoustic engines and refrigerators, and I'm fascinated by some of the latest developments in thermoacoustics: mixture separation via oscillating thermal diffusion [1, 2], self-excited oscillating heat pipes ("Akachi" or bubble-driven heat pipes) (See almost all of the papers in Sessions B6 and B7 in the Proceedings of the 11th International Heat Pipe Conference, 12–16 September 1999, Musashinoshi Tokyo, Japan, sponsored by the Japan Association for Heat Pipes and by Seikei University), and deliberate superposition of steady flow [3, 4]. At night I often dream of a future world in which thermoacoustics is widely practiced. One dream had linear-motor-driven thermoacoustic heat pumps atop the hot-water heaters in half the homes in Phoenix, pumping heat from room air into the hot water—the production of a little cooling in the homes was a nice by-product. Another dream featured a small thermoacoustic system next to the liquid-nitrogen and liquid-oxygen dewars in the back of our local hospital. This system had a thermoacoustic engine, heated by combustion of natural gas, driving several pulse-tube refrigerators, which provided the cooling necessary to liquefy air, to distill it to produce purified nitrogen and oxygen, and to reliquefy the pure gases for storage in the dewars. A third dream had hundreds of enormous combustion-powered thermoacoustic natural-gas liquefiers arrayed on an offshore platform, using the natural gas (methane) itself as the thermoacoustic working gas and filling a vacuum-insulated supertanker with cryogenic liquefied natural gas for transport to distant shores. Yet another dream showed an extensive thermoacoustic apparatus on Mars—a thermoacoustic engine driven by a small nuclear reactor produced 100 kW of acoustic power, which was piped to assorted thermoacoustic mixture separators and refrigerators, splitting atmospheric carbon dioxide and mined frozen water into pure H_2 and O_2 and liquefying these for use in fuel cells on each of the many robots scooting around building a colony for eventual human habitation.

The dreams are always different, but they have some features in common. First, they all feature low-tech hardware: big pipes, welded steel, conventional shell-and-tube heat exchangers, molded plastic, etc. Second, I know that this simplicity is deceptive, because the technical challenge of designing this easy-to-build hardware is extreme. Third, there are no people in these dreams, because I know so few people who are skilled in thermoacoustic engineering today. So I wake up, afraid that none of this will ever happen, afraid that integrated thermoacoustic process engineering is an opportunity that will never have a chance. So I get up and I write a few more paragraphs of this book, hoping to help newcomers learn basic thermoacoustics quickly, so they can go on to design, build, and debug wonderful thermoacoustic systems of all kinds.

This is an introductory book, not a full review of the current status of the field of thermoacoustics. It is evolving from the short course that I gave on this subject at the March 1999 Berlin acoustics meeting. The hardware examples used here to illustrate the elementary principles are thermoacoustics apparatus developed at Los Alamos or with our close collaborators, and the mathematical approach to the gas dynamics and power flows closely follows that pioneered by Nikolaus Rott. (Time pressure induces me to stick with topics most familiar to me! And, indeed, the Los Alamos approach to thermoacoustics has been quite successful.) Many aspects of thermoacoustics will be introduced, in an attempt to help the reader acquire both an intuitive understanding and the ability to design hardware, build it, and diagnose its performance.

At Los Alamos, we have found it most productive to stay focused on experimental and development hardware while maintaining several abstract points of view including phasor display of acoustic variables, an intuitively appealing picture of gas motion, and an entropy-generation perspective on the second law of thermodynamics. Intuition is important because it helps us humans organize our thoughts. Mathematics is unavoidable, because it is the common language with which scientists and engineers communicate and it allows us to interpolate and extend our knowledge quantitatively. But experiment is the source of all real truth, so the experiments are our most important and time-consuming activity at Los Alamos. Weaving intuition, mathematics, and experimental results together in this book, I will put a strong emphasis on the mathematics, because without this common vocabulary, we can get nowhere. Intuition gets second-highest emphasis, as appropriate for an introductory treatment. Experimental results get the least emphasis in this text. But please remember that the mathematical and intuitive discussions presented here are actually distillations of many experiments spanning many decades throughout the world.

Many readers will find that they have only part of the background needed to learn thermoacoustics. Mechanical engineers and chemical engineers may have insufficient acoustics background. They should study an introductory treatment like Chaps. 5–10 (and perhaps 14) in *Fundamentals of Acoustics* by Kinsler, Frey, Coppens, and Sanders [5]. Acousticians, on the other hand, may need to study something like the first half of *Fundamentals of Classical Thermodynamics* by Van Wylen and Sonntag [6]. Someone for whom the expression $i = \sqrt{-1}$ is unfamiliar must begin with a review of complex arithmetic in an engineering mathematics text such as *Advanced Engineering Mathematics* by Kreyszig [7]. There is much to learn, so be patient.

We are all deeply indebted to the Office of Basic Energy Sciences (OBES) in the US Department of Energy (DOE) for its steady, patient, dependable financial support of fundamental thermoacoustics research, dating back 20 years to when the late John Wheatley first led us in this direction. Working with the postdocs and students supported by OBES/DOE to study these oscillating thermodynamic systems at Los Alamos—Vince Kotsubo, John Brisson, Jeff Olson, Mike Hayden, Bob Reid, Phil Spoor, Bob Hiller, Scott Backhaus, Bart Smith, and Drew Geller— has taught me most of what I know about thermoacoustics and about how people learn this difficult subject. The engineering and technical skill and creativity of Bill Ward, David Gardner, and Chris Espinoza, and Bill's sophisticated computer skills, have been indispensable, to make rapid progress and to keep our understanding of thermoacoustics realistically grounded in experimental hardware.

Finally, thank you Sharon Dogruel, for your brainstorming, steady encouragement, and patience through a year of long days, no whole weekends, and too few holidays. I wish I could promise that it won't happen again.

Los Alamos, NM, USA   Gregory W. Swift
1999

Addendum, 2001: Now I must add thanks to the many individuals who have reported typos, math errors, and sources of confusion during the year the "fourth draft" has been widely available. My greatest debt by far is due to Robert Keolian and his 2000 thermoacoustics class at Penn State University, who suggested a couple hundred important improvements, including especially a total reorganization and great expansion of what are now the first three chapters.

Addendum, 2017: This new edition by ASA and Springer gives me the opportunity to fix a few small errors that were present in all previous printings. The thermal-mixing entry in the table in Sect. 6.1 was incorrect. (Thanks to Robert Keolian for finding that error.) Figures 4.5 and 4.15 and Eq. (7.81) had typos, and some signs in Fig. 4.22 were wrong. "Newton" was incorrectly abbreviated as Nt instead of N. Fonts in some figures have been improved, and some punctuation has been improved. Two good Stirling-engine historical references have been added. The DELTAE files that were in the Appendix have been upgraded to the newer DELTAEC format, without significant change in numerical results.

Minor changes in wording throughout have made line breaks and page breaks compatible with this new, smaller page format. Color figures have been added, and internal hyperlinks have been added in the e-book version. The accompanying computer animations, introduced in Sect. 1.3, have been upgraded for compatibility with modern operating systems. To obtain the animations, see Sect. 1.3.

Contents

1	**Introduction**	1
	1.1 Themes	1
	1.2 Length Scales	7
	1.3 Overview and Examples	8
	1.3.1 Standing-Wave Heat Engine	9
	1.3.2 Standing-Wave Refrigerator	14
	1.3.3 Orifice Pulse-Tube Refrigerator	17
	1.3.4 Thermoacoustic-Stirling Heat Engine	20
	1.4 Thermoacoustics and Conventional Technology	24
	1.5 Outline	26
	1.6 Exercises	28
2	**Background**	31
	2.1 Laws of Thermodynamics	31
	2.1.1 The First Law	31
	2.1.2 The Second Law	35
	2.2 Laws of Fluids	39
	2.2.1 Continuity (Mass)	40
	2.2.2 Momentum	41
	2.2.3 Energy	43
	2.2.4 Entropy	44
	2.3 Ideal Gases	45
	2.3.1 Thermodynamic Properties	45
	2.3.2 Transport Properties	48
	2.3.3 Shortcuts	49
	2.3.4 Mixtures	50
	2.4 Some Consequences of the Laws	50
	2.4.1 Carnot's Efficiency	50
	2.4.2 Maxwell Relations	52
	2.5 Exercises	54

3   Simple Oscillations.. 59
  3.1   The Harmonic Oscillator and Complex Notation 59
        3.1.1   Complex Notation 60
        3.1.2   Damped, Driven Harmonic Oscillator......................... 62
  3.2   Acoustic Approximations to the Laws of Gases 65
  3.3   Some Simple Oscillations in Gases.................................... 69
        3.3.1   The Gas Spring 69
        3.3.2   Simple Sound Waves 71
  3.4   Exercises 73

4   Waves.. 77
  4.1   Lossless Acoustics and Ideal Resonators............................. 78
  4.2   Viscous and Thermal Effects in Large Channels..................... 85
        4.2.1   Viscous Resistance 86
        4.2.2   Thermal-Relaxation Conductance 90
  4.3   Inviscid Boundary-Layer Thermoacoustics 94
  4.4   General Thermoacoustics 97
        4.4.1   The Math... 97
        4.4.2   The Ideas.. 102
  4.5   Exercises 114

5   Power 117
  5.1   Acoustic Power 118
        5.1.1   Acoustic Power Dissipation with $dT_m/dx = 0$ 121
        5.1.2   Acoustic Power with Zero Viscosity 124
  5.2   Total Power 130
        5.2.1   Traveling Waves.. 136
        5.2.2   Standing Waves... 137
  5.3   Some Calculation Methods 138
        5.3.1   Order-of-Magnitude Estimates 138
        5.3.2   Spreadsheet Calculations 139
        5.3.3   DELTAEC's Numerical Integration of Rott's Acoustic
                Approximation 140
        5.3.4   More Sophisticated Numerical Integration.................. 142
  5.4   Examples.. 143
  5.5   Exercises 146

6   Efficiency.. 149
  6.1   Lost Work and Entropy Generation.................................... 150
  6.2   Exergy 156
  6.3   Examples.. 162
  6.4   Exercises 165

7   Beyond Rott's Thermoacoustics 171
  7.1   Tortuous Porous Media... 174
  7.2   Turbulence 177
        7.2.1   Minor Losses 183

7.3 Entrance Effects and Joining Conditions 188
    7.3.1 Entrance Effects 189
    7.3.2 Joining Conditions 190
7.4 Mass Streaming 197
    7.4.1 Gedeon Streaming ("dc flow") 200
    7.4.2 Rayleigh Streaming 204
    7.4.3 Jet-Driven Streaming 209
    7.4.4 Streaming Within a Regenerator or Stack 211
    7.4.5 Deliberate Streaming 211
7.5 Harmonics and Shocks 219
7.6 Dimensionless Groups 222
    7.6.1 Insight 223
    7.6.2 Empirical Correlation 225
    7.6.3 Scale Models 225
7.7 Exercises 226

8 Hardware 231
8.1 Prelude: The Gas Itself 231
8.2 Stacks and Regenerators 233
    8.2.1 Standing Wave 233
    8.2.2 Traveling Wave 238
8.3 Heat Exchangers 240
    8.3.1 Common Arrangements 240
    8.3.2 Thermoacoustic Choices 242
8.4 Thermal Buffer Tubes, Pulse Tubes, and Flow Straighteners 245
8.5 Resonators 246
    8.5.1 Dissipation 246
    8.5.2 Size, Weight, and Pressure-Vessel Safety 248
    8.5.3 Harmonic Suppression 249
8.6 Electroacoustic Power Transducers 250
8.7 Exercises 255

9 Measurements 259
9.1 Easy Measurements 259
    9.1.1 Pressures and Frequency 260
    9.1.2 Mean Temperature 262
9.2 Power Measurements 263
    9.2.1 Acoustic Power 263
    9.2.2 Heat 267
9.3 Difficult Measurements 269
9.4 Points of View 269
    9.4.1 Natural Dependence 270
    9.4.2 Evidence 272
    9.4.3 Performance 274
    9.4.4 A Thermoacoustic Perspective 277
9.5 Exercises 287

A Common Pitfalls 291

B DELTAEC Files 293
    B.1   Standing-Wave Engine 294
    B.2   Standing-Wave Refrigerator... 296
    B.3   Orifice Pulse-Tube Refrigerator....................................... 299
    B.4   Thermoacoustic-Stirling Heat Engine 302

References 309

Author index 317

Subject index 321

List of Symbols

The equation numbers indicate the variable's definition, an early significant use of the variable, or the vicinity of the in-text definition of the variable. If a second equation number is given, the clarity or context may differ. MKS units are also given.

English Letters

A	Area, m^2, Eq. (4.1)
a	Sound speed, m/s, Eqs. (1.1), (2.56)
a	Acceleration, m/s^2, Eq. (9.8)
B	Bulk modulus, Pa, Eq. (2.48)
$\mathcal{B}$	Magnetic field, Tesla, near Eq. (8.7)
b	Exponent for T dependence of μ or k, Eqs. (2.63), (7.79)
b	Flow availability per unit mass, J/kg, near Eq. (6.16)
C	Compliance, m^3/Pa, Eq. (4.4)
C	A constant or function, Eq. (3.51)
C_V	Valve flow coefficient, weird units, Eq. (7.37)
c	Compliance per unit length, m^2/Pa, Eq. (4.77)
c	A constant or function, Eq. (7.15)
c	Heat capacity per unit mass, J/kg·K, Eq. (2.12)
c_p	Isobaric heat capacity per unit mass, J/kg·K, Eqs. (1.2), (2.58)
COP	Coefficient of performance, Eq. (2.80)
D	Diameter, m, Eq. (7.46)
$\mathcal{E}$	Energy, J, Eq. (2.1)
$\dot{E}$	Acoustic power, W, Eq. (5.2)
e	2.71828...
F	Force, N, Eqs. (2.24), (3.9)
$\mathcal{F}$	A function, Eq. (4.17)
f	Frequency, Hz, Eqs. (1.1), (3.3)
f	Spatially averaged thermoviscous function, Eqs. (4.54), (4.68)

f	Friction factor, Eq. (7.13)
g	Complex gain factor arising in continuity equation, Eq. (4.80)
g	Acceleration of gravity, m/s^2, Eq. (2.10)
$\dot{H}$	Rate at which total energy flows (total power), W, Eq. (5.24)
h	Thermoviscous function, Eqs. (4.53), (4.67)
h	Enthalpy per unit mass, J/kg, Eqs. (2.7), (5.18)
I	Electric current, Amp, Eq. (8.7)
i	$\sqrt{-1}$, Eq. (3.4)
J	Bessel function, Eq. (4.59)
j	Electric current density, Amp/m^2, Eq. (6.13)
K	Spring constant, N/m, Eqs. (2.10), (3.1)
K	Minor-loss coefficient, Eq. (7.31)
k	Thermal conductivity, W/m·K, Eqs. (1.2), (2.3)
k	Wave vector, m^{-1}, Eq. (3.51)
L	Inertance, kg/m^4, Eq. (4.7)
L	Electric inductance, Henry, Eq. (8.7)
l	Inertance per unit length, kg/m^5, Eq. (4.73)
l	Length, m, Eq. (3.54), below Eq. (8.7)
M	Mass, kg, Eqs. (2.5), (3.1)
$\dot{M}$	Mass flow, kg/s, Eq. (7.67)
m	Molar mass, kg/mol, Eq. (2.46)
m	Multiplier for turbulent viscous dissipation, Eq. (7.27)
$\dot{m}$	Mass-flux density, kg/s·m^2, Eq. (7.66)
N_R	Reynolds number, Eqs. (7.12), (7.14)
n	Screen mesh number, typically inch^{-1}, Eq. (7.11)
n	Mole fraction, Eq. (2.73)
p	Pressure, Pa, Chap. 2
Q	Heat, J, Eq. (2.1)
Q	Quality factor of resonance, Eq. (3.14)
$\dot{Q}$	Rate at which heat is moved (thermal power), W, Eqs. (2.3), (6.1)
q	Heat per unit mass, J/kg, Eq. (2.4)
R	Acoustic resistance, Pa·s/m^3, Eqs. (4.23), (4.36)
R	Electric resistance, Ohm, Eq. (8.7)
R	Mechanical resistance, kg/s, Eqs. (3.9), (8.8)
$\mathcal{R}$	Radius, m, Eq. (4.59)
$\mathfrak{R}$	Gas constant, J/kg·K, Eq. (2.44)
$\mathfrak{R}_{univ}$	Universal gas constant, J/mol·K, Eq. (2.46)
r	Acoustic resistance per unit length, Pa·s/m^4, Eqs. (4.74), (4.78)
r	Radius, m, Eq. (4.10)
r	Radial coordinate, m, Eq. (4.59)
S	Entropy, J/K, Eq. (2.14)
S	Surface area, m^2, Eq. (4.23)
$\dot{S}$	Rate of entropy generation, J/K·s, Eq. (6.10)
s	Entropy per unit mass, J/kg·K, Eq. (2.19)
s	Stress, Pa, Eq. (8.3)

T	Temperature, K, Chap. 2
t	Time, s
t	Wall thickness, m, Eq. (8.3)
U	Volume flow rate, m^3/s, Eq. (4.1)
u	x component of velocity, m/s, Eqs. (2.10), (2.22)
V	Volume, m^3, Eq. (2.2)
V	Voltage, Volt, Eq. (8.7)
v	Vector velocity, m/s, Eqs. (2.4), (2.25)
v	y component of velocity, m/s, near Eqs. (2.22), (7.81)
W	Work, J, Eq. (2.1)
W	Weight, N, Eq. (8.5)
$\dot{W}$	Rate at which work is done (mechanical power), W, Eqs. (3.13), (6.1)
w	z component of velocity, m/s, near Eq. (2.22)
w	Work per unit mass, J/kg, Eq. (2.4)
X	Exergy, J, Eq. (6.20)
X	Electric reactance, Ohm, Eq. (8.7)
X	Mechanical reactance, kg/s, Eq. (8.8)
$\dot{X}$	Rate at which exergy flows, W, Eqs. (6.20), (7.9)
x	Coordinate along sound-propagation direction, m, Eq. (3.45)
x	Position, m, Eq. (2.10)
Y	Neumann function, Eq. (4.64)
y	Coordinate perpendicular to sound-propagation direction, m, Eq. (4.12)
Z	Acoustic impedance, $Pa \cdot s/m^3$, above Eq. (4.8), Eqs. (4.20), (4.33)
z	Coordinate perpendicular to sound-propagation direction, m, Eq. (4.47)
z	Specific acoustic impedance, $kg/m^2 \cdot s = Pa \cdot s/m$, Eq. (4.84)

Greek Letters

α	Variable defined below Eq. (4.65)
α	Electric conductivity, $(Ohm \cdot m)^{-1}$, Eq. (6.13)
β	$= -(1/\rho)(\partial \rho/\partial T)_p$, thermal expansion coefficient, K^{-1}, Eq. (2.47)
Γ	$= (dT_m/dx)/\nabla T_{crit}$, Eq. (5.37)
γ	Ratio of isobaric to isochoric specific heats, Eq. (2.45)
Δ	Big difference, Eqs. (2.10), (4.1)
δ	Penetration depth, m, Eqs. (1.2), (1.3)
δ	Small difference, Eq. (6.9)
δ_{ik}	The unit tensor, equal to unity for $i = k$ and zero for $i \neq k$, Eq. (2.43)
ε	Surface roughness, Eq. (7.28)
ϵ	Internal energy per unit mass, J/kg, Eq. (2.4)
ζ	Bulk viscosity, $kg/m \cdot s$, Eq. (2.26)
η	Efficiency, Eq. (2.79)
θ	Phase angle, Eqs. (7.52), (7.79)
κ	Thermal diffusivity, m^2/s, Eq. (1.2)

Λ	$= 1 - \delta_\nu/r_h + \delta_\nu^2/2r_h^2$, Eq. (5.38)
λ	Wavelength, m, Eq. (1.1)
μ	Dynamic viscosity, kg/m·s, Eqs. (1.3), (4.15)
ν	Kinematic viscosity, m²/s, Eq. (1.3)
ξ	Displacement of gas along x, m, Eqs. (1.5), (3.10), below Eq. (4.16)
ξ	Position of mass, m, Eqs. (3.1), (3.10)
Π	Perimeter, m, Eq. (4.10)
π	3.14159 . . .
ρ	Density, kg/m³, Eq. (1.2), Chap. 2
σ	Prandtl number, Eq. (1.4)
σ_{fric}	Rubbing-friction coefficient, N/m², table in Sect. 6.1
σ'	Nine-component viscous stress tensor, Pa, Eq. (2.26)
τ	Transduction coefficient, N/Amp = Volt/(m/s), Eqs. (8.7), (8.8)
Φ	Viscous dissipation function, s⁻², Eq. (2.43)
ϕ	Phase angle, Eq. (3.2)
ϕ	Volumetric porosity, Eq. (7.10)
ψ	Rott's joining function, m³/s², Eq. (7.49)
ω	$= 2\pi f$, angular frequency, s⁻¹, Eqs. (1.2), (3.3)

Subscripts

a	Amplitude, Eq. (3.2)
C	Cold, Eq. (2.76)
H	Hot, Eq. (2.75)
h	Hydraulic, Eq. (4.10) (i.e., r_h is hydraulic radius)
L	Left, Fig. 7.11
M	Moody, Eq. (7.22)
m	Mean, Eq. (3.18)
p	Parallel, Eq. (4.24)
R	Reynolds, Eq. (7.14) (i.e., N_R is Reynolds number)
R	Right, Fig. 7.11
s	Series, Eq. (4.24)
x	Exchanger, Fig. 7.11
κ	Thermal, Eq. (1.2)
ν	Viscous, Eq. (1.3)
0	"Environment" or "Ambient," Eq. (2.75)
0	Independent of time, Eq. (7.65)
1	First order, Eqs. (3.4), (3.18), usually a complex amplitude
2	Second order, Eqs. (5.2), (7.65)
acoust	Acoustic, Eq. (8.10)
cont	Continuity, Eq. (5.42)
crit	Critical, Eq. (4.44) (i.e., ∇T_{crit})

dimless	Dimensionless, Eq. (7.112)
elec	Electric, Eq. (8.7)
entr	Entrance, Eq. (7.45)
fb	Feedback, Fig. 1.23
fFc	Fundamental Fourier component, Eq. (7.57)
fric	Friction, table in Sect. 6.1
gen	Generated, Eqs. (2.15), (6.10)
max	Maximum, Eq. (2.11)
mech	Mechanical, Eqs. (3.9), (8.8)
meth	Methane, Fig. 9.14
mic	Microphone, Fig. 9.12
mom	Momentum, Eq. (5.41)
ml	Minor loss, Eq. (7.76)
pow	Power, Eq. (5.43)
pt	Pulse tube, Fig. 4.21
reg	Regenerator, Eq. (7.72)
res	Resonator, Fig. 1.23b
trans	Transducer, Eq. (8.13)
turb	Turbulent, Eq. (7.26)
univ	Universal, Eq. (2.46) (i.e., molar gas constant $\Re_{univ} \simeq 8.3$ J/mol·K)

Special Symbols

Im[ ]	Imaginary part of, Eq. (4.74)
Re[ ]	Real part of, Eq. (3.4)
$\langle\,\rangle$	Spatial average perpendicular to x, Eq. (4.17)
||	Magnitude of complex number, Eq. (3.5)
overdot	Time derivative, time rate, Eqs. (2.3), (5.2), (7.67); Exercise 6.15
overbar	Time average, above Eq. (3.7), Eq. (5.7)
tilde	Complex conjugate, Eqs. (3.8), (5.1)

About the Author

Gregory W. Swift received his PhD in physics from the University of California at Berkeley in 1980 and has worked in the Condensed Matter and Thermal Physics Group at Los Alamos National Laboratory (LANL) ever since. He is a Fellow of the Acoustical Society of America, of the American Physical Society, and of LANL. He received the Acoustical Society's Silver Medal in Physical Acoustics in 2000, an award that has been given, on average, only every three years. He received the US Department of Energy's E.O. Lawrence Award in 2004, in the category of Environmental Science and Technology. The main focus of Greg's research has been the invention and development of novel energy-conversion technologies. He enjoys the thermodynamics of heat engines and refrigerators, the thermodynamics of non-ideal-gas fluids, physical acoustics, hydrodynamics, and low-temperature physics. He hopes that thermoacoustic engines and refrigerators will play a meaningful role in the energy economy of the twenty-first century.

Chapter 1
Introduction

This book brings a unifying thermoacoustic perspective to heat engines and refrigerators that depend on powerful oscillating pressure and oscillating flow along a temperature gradient. Applying a thermoacoustic point of view to such engines and refrigerators enables clear, easy analysis of their operation, and it leads to simpler hardware through the elimination of moving parts.

Readers who are particularly interested in only one type of engine or refrigerator should supplement this book with one or more additional books, review articles, or other material. Standing-wave engines and refrigerators are reviewed in [8–13]. Stirling engines and/or refrigerators are covered in [14–18] and current status is summarized in the proceedings of the annual Intersociety Energy Conversion Engineering Conferences and the less frequent International Stirling Engine Conferences. The short quarterly Stirling Machine World (B.A. Ross, Stirling Machine World. A quarterly newsbulletin. Los Olivos CA. Contact stirmach@juno.com or john.corey@chartindustries.com) provides a succinct summary of the latest conferences and publications related to Stirling technology. Stirling cryocoolers and pulse-tube refrigerators are reviewed in [19–21], and current status is reported in the proceedings of the International Cryocooler Conferences (available at cryocooler.org/past-proceedings/) and the Cryogenic Engineering Conferences. The traveling-wave description of Stirling thermodynamics is summarized in [22–24].

This chapter summarizes the common themes of many advances that have been made in this class of engines and refrigerators during the past few decades, introduces four examples that will be examined in greater detail in later chapters, and presents the important length scales of thermoacoustics.

1.1 Themes

Figure 1.1 shows an arrangement typical of Stirling engines of the late twentieth century. There are many moving parts: a rotating crankshaft, moving connecting rods, and reciprocating pistons. These components conspire to take the working

© Acoustical Society of America 2017
G.W. Swift, *Thermoacoustics*, DOI 10.1007/978-3-319-66933-5_1

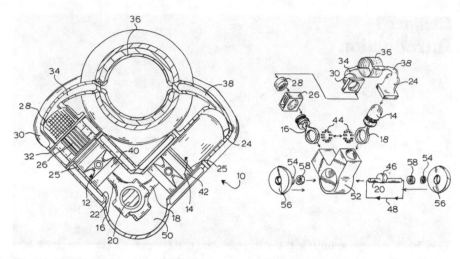

Fig. 1.1 Two views of a typical mechanical Stirling engine of the late twentieth century [25]. The thermal parts are the hot heat exchanger 36, the regenerator 28, and the ambient heat exchanger 26. The hot piston 14 is elongated so that its lower end can be near ambient temperature. The relative positions of the hot piston 14 and the ambient piston 25 are determined by links 16, 18 and crankshaft 20

gas in the engine through a periodic cycle, as the two pistons oscillate smoothly with approximately 90° difference between the time phases of their motions. The pressure of the working gas as a function of time is determined by the time-dependent total volume in which the gas is sealed and by the time-dependent average temperature of the gas. If the two pistons were to move 180° out of phase—one moving up whenever the other moved down—then the total volume available to the gas between them might not change, but the average temperature of the gas would change as the gas was displaced toward hot or toward ambient through the thermal components. On the other hand, if the two pistons were to move in phase—moving up together and moving down together—then the average temperature of the gas would not change significantly while the total volume would change dramatically. As will be discussed in great detail later, the desired, properly phased motions of the pistons approximately 90° apart cause more complicated behavior—thermal expansion of the gas while the pressure is high and thermal contraction while the pressure is low—which delivers net work to the pistons. This work keeps the parts moving and can do something useful: turn industrial machinery or pump water or circulate air. Meanwhile, heat must be supplied to the heater to keep it hot, and heat must be removed from the ambient heat exchanger to keep it cool.

Before the advent of the computer, the analysis of Stirling engines used the most powerful computational tools available at the time: paper and ink. Given that severe limitation, extreme approximations were made to simplify the analysis. In the most extreme simplification, it was assumed that the pistons moved abruptly instead of sinusoidally, so that the gas was carried through a cycle of four discrete steps: compression, temperature change via constant-volume displacement one way through the thermal components, expansion, and temperature change via displacement the other way. It was also assumed that the volume of gas entrained in the heater, regenerator, and cooler passages was negligible, that no space remained above each piston when it was at the top of its motion, and that when the gas was between a heat exchanger and the adjacent piston it was always at that heat exchanger's temperature. With these approximations, the entire mass of working gas would be carried through a four-step cycle consisting of two constant-volume temperature changes and two constant-temperature volume changes. This idealized process is known as the Stirling cycle.

Neither the Stirling cycle nor any other textbook thermodynamic cycles will be examined in detail in this book, because real Stirling engines and refrigerators [17, 20, 26] are so complex that the Stirling *cycle* itself is of only academic interest. A typical parcel of gas in a real regenerator moves, sinusoidally, only a fraction of the length of the regenerator, so its cycle cannot be analyzed by direct reference to a textbook cycle of four discrete steps covering the entire temperature span of the device.

Fortunately, today's computers allow easy, reasonably realistic analysis of these devices, including the simultaneous time dependences and position dependences of the pressure changes and gas motion, which require proper accounting for non-negligible gas volumes everywhere. Well suited to personal computers, the particular style of analysis that is taken in this book will be called the *thermoacoustic* approach, because it brings techniques of acoustics to the analysis of the coupled position-dependent oscillations in these thermodynamic devices.[1] Ordinarily, sound waves are treated as *small* coupled oscillations of pressure and velocity: Even a painfully loud sound is a "small" oscillation, because at the threshold of pain the oscillating pressure amplitude is only about 2×10^{-4} times atmospheric pressure. Fortunately, the acoustic techniques remain usefully accurate even for the "large" oscillations encountered in these engines and refrigerators, where pressure amplitudes are often more than 1 atm with mean pressures more than 10 atm. These acoustic techniques rely heavily on the pioneering work of Nikolaus Rott, who was the first to derive correct expressions for motion and pressure [12] and time-averaged energy transport [13] in a channel with small, sinusoidal oscillations and with a temperature gradient.

In the typical Stirling engine of a century ago, the mechanical parts dominated the thermal parts—in volume, weight, and visual impact [27]. Since then, engineers have sought to simplify such heat engines and refrigerators by the elimination of moving parts. In the 1960s, Cooke-Yarborough and collaborators, and, independently, Beale, were thinking about the gas-pressure forces acting on moving pistons, and realized that under the correct circumstances the forces on the connecting rods were small while the pistons continued to move correctly. This invention

---

[1]In his 1980 review "Thermoacoustics," Nikolaus Rott [8] declared the meaning of the word to be "rather self-explanatory," a blend of thermodynamics and acoustics.

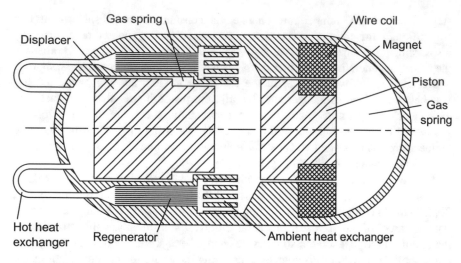

Fig. 1.2 Schematic of a typical free-piston Stirling engine. Gas-spring spaces change volume and exert force when the piston and displacer move

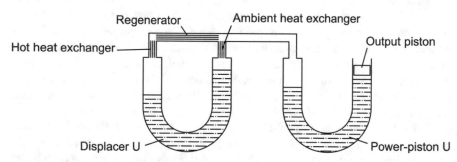

Fig. 1.3 The Fluidyne (liquid piston) Stirling engine [29]. Water or other liquid in the two U tubes oscillates

led to what are called free-piston [15, 17, 28] Stirling engines and refrigerators, as illustrated in Fig. 1.2, in which the massive moving pistons bounce against gas springs in resonance, with other moving parts such as connecting rods and crankshafts eliminated.[2] If the gas springs, masses, etc. are designed correctly, the pistons will indeed naturally move with the desired relative amplitudes and time phasing.

Another method to eliminate moving parts is the liquid-piston Stirling engine, also known as the Fluidyne engine [29], illustrated in Fig. 1.3. In Fluidyne engines,

---

[2]Some Stirling mechanical arrangements (e.g., Fig. 1.2) use a piston having *both* ends exposed to the working gas. Such a piston is called a displacer. The desired sequence of four steps in the thermal components—compression, displacement, expansion, displacement—is the same as for the arrangement of Fig. 1.1.

liquid in two U tubes, oscillating naturally with the same frequencies but different phases, serves the function of the pistons of the ordinary Stirling engine, confining and moving the working gas through the regenerator and heat exchangers appropriately. The Fluidyne's resonance phenomena are similar to those of the free-piston Stirling machines, but with gravity and liquid density contributing to the effective spring constant of the resonance.

In cryogenic refrigeration, the accidental discovery and subsequent development of "pulse-tube" refrigeration by Gifford and Longsworth [30], in which pulses of pressure in tubes of certain geometry produced refrigeration, and the dramatic improvement achieved with the addition of the "orifice" by Mikulin et al. [31], preceded the realization that what is now called the orifice pulse-tube refrigerator is essentially a Stirling cryocooler in which the cold piston has been replaced by non-moving components. Whatever the history, the orifice pulse-tube refrigerator's elimination of the cold sliding or flexing seal from the Stirling cryocooler represents a dramatic simplification in hardware.

More recently, Peter Ceperley [23, 24] realized that the phasing between pressure and velocity in the thermodynamic elements of Stirling machines is the same as the phasing between pressure and velocity in a traveling acoustic wave, so he proposed eliminating everything but the working gas itself, using acoustics to provide all the control of the gas motion and gas pressure. Figure 1.4 shows such a device, with two heat exchangers and a regenerator but no moving parts whatsoever. Shortly thereafter, the Los Alamos group began developing what are now called standing-wave thermoacoustic engines and refrigerators, relying on intrinsically irreversible heat transfer in a type of heat exchanger known as a "stack" and using a different

Fig. 1.4 The traveling-wave engine concept. Tag lines *2* and *3* are heat exchangers; *1* is a regenerator. The complicated shape of the enclosure ensures the correct relative amplitudes and phasing of the oscillating pressure and gas motion. Reproduced from one of Ceperley's patents [32, 33]

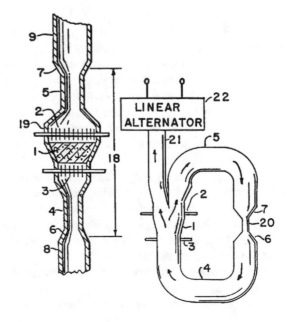

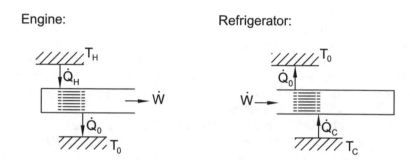

Fig. 1.5 The types of devices under consideration in this book. The *upper-left sketch* is an idealized, textbook heat engine, a device producing work at a rate $\dot{W}$ while absorbing heat at a rate $\dot{Q}_H$ from a source at high temperature T_H and rejecting heat at a rate $\dot{Q}_0$ to a lower temperature. The *lower sketches* show idealized thermoacoustic versions of such an engine and a similar refrigerator. Within the engine or refrigerator are three heat-exchange components. Two heat exchangers transfer heat to or from the external thermal reservoirs. Between them is a third heat-exchange component, called a stack in standing-wave devices and a regenerator in traveling-wave (e.g., Stirling) devices, with no connection to an external thermal reservoir

time phasing from that of Ceperley's ideas and Stirling machines, which strive for reversible heat transfer in the regenerator.

This book will attempt to integrate these themes together into a coherent picture of such heat engines and refrigerators, with physical, mathematical, and intuitive foundations. The book will use the kind of mathematics that Rott [8, 12, 13] developed, and it will emphasize hardware with the features of the work of Cooke-Yarborough, Beale, West, and especially Ceperley—the elimination of moving parts whenever possible to achieve simplicity and reliability, while maintaining the highest possible efficiency.

As suggested in Fig. 1.5, this integration will often involve devices in which the gas-enclosing cavity comprises an acoustic resonator that determines the relationship between gas motion and pressure, in which heat exchangers exchange heat with external heat sources and sinks, and in which a stack or a regenerator between the heat exchangers makes the device function as a thermoacoustic standing-wave device or a Stirling device.

In one respect, thermoacoustic devices are more interesting than crankshaft-based systems, e.g, as shown in Fig. 1.1, where hardware motion was determined by design and exerted direct control on gas motion. In thermoacoustic systems, including linear-motor free-piston Stirling systems, oscillating pressure gradients cause the oscillating motion, and the oscillating motion causes the oscillating pressure gradients. Analyzing a dynamical system having such interdependent causes and effects is inherently challenging.

1.2 Length Scales

The important length scales in thermoacoustic engines and refrigerators are illustrated in Fig. 1.6.

Along the wave-propagation direction x (the direction of motion of the gas), the wavelength of sound, λ, is an important length scale, given by

$$\lambda = a/f, \tag{1.1}$$

where a is the speed of sound and f is the oscillation frequency. When gas inertia contributes to the resonance behavior, the whole length of the apparatus may typically be a half wavelength or a quarter wavelength. However, when massive mechanical components participate meaningfully in the resonance behavior, as in free-piston Stirling systems, the size of the system is typically much smaller than the wavelength. In all cases, the lengths of heat-exchange components are much shorter than the wavelength.

Another important length scale in the direction of motion of the gas is the gas displacement amplitude $|\xi_1|$, which is the velocity amplitude $|u_1|$ divided by the angular frequency $\omega = 2\pi f$ of the wave. This displacement amplitude is often a very large fraction of the stack length or regenerator length, and may be larger than the lengths of the heat exchangers at either end of the stack or regenerator. The displacement amplitude is always shorter than the wavelength.

Perpendicular to the direction of motion of the gas, two characteristic lengths are the thermal penetration depth

$$\delta_\kappa = \sqrt{2k/\omega\rho c_p} = \sqrt{2\kappa/\omega} \tag{1.2}$$

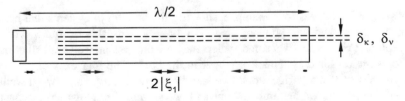

Fig. 1.6 Important length scales in a standing-wave thermoacoustic device

and the viscous penetration depth

$$\delta_v = \sqrt{2\mu/\omega\rho} = \sqrt{2\nu/\omega},\tag{1.3}$$

where k and κ are the thermal conductivity and diffusivity of the gas, μ and ν are its dynamic and kinematic viscosities, and c_p is its specific heat per unit mass at constant pressure. These characteristic lengths tell us how far heat and momentum can diffuse laterally during a time interval of the order of the period of the oscillation divided by π. At distances much greater than these penetration depths from the nearest solid boundary, the gas feels no thermal contact or viscous contact with the solid boundaries. In parts of the apparatus whose lateral dimensions are of the order of the viscous and thermal penetration depths, the gas does feel both thermal and viscous effects from the boundaries. Clearly the heat exchange components in thermoacoustic systems must have lateral dimensions of the order of δ_κ in order to exchange heat with the working gas (but turbulence can relax this requirement).

In the square of the ratio of these two penetration depths,

$$\left(\frac{\delta_v}{\delta_\kappa}\right)^2 = \frac{\mu c_p}{k} = \sigma \lesssim 1,\tag{1.4}$$

the frequency and the density cancel, leaving simply the ratio of viscosity times heat capacity to thermal conductivity. This ratio, σ, is called the Prandtl number. The Prandtl number is close to unity for typical gases, so viscous and thermal penetration depths are comparable. Hence, thermoacoustic engines and refrigerators typically suffer from substantial viscous effects.

In ordinary audio acoustics, the displacement amplitude of the gas is much smaller than the thermal and viscous penetration depths, which in turn are much smaller than the wavelength. In thermoacoustic engines and refrigerators typically the gas displacement amplitudes $|\xi_1|$ are much larger than the penetration depths, but still much smaller than the acoustic wavelength:

$$\delta_v, \; \delta_\kappa \ll |\xi_1| \ll \lambda.\tag{1.5}$$

1.3 Overview and Examples

This section introduces four examples showing the breadth of systems that benefit from the thermoacoustic point of view. For brevity here, this overview will use concepts and vocabulary that some readers may not understand well until all of these concepts are revisited systematically later in the book. The four examples include two engines and two refrigerators; two standing-wave systems and two traveling-wave systems; and two research systems and two systems bringing thermoacoustics toward a more practical stage of development. These same four examples will be revisited many times, as various aspects are examined in detail at appropriate places in subsequent chapters.[3]

The computer animation "Ani. Standing" discussed below is the first of many animations to be used in this book. Watching the animations while reading the text is vital—studying this book without the animations would be as frustrating and ineffective as studying an engineering textbook without looking at the figures. As of the 2015 printing, the animations are available for Windows 95, 98, NT, 2000, Vista, and 7, and for Mac OSX 10.5 through 10.10. Download and unzip the file appropriate for your computer from either http://www.lanl.gov/thermoacoustics/ movies.html or https://www.springer.com/us/book/9783319669328 (click on "Supplementary Material" about halfway down the page). Open the readme file for the latest instructions for your operating system. Expect minor differences between the descriptions in this book and how the animations are actually accessed; for example, what is called "Ani. Wave /s" here is accessed via "Wave_s (Standing)" in the Windows shortcut subfolder.

1.3.1   Standing-Wave Heat Engine

The purpose of a heat engine is to produce work from high-temperature heat, and one means to do this is with a standing-wave thermoacoustic heat engine. Animation Standing /e illustrates the most important features, with a moving view resembling Fig. 1.7.

An acoustic resonator is shown across the top, with its solid walls heavily filled. The standing-wave resonance in the gas trapped inside the resonator is indicated by the moving blue vertical lines, which can be imagined as massless sheets of cellophane or thin clouds of smoke moving with the gas. When the lines move to the left, crowding together near the left end of the resonator, the density and pressure rise there. The left end of the resonator contains three heat exchangers through which the oscillating gas passes rather freely. From left to right, these are the hot heat exchanger, the "stack," and the ambient heat exchanger. The hot and ambient heat exchangers are connected (by means not shown) to thermal reservoirs outside the resonator, so that they can maintain desired temperatures at the ends of the stack.

The conversion of heat to work takes place inside the stack. The bottom 2/3 of the animation shows the details of that process qualitatively, by following a single imaginary "parcel" of the gas—a mass element—as it oscillates. The parcel is shown as a blue rectangle, oscillating right and left between two parallel plates of the stack. This is a magnified view of the portion of the stack that is enclosed in the yellow ellipse (in which the same moving parcel is represented by a small moving blue

---

[3] You may enjoy re-reading this section as you finish each of the subsequent chapters. Mark these pages (with clips or tape or . . .) so you can refer back here easily.

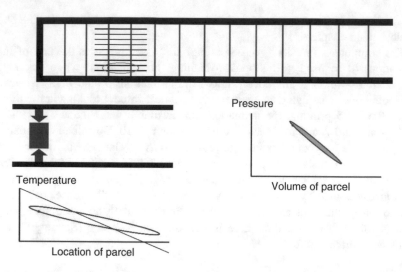

Fig. 1.7 Running Ani. Standing /e on a personal computer should produce a moving display like this, illustrating some of the important phenomena in standing-wave thermoacoustic engines

speck). The time-dependent size of the rectangle indicates the volume of the parcel, and the red arrows show heat transfer between the parcel and the adjacent plates of the stack. The graphs below and right of the moving parcel show the temperature, location, pressure, and volume oscillations of the parcel as functions of time.

In the temperature–location graph, the white diagonal line (white on the computer screen, but black in Fig. 1.7) indicates the temperature of the solid plates as a function of position, and the moving blue dot traces the temperature of the gas parcel as a function of position and time. When the parcel is left of center, it is cooler than the surrounding plates, so it absorbs heat from the plates; when the parcel is right of center, it is warmer than the surrounding plates, so it rejects heat to the plates. Hence, the parcel experiences thermal expansion while it is left of center, and thermal contraction while it is right of center. Meanwhile, the time phasing between motion and pressure, imposed by the standing wave, is crucial: When the parcel is left of center, its pressure is high; when it is right of center, its pressure is low. Hence, the thermal expansion of the parcel occurs while the pressure is high, and the thermal contraction occurs while the pressure is low. The pressure–volume graph shows the resulting pressure and volume of the parcel as functions of time. Absent thermal expansion and contraction, this graph would simply show an oscillation along a diagonal line, but the thermal expansion and contraction swell that line into a narrow ellipse. The area of the ellipse, which is the cycle integral of the pressure with respect to the volume, is the net work done per cycle by the parcel of gas on its surroundings. The sum of such works by all parcels in the stack is the work produced by the engine. This work is produced at the acoustic frequency of the standing wave, so it is acoustic power.

Note that the thermal contact between the parcel and the adjacent plates must be neither too weak nor too strong. If the thermal contact were too weak, no heat would be transferred between the parcel and the plates, so no thermal expansion/contraction or work would occur. More subtly, if the thermal contact were too strong, the parcel's temperature would trace an oscillating line exactly on top of the solid's local temperature in the temperature–location graph, which would shift the time phasing of the thermal expansion and contraction by roughly 90°. With such phasing, the pressure–volume graph would become a reciprocating line instead of a narrow ellipse, and again no work would occur. Successful operation of a standing-wave engine requires deliberately imperfect thermal contact between the gas and the stack, which is obtained when the spacing between the plates is roughly a few δ_κ.

Summarizing this discussion: A standing-wave thermoacoustic engine has a "stack" with gaps of the order of δ_κ through which acoustic oscillations with standing-wave time phasing occur. If the heat exchangers maintain a sufficiently large temperature difference across the stack, the gas in the stack produces work, because the gas in the stack experiences thermal expansion when the pressure is high and thermal contraction when the pressure is low. Thus the gas in the stack pumps acoustic power into the standing wave. The standing wave in turn provides the oscillating pressure, and the oscillating motion that causes the gas in the stack to experience the oscillating temperature responsible for the thermal expansion and contraction. The velocity of the gas along the stack's temperature gradient is 90° out of phase with the oscillating pressure, so imperfect thermal contact between gas and stack is required to enable the thermal expansion and contraction steps to be in phase with the oscillating pressure. These complex, coupled oscillations appear spontaneously whenever the temperature difference across the stack is high enough.

An example of such a standing-wave heat engine is one of the two identical engines in the heat-driven cryogenic refrigerator [34] assembled at Tektronix in 1994 and shown in Figs. 1.8 and 1.9. Building this system was the second step in a development aimed at providing a small amount of cryogenic refrigeration for advanced electronic components. Helium at a pressure of 30 atm filled the system. A two-stage orifice pulse-tube refrigerator (not discussed here) provided the refrigeration, driven by oscillating pressure and oscillating motion—i.e., by acoustic power. The acoustic power was created from heat, in the two standing-wave thermoacoustic engines, each heated by a red-hot electric heater. We hoped eventually to make such a system smaller, and sufficiently efficient that the necessary electric power could be supplied from an ordinary wall plug.

The resonator lies horizontally in the figures, with the standing-wave thermoacoustic engines on each end, and an upward side branch near the left end connecting to the two-stage orifice pulse-tube refrigerator. The sound speed of helium is about 1000 m/s at room temperature, and the horizontal length of the resonator in Figs. 1.8 and 1.9 was about 1 m, so one would expect that the half-

Fig. 1.8 Standing-wave engine example: Tektronix researchers Ying Ki Kwong, Kim Godshalk, and Ed Hershberg with a heat-driven refrigerator, 1995. The resonator, horizontal, has standing-wave engines on each end. A two-stage orifice pulse-tube refrigerator stands vertically near the *left end*. Photo courtesy of Kim Godshalk

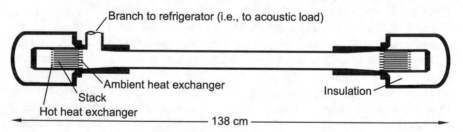

Fig. 1.9 Scale drawing of the Tektronix resonator and its standing-wave engines. The annular spaces around each end were packed with thermal insulation, and did not experience the acoustic oscillations. The spacing in stacks and heat exchangers is not to scale. The upward branch near the *left end* is the connection to the two-stage pulse-tube refrigerator

wavelength resonance in this system would be near 500 Hz. Indeed, the system operated near[4] 370 Hz.

Water cooling lines, seen extending through the bottom edge of Fig. 1.8, removed waste heat from the ambient heat exchangers. A hot heat exchanger is shown in Fig. 1.10, and a stack is shown in Fig. 1.11. The "open" nature of the hot heat exchanger is apparent, and even the stack's cross-sectional area is 83% open

---

[4]Frequency varied ±20 Hz, depending on acoustic load, water temperature, . . .

Fig. 1.10 Electric-resistance hot heat exchanger for the standing-wave engine of Figs. 1.8 and 1.9. A 6-mm-wide, 25-μm-thick ribbon of NiCr winds up and down in the figure, barely visible because it reflects the white ceramic and dark background like a mirror. Slits in the ceramic "combs" that maintain the spacing between adjacent parts of the NiCr ribbon are clearly visible; the center-to-center spacing of these slits is 1.6 mm. Three ceramic ribs, perpendicular to the combs, interlock with the ribs on the far side to stabilize the longest ribs

Fig. 1.11 The stack used in one of the standing-wave engines of Figs. 1.8 and 1.9. The overall diameter is 6.3 cm. The spiraling stainless-steel sheet is 50 μm thick; the gap between layers of the spiral is 250 μm. Each end is held by 24 radial ribs; the intersections between each rib and each turn of the spiral were welded

space despite its "solid" appearance. At 30 atm and 350 Hz, the thermal penetration depth in helium is approximately 0.1 mm. Correspondingly, to ensure the necessary imperfect thermal contact, the spacing between layers in the stack was 0.25 mm.

In operation, electric heat was applied to the hot heat exchanger, and the wave would appear, at 350 Hz, as soon as the hot temperature was high enough. Increasing the heater power then increased the oscillation amplitude. These heat-driven engines and their resonator could deliver up to 1 kW of acoustic power to the refrigerator. At one favorite operating point, they delivered 500 W to the pulse-tube refrigerator with an efficiency of 23% of the Carnot efficiency. (Here, efficiency is the acoustic power delivered to the pulse-tube refrigerator divided by the total electric power supplied to the two engines' heaters, and the temperatures used in the Carnot factor are those of the engines' cooling water and of the cases around the hot heat exchangers.)

1.3.2   Standing-Wave Refrigerator

The purpose of a refrigerator is to remove heat from a low temperature, necessarily consuming work and rejecting waste heat to a higher temperature. One means to do this is with a standing-wave thermoacoustic refrigerator. Animation Standing /r illustrates the most important features, with a moving display resembling Fig. 1.12. An acoustic resonator is shown across the top of the animation (not included in Fig. 1.12), similar to that of the standing-wave engine discussed above. Some means to drive the wave, not shown in the animation, is required. The left end of the resonator contains three heat exchangers through which the oscillating gas passes freely. From left to right, these are the ambient heat exchanger, the stack, and the cold heat exchanger. The ambient and cold heat exchangers are connected (by means

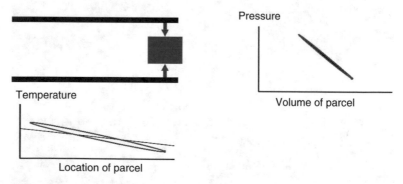

Fig. 1.12 Running Ani. Standing /r on a personal computer should produce a moving display including this view in its lower half, illustrating some of the important phenomena in standing-wave thermoacoustic refrigerators. The plate temperature gradient is shallower than in Fig. 1.7, resulting in a change of sign of the heat flows and net work

not shown) to thermal reservoirs outside the resonator. Removal of heat from the cold reservoir is the purpose of the refrigerator.

The pumping of heat up the temperature gradient, from right to left, takes place inside the stack. The bottom 2/3 of the animation shows the details of this process qualitatively, by following a single imaginary "parcel" of the gas as it oscillates between two parallel plates of the stack. This is a magnified view of the portion of the stack that is enclosed in the yellow ellipse. The time-dependent size of the blue rectangle indicates the volume of the parcel, and the red arrows show heat transfer between the parcel and the adjacent plates of the stack. The graphs below and right of the moving parcel show the temperature, location, pressure, and volume of the parcel as functions of time.

In the temperature–location graph, the white diagonal line (black in Fig. 1.12) indicates the temperature of the solid plates as a function of position, and the moving blue dot traces the temperature of the gas parcel as a function of position and time. The time phasing between motion and pressure, imposed by the standing wave, is crucial: When the parcel is left of center, its pressure is high; when it is right of center, its pressure is low. To the extent that the parcel's thermal contact with the plates is poor, adiabatic temperature oscillations accompany the pressure oscillations, so the parcel's temperature tends to rise adiabatically as it moves left and tends to fall adiabatically as it moves right. However, the plates' temperature gradient is relatively small, so when the parcel is right of center, it is cooler than the surrounding plates and can absorb heat from the plates; when the parcel is left of center, it is warmer than the surrounding plates, so it rejects heat to the plates. Hence, heat is pumped up the temperature gradient along the plates. The combined action of all parcels in the stack removes heat from the cold heat exchanger at the right end of the stack and rejects heat at the ambient heat exchanger at the left end of the stack.

Just as for the standing-wave engine described above, note that the thermal contact between the parcel and the adjacent plates must be neither too weak nor too strong. If the thermal contact were too weak, no heat would be transferred between the parcel and the plates, so no heat pumping would occur. More subtly, if the thermal contact were too strong, the parcel's temperature would trace an oscillating line exactly on top of the solid's local temperature in the temperature–location graph, which would shift the time phasing of the heat transfer by roughly $90°$. With such phasing, the net heat transfer from the parcel to any particular location on the stack would be zero, and again no heat pumping would occur. Successful operation of a standing-wave refrigerator requires deliberately imperfect thermal contact between the gas and the stack, which is obtained when the spacing between the plates of the stack is roughly a few δ_κ. This imperfect thermal contact affects both the magnitude and the time phasing of the temperature oscillations throughout the stack.

Summarizing: A standing-wave thermoacoustic refrigerator has a "stack" with gaps of the order of δ_κ through which acoustic oscillations with standing-wave time phasing occur. The key idea is that the gas in the stack experiences partly adiabatic temperature oscillations, in phase with the standing-wave pressure oscillations, and

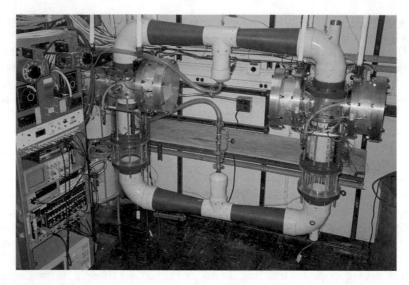

Fig. 1.13 Standing-wave refrigerator example: Loudspeaker-driven research apparatus for investigating the superposition of deliberate steady flow on the oscillating thermoacoustic flow

simultaneous oscillating standing-wave displacements in phase with the pressure oscillations and hence in phase with those temperature oscillations. Thus the gas in the stack cools a little as it is displaced in one direction and warms a little as it is displaced in the other direction. Imperfect thermal contact between gas and stack enables the required heat transfer at the extremes of the motion.

The system shown in Figs. 1.13 and 1.14 is an example of such a standing-wave refrigerator. It was built by Bob Reid to investigate the deliberate superposition of steady flow [3, 4, 35] with the oscillating flow of the standing-wave thermoacoustic refrigeration process. The superimposed steady-flow velocity was much smaller than the standing-wave oscillating velocity. The refrigerator was driven by four loudspeakers, enclosed in the obvious aluminum housings shown in Fig. 1.13. The left and right pairs of loudspeakers were driven 180° out of phase, to produce a standing wave in the toroidal resonator having essentially one full wavelength around the torus, with pressure nodes centered in the top and bottom sections of the resonator and pressure maxima near the loudspeakers. With the resonator filled with a mixture of 92% helium and 8% argon at 3 atm of pressure, this resonance occurred at 94 Hz. The branches at the pressure nodes allowed injection and removal of steady flow without disturbing the standing wave. Slow steady flow from top to bottom through the heat-exchange components, superimposed on the thermoacoustic oscillating flow, produced cold gas leaving the resonator at the bottom pressure node. The spacing in the heat-exchange components was about 1 mm, a few times the thermal penetration depth in this gas at this frequency. The stacks were made of flat fiberglass sheets. The ambient heat exchangers were of the finned, water-cooled variety, similar to the one shown in Fig. 8.10.

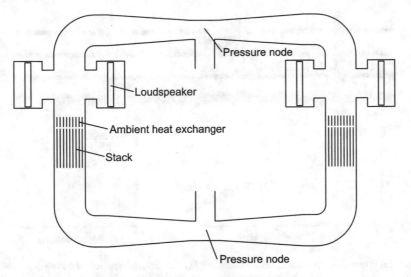

Fig. 1.14 Schematic of the standing-wave refrigerator example

1.3.3 Orifice Pulse-Tube Refrigerator

Traveling-wave devices operate much differently than standing-wave devices. An illustration of the heart of a traveling-wave refrigerator—its "regenerator"—is shown in Ani. Ptr /r (and Fig. 1.15), which should be compared with Ani. Standing /r. The oscillating velocity and oscillating pressure are in phase in the traveling-wave refrigerator, whereas the position and pressure are in phase in the standing-wave refrigerator. Furthermore, in contrast to the standing-wave refrigerator's deliberately imperfect thermal contact, the thermal contact between the parcel and the solid plate (or, more commonly, solid matrix) in the regenerator is as good as can be: The parcel's temperature as a function of time always lies atop the local solid temperature in the temperature–location graph. Qualitatively, heat pumping up the temperature gradient occurs in the traveling-wave refrigerator because the parcel experiences expansion near the right end of its travel, absorbing heat from the solid, and experiences compression near the left end of its travel, rejecting heat to the solid. (The picture is complicated by the superimposed heat exchange occurring as the parcel moves along the temperature gradient, but the net heat transfer between the parcel and each portion of the solid due to this effect is zero.)

The orifice pulse-tube refrigerator, shown in Ani. Ptr (after typing m and o from within the animation) and Fig. 1.16, provides one way to achieve the required time phasing between pressure and velocity. An oscillating driver piston at the left end supplies work to the system, and an acoustic impedance network at the right end dissipates power and sets the relative amplitudes and time phasing of the pressure and velocity. An ambient heat exchanger left of the regenerator and a cold

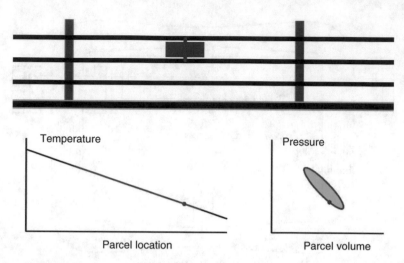

Fig. 1.15 Animation Ptr /r produces a moving display like this, qualitatively showing details of the position, pressure, temperature, and volume of a typical parcel of gas in the heart of the regenerator of a traveling-wave refrigerator

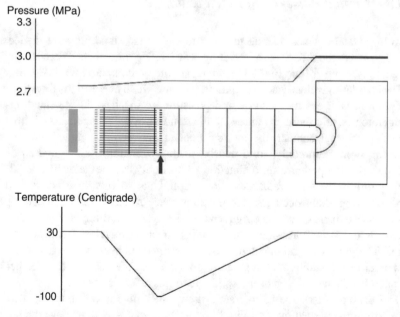

Fig. 1.16 Running Ani. Ptr (and then typing m and o) shows a moving view of pressure and gas motion in an orifice pulse-tube refrigerator. Work is supplied from the *left end* by a piston (the moving *green rectangle*), and the impedance of the components at the *right end* helps set the amplitudes and phases of motion and pressure in the regenerator

heat exchanger right of the regenerator transfer heat between the working gas and external thermal reservoirs.

The desired cooling effect at the cold heat exchanger in this and other Stirling, pulse-tube, and traveling-wave refrigerators can be understood in terms of the heat pumping in the regenerator described in the first paragraph of this subsection. From this point of view, the heat pumping in the regenerator removes heat from the cold heat exchanger, carrying it into and through the regenerator, while nothing on the other side of the cold heat exchanger replenishes that heat.[5] Alternatively, focusing on the other side of the cold heat exchanger by typing c within Ani. Ptr, the cooling effect can be viewed as being caused by gas that is displaced out of the cold heat exchanger and into the open space away from the regenerator, adiabatically cooled, and displaced back into the cold heat exchanger where its low temperature lets it absorb heat from the cold heat exchanger. Action on the other side of the heat exchanger rejects heat into the closest portion of the regenerator, where it can be picked up by gas behaving as in Ani. Ptr /r. In either of these two points of view, the wave must provide oscillating pressure and oscillating displacement, with the correct time phasing.

In Stirling refrigerators, as illustrated in Ani. Ptr by typing s, the open space is adjacent to a cold piston. The technical challenge of sealing around such a piston at cryogenic temperatures has been the main motivation for the development of orifice pulse-tube refrigerators as an alternative to Stirling cryocoolers. In pulse-tube refrigerators, the cold piston is replaced by the aforementioned acoustic impedance network and a nearly adiabatic, thermally stratified column of moving gas—the "pulse tube" itself. The pulse tube must provide a thermal barrier between the acoustic impedance network and the cold heat exchanger.

The orifice pulse-tube refrigerator shown in Figs. 1.17 and 1.18, an example of a traveling-wave refrigerator, was built at Cryenco (now a part of Chart Industries) for liquefaction of natural gas [36]. At about 3 atm, natural gas (methane) liquefies at 120 K. With a few modifications made after the photograph was taken, this refrigerator provided 2 kW of refrigeration at that temperature, with a *COP* as high as 23% of the Carnot *COP*. (Here, *COP* is the heat removed from the methane stream divided by the acoustic power incident on the aftercooler from below, and the temperatures used in the Carnot factor are those of the liquefied methane and of the cooling water.) The working gas was helium at a pressure of 30 atm, driven at 40 Hz ($\lambda = 25$ m) by a standing-wave thermoacoustic engine through a resonator, not shown in the figures.

The regenerator was a pile of fine stainless-steel screen, with pores much smaller than δ_κ, and the flow straighteners were shorter piles of coarser screen. The heat exchangers were cross-flow "shell and tube," with the helium oscillating through the tubes while the water or methane flowed steadily around and between the tubes. This style of heat exchanger is illustrated schematically in Fig. 1.19.

---

[5]John Wheatley used to say that a heat exchanger with such imbalanced surroundings experienced "broken thermodynamic symmetry."

Fig. 1.17 Traveling-wave refrigerator example: the Cryenco 2-kW orifice pulse-tube refrigerator

The acoustic impedance network comprised a compliance (the large tank in the lower-right quadrant of the photograph), two valves serving as variable flow resistances (at the top of the photograph), and pipes connecting them.

1.3.4 Thermoacoustic-Stirling Heat Engine

The key idea in Stirling engines and other traveling-wave engines is that the gas in the regenerator experiences thermal expansion when the pressure is high and thermal contraction when the pressure is low, as illustrated in Ani. Tashe /r and Fig. 1.20. Thus the gas in the regenerator does work every acoustic cycle, pumping acoustic power into the gas. The oscillating gas motion and pressure are caused by moving pistons in traditional Stirling engines, but they can also appear spontaneously without pistons, whenever the temperature gradient through the regenerator is high enough, if the regenerator and attendant heat exchangers are in a suitable acoustic network such as that illustrated in Ani. Tashe /t and Fig. 1.21. The velocity of the gas along the regenerator's temperature gradient is substantially in phase with the oscillating pressure, so good thermal contact between gas and

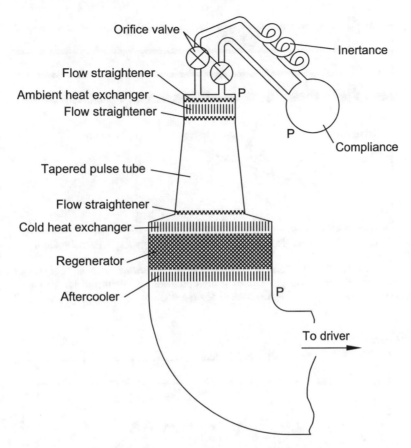

Fig. 1.18 Schematic of the Cryenco orifice pulse-tube refrigerator. "P" indicates the location of a pressure sensor. The ambient heat exchanger and the "aftercooler" are held at ambient temperature by flowing water

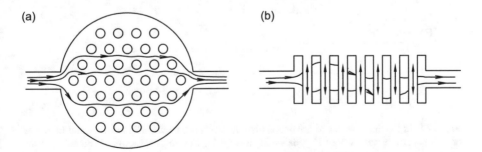

Fig. 1.19 Two perpendicular central cross sections through a cross-flow shell-and-tube heat exchanger. In some thermoacoustic applications, the helium or other working gas oscillates through the tubes, and the water or other thermal-reservoir exchange fluid passes unidirectionally though the shell, around and between the tubes. See also Fig. 1.24

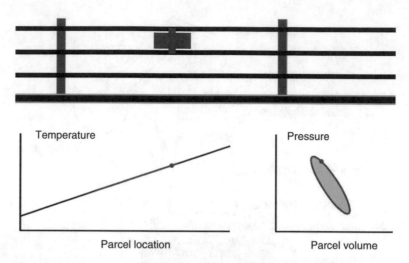

Fig. 1.20 As shown in Ani. Tashe /r, the gas motion, temperature gradient, heat transfer, and pressure combine to make the gas expand at high pressure and contract at low pressure in the regenerators of traveling-wave engines such as Stirling engines

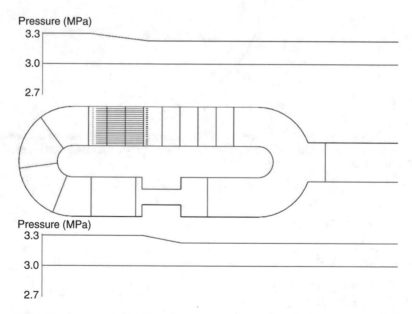

Fig. 1.21 In the thermoacoustic-Stirling engine, a short toroidal acoustic network and long side branch support a wave whose impedance is suitable for traveling-wave work production at the location of the regenerator and heat exchangers

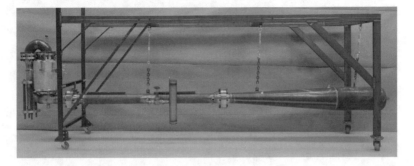

Fig. 1.22 Traveling-wave engine example: a research apparatus for initial Los Alamos studies of the performance of a thermoacoustic-Stirling heat engine

regenerator is required to cause the thermal expansion and contraction steps to be in phase with the oscillating pressure. This thermal contact is achieved by making the channel size in the regenerator much smaller than the thermal penetration depth. In contrast, in the standing-wave engine discussed above, the thermal contact in the stack is not good, and the displacement—not velocity—of the gas is in phase with the pressure. Either of these situations can serve as the heart of a heat engine.

As an example of a traveling-wave engine, the thermoacoustic-Stirling heat engine shown in Figs. 1.22 and 1.23 was built by Scott Backhaus to demonstrate Stirling-cycle engine thermodynamics with no moving parts—a hybrid thermoacoustic-Stirling heat engine—and to investigate issues affecting its efficiency [37, 38]. Heat was supplied to the engine electrically, waste heat was removed by a water stream, and acoustic power was thereby created and delivered to the resonator. Some of that acoustic power was dissipated in the resonator, but most was dissipated in a variable acoustic load, the small vertical cylinder and the valve near the center of Fig. 1.22. This engine delivered acoustic power $\dot{E}_{res}$ with an efficiency as high as 42% of the Carnot efficiency. (Here, efficiency is the acoustic power delivered into the resonator to the right of the junction labeled in Fig. 1.23, divided by the electric power supplied to the engine's heater; the temperatures used in the Carnot factor are those of the gas just below the hot heat exchanger and of the engine's cooling water.) Suppression of steady flow around the torus containing the heat exchangers was crucial for this high efficiency, as discussed in Chap. 7. Thirty-atmosphere helium filled the system, oscillating at 80 Hz. The resonator was essentially half-wavelength, with the pressure oscillations in the fat portion on the right end of Fig. 1.23a 180° out of phase from those in the small torus containing the heat exchangers on the left. The highest velocity occurred in the center of the resonator, at the small end of the long cone. The highest pressure amplitude occurred in the main ambient heat exchanger. The regenerator was a pile of fine-mesh screen and the hot heat exchanger was similar to that shown in Fig. 1.10. The welded stainless-steel shell-and-tube main ambient heat exchanger is shown in Fig. 1.24.

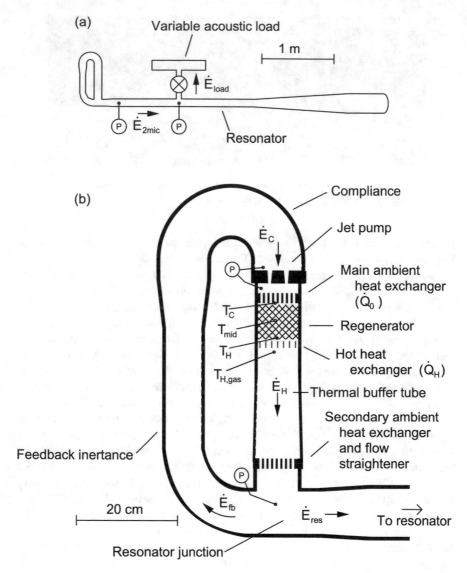

Fig. 1.23 The thermoacoustic-Stirling heat engine. (**a**) Overview, as in the photo. (**b**) The heart of the engine

1.4 Thermoacoustics and Conventional Technology

Keeping in mind the four examples of thermoacoustic devices described above, consider what the advantages and disadvantages of thermoacoustic systems might be in comparison to the energy conversion systems that are in widespread use, such as the internal combustion engine, the steam turbine, the reverse-Rankine or vapor-compression refrigeration cycle, the gas turbine, etc.

Fig. 1.24 Looking down on the top ends of the tubes of the main ambient heat exchanger in the thermoacoustic-Stirling engine. The helium oscillated through these tubes. Three water pipes at the *top of the picture* fed ambient water into the shell of the heat exchanger; three pipes at the *bottom of the picture* carried the water away

These conventional systems have moving parts. From that point of view one might think of them as being less reliable and more expensive than thermoacoustic devices. However, decades of investment in engineering development have given these conventional devices high reliability, and decades of manufacturing experience have given them low cost. Thermoacoustic devices seem to have the immediate potential to have comparably high reliability and low cost, because they use no moving parts, no exotic materials, and no close tolerances. Realizing this potential is one of the challenges of thermoacoustics research.

Those decades of experience and investment in conventional energy-conversion devices have also given them very high efficiencies. Today's automobile engines convert the fuel's heat of combustion to shaft work with 30% efficiency; large diesel engines achieve 40%. Vapor-compression refrigeration systems with COP's above 50% of Carnot's COP are common. Thermoacoustic devices have not yet exceeded such high efficiencies; increasing thermoacoustic device efficiency into this range is another important current challenge.

However, it is important to realize that a new technology can be commercially successful even if its efficiency does not exceed that of older technologies. This point is illustrated in Fig. 1.25. The Stirling refrigerator is quite old, and in principle it can have Carnot's COP. However, high COP can only be achieved at great expense, as the mass of heat exchange metal, quality of lubricants, etc. is made exceedingly high. For instance, qualitatively, the efficiency vs capital cost of a Stirling refrigerator might resemble the Stirling curve in Fig. 1.25, with very high efficiency possible if one is willing to spend an exponentially large amount of money building the refrigerator. However, in the commercial world, energy costs are currently low enough that no one chooses to invest so much capital to achieve such high efficiency. Instead, vapor-compression refrigeration is used for almost all commercial refrigeration systems, because at useful efficiencies it has much

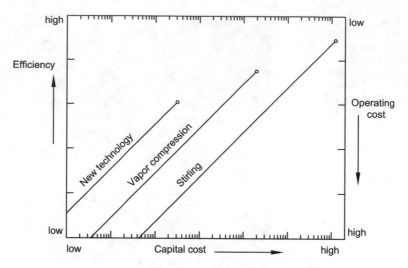

Fig. 1.25 Different technologies exhibit different trade-offs between cost and efficiency

lower capital cost than Stirling refrigeration. It's not important that the *COP* of vapor-compression refrigeration can in principle never reach the ultimate *COP* of the Stirling cycle. Furthermore, each application requires a different compromise between construction costs and efficiency. For example, the vapor-compression air-conditioning system for a large office building in Phoenix might lie higher on the efficiency–cost curve than the vapor-compression chiller in a soft-drink vending machine in Seattle. If thermoacoustics turns out to have an efficiency–cost curve somewhere to the left of existing technologies, it will succeed in some commercial markets. Whether the terminus of the thermoacoustics curve is higher or lower than that of older technologies is not as important as whether the capital costs will be lower at modest efficiencies.

(For a broad discussion of issues surrounding the commercialization of nascent technologies, see [39].)

1.5 Outline

This book promotes a unifying description allowing an understanding of both engines and refrigerators, both traveling-wave and standing-wave. This description is based firmly on both acoustics and thermodynamics. For brevity, I will refer to the devices of interest as "thermoacoustic engines and refrigerators," although I really mean "cyclic engines and refrigerators—including Stirling engines, Stirling refrigerators, pulse-tube refrigerators, and intrinsically irreversible standing-wave engines and refrigerators—studied using the unifying thermoacoustic approach." In other words, thermoacoustics really refers to a mental framework, not a class of devices, in this book.

The thermoacoustic description is so complicated that it is difficult for the human mind to hold all of its aspects simultaneously, even for a single device. Changing point of view from one manageable aspect to another quickly and effortlessly within the thermoacoustic description is the best I've ever been able to do. Hence, I've organized the book according to aspects—six chapters, 4 through 9—each of which is relatively small and manageable.

Only the exceptional reader is prepared to understand all the aspects of thermoacoustics immediately: Most readers lack necessary experience with one half or the other, either "thermo-" or "-acoustics." Hence, Chap. 2, **Background**, and Chap. 3, **Simple Oscillations**, provide brief reviews of relevant thermodynamics, fluid mechanics, acoustics, and complex-variable notation.

Chapter 4, **Waves**, examines the oscillating pressure and velocity. This includes the relevant fundamentals of ordinary acoustics, resonance phenomena in resonators, and viscous damping and thermal contact to the side walls of the channel in which the wave propagates. After studying the Waves chapter, you should be able to visualize and understand the overall wave behavior of a thermoacoustic apparatus, based on a drawing or description. This process involves thinking about all the dimensions of the apparatus in comparison to λ, δ_κ, and δ_ν, locating the nodes and antinodes of oscillating pressure and motion, identifying which parts of the wave are inertial and which are compliant, identifying which parts are affected by viscous flow resistance or by thermal relaxation effects, and predicting approximately the relative phases of pressure and motion throughout the apparatus. At a coarse, qualitative level, this process can be done by eye and by hand, but quantitative results can be obtained easily by numerical integration of the momentum and continuity equations.

The approach introduced in Chap. 4 (and used in subsequent chapters) depends on some choices. I chose $e^{i\omega t}$ rather than $e^{-i\omega t}$ for the time dependence. I chose oscillating pressure p_1 and oscillating volume flow rate U_1 as the primary dynamic variables, although $\dot{M}_1 = \rho_m U_1$ or $\langle u_1 \rangle = U_1/A$ would serve just as well as U_1. I chose an Eulerian formulation, instead of Lagrangian [14]. Following Rott, I chose to use f_ν and f_κ to keep track of complex viscous and thermal-relaxation effects; however, $F = 1 - \tilde{f}$ is also used in the literature [40], and there are many other valid approaches involving terminology like complex density [40], complex viscosity, and complex Nusselt number [41]. I chose to concentrate on differential equations instead of, for example, a formulation using transfer-function matrices [42–44]. I chose to draw impedance diagrams using the electrical analogy, not the equally valid mechanical (see Exercise 4.12) or acoustical analogies [5]. Any of these alternatives can accurately capture the essential physics of thermoacoustics. I simply decided to stick with one set of choices and get on with more advanced issues.

The power—both thermal and acoustic—is of greatest interest in heat engines and refrigerators, so Chap. 5, **Power**, adds these concepts to the pressure and velocity discussion. Acoustic power, the time-averaged product of pressure and volume flow rate, is produced in engines and consumed by refrigerators. Total power, which is the power that is subject to the first law of thermodynamics, is of even greater importance. After studying the Power chapter, you should be able to understand thermoacoustic engines and refrigerators in terms of energy conservation, power flows, and acoustic-power dissipation and production. Quantitative results can be obtained as desired from products of the oscillatory variables of Chap. 4.

The second law of thermodynamics, which puts bounds on the performance of all engines and refrigerators, forms the basis of Chap. 6, **Efficiency**. Key concepts are entropy generation, which can be subdivided according to location and mechanism, and exergy, which provides a quantitative figure of merit for thermodynamic devices and their interactions with their environments. Together, these concepts provide a standard, formal accounting method for sources of inefficiency. After studying this chapter, you should have an understanding of thermodynamic efficiency that goes beyond what is taught in standard physics courses and undergraduate engineering courses. You should be able to visualize and understand the locations and mechanisms of irreversibility in a thermoacoustic apparatus, and account quantitatively for the irreversibilities.

Chapter 7, **Beyond Rott's Thermoacoustics**, introduces many issues that go beyond the low-amplitude, monofrequency approximation implicit in Rott's pioneering work. These issues—of great concern in real devices, which usually operate at high amplitude in order to achieve high power per unit volume—include turbulence, minor losses, entrance effects, joining conditions, superimposed steady flow (streaming), and harmonics. While Chaps. 4 through 6 stand on a well-understood foundation, these high-amplitude issues are at the frontier of current knowledge. As in other branches of engineering, the use of dimensionless groups of variables helps bring organization to these complicated phenomena.

Chapter 8, **Hardware**, presents a few construction techniques, for stacks, regenerators, heat exchangers, pulse tubes and thermal buffer tubes, resonators, and electroacoustic power transducers, with emphasis on techniques that we currently use at Los Alamos.

Chapter 9, **Measurements**, outlines the simplest measurement techniques, suitable for routine use, and discusses some points of view that enable meaningful conclusions from such simple measurements.

The first **Appendix** lists some pitfalls I wish I had not had to discover "the hard way." The second **Appendix** gives four DELTAEC files (DELTAE in the First Edition) describing the four hardware examples presented earlier in this chapter.

A **List of Symbols** is included at the beginning of the book, and the **Bibliography**, **Author Index**, and **Subject Index** are at the end.

1.6 Exercises

1.1 Can you think of any other length scale in oscillations in an ideal gas, besides λ, $|\xi_1|$, δ_ν, and δ_κ? If you can, when would it be important in thermoacoustics? How can you be sure that no important length has been forgotten?

1.2 Do you know roughly what a kitchen refrigerator costs, and what is the cost of the electricity it consumes each year? If so, would you expect most people to choose to pay 25% more to buy a refrigerator that uses 50% less electricity? Would you do so? Why or why not?

1.3 As Earth rotates under the Sun, the surface of the ground experiences a daily temperature oscillation. What is the penetration depth of the resulting diffusion wave into the soil? (Assume that the thermal diffusivity of soil is 0.01 cm^2/s.) Seasonal variation in solar heating of the ground causes an annual temperature oscillation at greater depths. What is the penetration depth of the annual phenomenon? How deeply should water pipes be buried where you live?

Chapter 2
Background

This chapter provides a brief summary of thermodynamics and fluid mechanics,[1] on which thermoacoustics relies. The approach is a blend of Bejan's [45] and Landau's [46]. Thermodynamics [6, 45, 47–49] comes first, beginning with the first and second laws. Each law has two versions, one for fixed-mass "closed" systems and another for "open" systems into and out of which mass can flow. Though more difficult to derive, the "open" versions are more useful for gases and liquids. Next, the laws of fluid mechanics [46, 50–53] are outlined, including the fluid-mechanical expression of the second law of thermodynamics, which is seldom encountered in elementary fluid mechanics. These laws are described in considerable generality here, even though later chapters will only use ideal gases and usually be limited to Rott's acoustic approximation (see Chap. 3).

The chapter ends with a summary of the properties of ideal gases and derivations of a few useful thermodynamic results that will be needed later.

2.1 Laws of Thermodynamics

2.1.1 The First Law

The first law of thermodynamics expresses conservation of energy: Energy can be converted from one type to another, or moved from one place to another, but it cannot be created or destroyed. Conservation of "mechanical" energy as it is converted among gravitational potential energy, kinetic energy, and energy in a

---

[1] Some readers (e.g., chemical engineers, mechanical engineers) should find this material familiar, and can regard this chapter as a quick review to establish the notation we will use throughout the book. Readers who find this material less familiar (e.g., acousticians) can regard this as an outline of relevant topics in thermodynamics and fluid mechanics, which should be learned in greater detail elsewhere.

© Acoustical Society of America 2017
G.W. Swift, *Thermoacoustics*, DOI 10.1007/978-3-319-66933-5_2

stored spring is a familiar example. The first law extends this conservation principle to include heat energy.

The mathematical expression of the first law for a macroscopic system of fixed mass is usually written

$$d\mathcal{E} = dQ - dW, \tag{2.1}$$

where $d\mathcal{E}$ is the small change of the energy *of* the system, dQ is the small heat *added to* the system, and dW is the small work *done by* the system. These sign conventions for W and Q reflect the historical emphasis on engines, not refrigerators, while thermodynamics was being developed. This process is illustrated schematically in Fig. 2.1a. The first law says that a change in the energy of the system can occur only through energetic interaction of the system with its surroundings—the exchange of work or heat (or both) between the system and its surroundings.

$\mathcal{E}$ measures energy stored in the system, by means of internal storage mechanisms such as flywheels or springs or massive objects that are lifted in a gravitational field, by means of the system's overall kinetic energy of motion through space, and also usually including mechanisms that are beyond the scope of classical physics: molecular kinetic energy, molecular rotational energy, and molecular internal vibrational energy. The energy of a system is a unique function of the state of the system, independent of the history by which the system was brought to that state.

The work dW includes all kinds of mechanical, electrical, and magnetic work, but of most interest in thermodynamics is pressure–volume work

$$dW = p\,dV \tag{2.2}$$

done by the system when its volume increases by dV against a pressure p. Work is not a state variable of the system, i.e., it makes no sense to try to talk about the "work content" of the system, and the work spent taking the system from one state to another state depends on the details of the path followed by the process—not simply on the beginning and ending states of the process.

Heat dQ is the energy that flows in response to temperature differences. In the simplest case of conduction through an isotropic material, the rate of heat transfer $dQ/dt \equiv \dot{Q}$ is given by

$$\dot{Q} = -kA\frac{dT}{dx}, \tag{2.3}$$

where dT/dx is the temperature gradient (assumed to be parallel to the x direction here), k is the thermal conductivity of the material, and A is its cross-sectional area perpendicular to x. Heat, like work, is not a state variable of the system, i.e., it makes no sense to try to talk about any sort of unique "heat content" of a given state of the system.

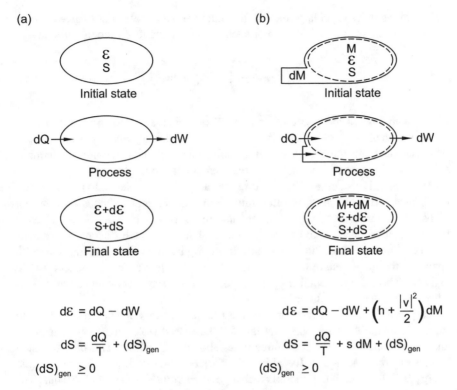

Fig. 2.1 One important key to understanding the laws of thermodynamics is careful definition of the system under consideration, its boundary, and what crosses that boundary. (**a**) A closed system: A system of fixed mass, including such disparate examples as 1 kg of air contained in a cylinder, 1 kg of water (partly liquid, partly ice) contained in a glass, and a television set. The *solid line* defines the boundary of the system. (**b**) An open system: A system whose boundary, marked by the *dashed line*, allows mass dM to enter (or leave), including examples such as a rocket engine and a living body. As described in the text, knowledge of the laws for the fixed-mass system bounded by the *solid line* yields the laws for the open system indicated by the *dashed line*. For the first law, ignore S; for the second law, ignore $\mathcal{E}$

History confirms the subtlety of these concepts. As recently as 200 years ago, it was believed that heat was essentially an invisible conserved fluid, called "caloric," whose flow between bodies was driven by temperature differences, with the "heat content of a body" a meaningful quantity. In those days, it would have been difficult to invent Eq. (2.1), because caloric and energy were believed to be such unrelated properties that they were not even quantified with the same units. But experiments eventually showed that heat was actually a type of energy.

Sometimes the first law must be applied to a "system" consisting of a microscopic mass element of fluid. (A fluid can be either a gas or a liquid. This book focuses on gases.) A microscopic mass element is an amount of fluid that is small enough that all of it can be regarded as having essentially the same temperature T, pressure p, and vector velocity $\mathbf{v}$, yet large enough to contain enough molecules to give meaning to these "macroscopic" variables. For such a small element of fluid, the first law is

$$d\left(\epsilon + \frac{|\mathbf{v}|^2}{2}\right) = dq - dw. \tag{2.4}$$

Here the energy[2] of the "system" consists of only its internal energy per unit mass ϵ plus its kinetic energy per unit mass $|\mathbf{v}|^2/2$. The internal energy per unit mass ϵ, which accounts for the energy stored in molecular motions, is a unique function of p and T, independent of the history of the element of fluid.

Thus far this discussion has focused on a system of fixed mass, but in fluid mechanics and acoustics it is often much more convenient to focus on a system whose boundaries are fixed in space and through which mass flows. Thus, it is important to derive expressions of the first law for such "open" systems, as illustrated in Fig. 2.1b. The dashed line in the figure shows the boundary, fixed in space, of the open system under consideration. Thermodynamic engineers usually call such a boundary a "control surface," and the open system contained within such a boundary a "control volume."

The derivation of an expression for the first law for this open system starts with Eq. (2.1), which expresses the first law for the closed (fixed-mass) system that is enclosed by the solid line at all three times shown in Fig. 2.1b. This closed system always has mass $M + dM$. Its initial energy is $\mathcal{E} + (\epsilon + |\mathbf{v}|^2/2)\,dM$, and its final energy is $\mathcal{E} + d\mathcal{E}$, so it experiences a change in energy $d\mathcal{E} - (\epsilon + |\mathbf{v}|^2/2)\,dM$. This change in energy is accomplished by a process involving addition of heat dQ and two work transfers, one being the unspecified work dW done by the system and the second being the pressure–volume work $p\,dM/\rho$ associated with pushing the mass dM, whose volume is dM/ρ, into the open system. Hence, Eq. (2.1) for the closed system shows that the first law for the closed system is

$$d\mathcal{E} - \left(\epsilon + \frac{|\mathbf{v}|^2}{2}\right) dM = dQ - \left(dW - p\frac{dM}{\rho}\right). \tag{2.5}$$

To obtain an expression for the change in energy $d\mathcal{E}$ of the open system, Eq. (2.5) is rearranged with $d\mathcal{E}$ alone on the left side, yielding

$$d\mathcal{E} = dQ - dW + \left(\epsilon + \frac{p}{\rho} + \frac{|\mathbf{v}|^2}{2}\right) dM, \tag{2.6}$$

which is the first law for an open system. Note that the energy added to the open system when mass dM flows into the open system is *not* simply the energy $(\epsilon + |\mathbf{v}|^2/2)\,dM$ contained *within* the mass dM. The p/ρ term shows that addition of the

---

[2]Gravity will usually be neglected in this book.

mass dM adds even more energy, which comes from the pressure–volume work that must be done on the system to push the mass dM into the system.

The combination

$$h = \epsilon + p/\rho, \tag{2.7}$$

which is called the enthalpy per unit mass of the fluid, occurs frequently in thermodynamics and fluid mechanics. Hence, the first law for macroscopic open systems is usually written

$$d\mathcal{E} = dQ - dW + \left(h + \frac{|\mathbf{v}|^2}{2} \right) dM, \tag{2.8}$$

showing how the energy change within a control volume results from flows of heat, work, enthalpy, and kinetic energy.

2.1.2   The Second Law

In principle, mechanical energy can be converted freely among gravitational potential energy, kinetic energy, rotational kinetic energy, and the energy stored in a spring, with the only fundamental constraint being the first law: The total energy $\mathcal{E}$ is constant. The same complete freedom does not exist when heat energy is also involved. The second law of thermodynamics constrains the conversion between heat and other forms of energy.

These points are illustrated in Figs. 2.2 and 2.3. Figure 2.2 shows a purely mechanical system in which energy is exchanged among various forms, with the total energy constant. Initially, the spring is held compressed (by a latch, not shown) a distance Δx and contains all the energy, $\mathcal{E}_{\text{initial}} = \frac{1}{2} K (\Delta x)^2$, in the system. When the latch is released, it launches the ball of mass M upward. As the ball flies upward, its velocity u at any height x can be calculated using conservation of energy,

$$\mathcal{E}(x) = \mathcal{E}_{\text{initial}} , \tag{2.9}$$

$$\frac{1}{2} M u^2 + M g x = \frac{1}{2} K (\Delta x)^2 , \tag{2.10}$$

where g is the acceleration of gravity. The height x_{max} at which the ball comes momentarily to rest is given by

$$M g x_{\text{max}} = \frac{1}{2} K (\Delta x)^2 . \tag{2.11}$$

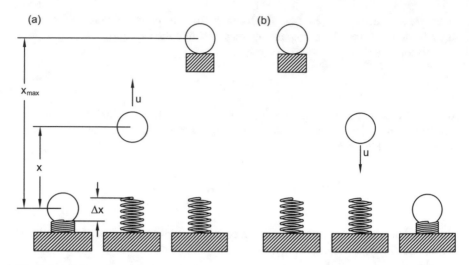

Fig. 2.2 (**a**) When the latch (not shown) holding a compressed spring K beneath a ball with mass M is released, the ball is launched upwards, with speed u depending on height x. When the ball comes to rest at height x_{max} at the peak of its trajectory, it is caught by a shelf. The total energy $\mathcal{E}$ begins as energy of compression $\frac{1}{2}K(\Delta x)^2$, later comprises the sum of kinetic and gravitational potential energies $\frac{1}{2}Mu^2 + Mgx$, and ends up stored as gravitational potential energy Mgx_{max}. (**b**) The process can be reversed, so that the gravitational potential energy is converted first to kinetic energy and finally to energy of compression in the spring

If the mass is caught and held at that height by quickly sliding a shelf under it, all of the energy that was initially stored as the spring's energy of compression is stored instead as gravitational potential energy. All these energy conversions are reversible, as illustrated in Fig. 2.2b: Releasing the mass from height x_{max} results in the conversion of gravitational potential energy first into kinetic energy and finally into energy stored as compression in the spring. Although real hardware might suffer from imperfections (e.g., air resistance, or friction in the latch that holds the spring compressed) preventing perfect conversion of energy among these three forms, there are no *fundamental* limitations to this "reversible" conversion other than the first law.

When heat is involved, the situation is fundamentally different, as illustrated in Fig. 2.3. The complete conversion of mechanical energy to heat is commonly observed, as in the example of Fig. 2.3a, which illustrates work being expended trying to drill a cannon barrel with a hopelessly dull drill. For $d\mathcal{E}/dt = 0$, the first law becomes $\dot{W} = \dot{Q}$, i.e.,

$$M_{weight}gu = \dot{M}_{water}c_{water}(T_H - T_0),\qquad(2.12)$$

which gives the increase $T_H - T_0$ in water temperature as a function of the water's mass flow rate $\dot{M}_{water}$, its heat capacity per unit mass, c_{water}, and the speed of descent u of the falling weight $M_{weight}g$. The first law would allow the reverse process, the

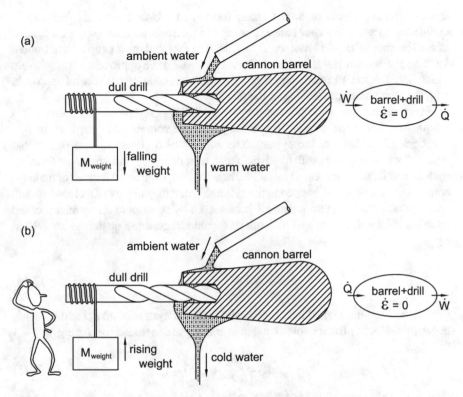

Fig. 2.3 (**a**) Count Rumford thought about heat and work while watching a cannon barrel being drilled. In today's terminology, consider "the system" to be the metal parts: the cannon barrel plus the drill bit. A source of work turns the drill bit, delivering work to the system, and a stream of water carries heat away from the system, keeping the system at a constant temperature. When the drill is very dull, it cuts very little, so the system changes negligibly with time; in particular, the energy $\mathcal{E}$ of the system is essentially constant. Hence, the process can be represented schematically by the diagram at the right, with the first law showing that $\dot{Q} = \dot{W}$. (**b**) The opposite process, in which a system simply absorbs heat from a water stream and produces work, has never been observed, even though it is allowed by the first law. It is forbidden by the second law; it would require reversing an irreversible process

direct conversion of heat into mechanical energy illustrated in Fig. 2.3b, to occur as well, subject only to

$$M_{\text{weight}} g u = \dot{M}_{\text{water}} c_{\text{water}} (T_0 - T_C), \tag{2.13}$$

but such a process has never been observed: Heat cannot be converted simply and directly into mechanical energy, whether by a dull drill and cannon barrel or by any other conceivable system. The second law describes this limitation quantitatively.

Introductory textbook discussions of the second law tend to follow any of three approaches: historical, beginning with something like Fig. 2.3 and proceeding through the arguments of Sadi Carnot; formal, in which a minimal postulate is claimed and its consequences are explored; and microscopic, based on a discussion of the statistics of random molecular motion. The fact that most people have trouble learning the second law indicates that none of these three approaches is intuitively satisfying. The brief review here follows a casual version of the second approach, with the next paragraph serving as the postulate.[3]

Every system has an important property called its entropy S. The entropy is a unique function of the state of the system, just as the volume and energy are unique functions of the state of the system. The entropy of a closed system (recall that "closed" means a system with fixed mass) can be reduced by removing heat from the system, and it can be increased by adding heat to the system or by letting (or making) something "irreversible" happen to the system. When the entropy of a closed system with spatially uniform temperature T is increased by an amount dS by means of the addition of heat dQ to the system from its surroundings, these quantities are related by

$$dS = \frac{dQ}{T}. \tag{2.14}$$

(The same formula holds if heat is removed from the system, with dQ and dS being negative numbers.) Irreversible changes in entropy are included by writing

$$dS = \frac{dQ}{T} + (dS)_{\text{gen}}, \tag{2.15}$$

$$(dS)_{\text{gen}} \geq 0, \tag{2.16}$$

where $(dS)_{\text{gen}}$ is entropy generated by irreversible processes. Equations (2.15) and (2.16) are the second law for a closed system with spatially uniform temperature T.

Figure 2.3a is called an irreversible process, because it never goes in reverse as in Fig. 2.3b. The irreversible process involved in Fig. 2.3a is friction between the dull drill and the cannon barrel. Friction is irreversible because it turns work into heat, but never turns heat into work. Viscous flow of a fluid is another example of an irreversible process. The system comprising the cannon barrel and the dull drill

---

[3] We all have difficulty learning the second law, because the concept of entropy is so non-intuitive, even though an object's entropy is just as well defined as an object's weight or altitude or volume. Humans live in a three-dimensional, mostly visual world, in which that property of a system called its volume has "obvious" meaning. As we become educated, learning to calculate the volume of objects first by multiplying three linear dimensions, and later by integral calculus, we continue to rely on our natural, intuitive feeling for volume. Energy is less intuitive, but early education with the various forms of mechanical energy, and play with bicycles, swings, etc., impart some energy intuition to us, providing a background for incorporating heat energy into our mental picture relatively easily. Entropy, however, must be learned without benefit of prior hands-on or visual experience.

is described in the vocabulary of Eqs. (2.15) and (2.16) by saying that the state of the system does not change in time, so its entropy S is constant and $\dot{S} \equiv dS/dt = 0$. The temperature of the system is T_0. Heat is removed from the system by the water stream at a rate dQ/dt equal to $M_{\text{weight}}gu$ according to Eq. (2.12). The system also experiences an irreversible rate of increase of entropy $\dot{S}_{\text{gen}}$ due to friction. Dividing the second law in Eqs. (2.15) and (2.16) by dt relates these quantities:

$$\dot{S}_{\text{gen}} = \frac{dQ/dt}{T_0} = \frac{M_{\text{weight}}gu}{T_0} \geq 0, \tag{2.17}$$

with the inequality showing that the process of Fig. 2.3b can never occur.

Thermally insulated systems, including the biggest system of all—the entire universe—have $dQ = 0$, so that the grandest expression of the second law is simply

$$\dot{S}_{\text{universe}} \geq 0. \tag{2.18}$$

Clausius, who coined the word "entropy," despaired [54] at this discovery in the 1860s, because it shows that the time evolution of the entire universe is and must be irreversible. Overall, the universe is aging irrevocably, due to irreversible processes.

An open system can experience additional entropy changes by means of the entropy carried by mass entering or leaving the system. The second law for an open system can be derived in the same way as the first law for an open system was derived above, by considering the injection of a small mass dM into the open system illustrated in Fig. 2.1b and changing the point of view from the closed system (which includes the small mass throughout the process) to the open system (which includes the small mass only at the end of the process). Applying Eqs. (2.15) and (2.16) to this composite closed system yields

$$dS = \frac{dQ}{T} + s\,dM + (dS)_{\text{gen}}, \tag{2.19}$$

$$(dS)_{\text{gen}} \geq 0, \tag{2.20}$$

where s is entropy per unit mass. These equations express in symbols the fact that the entropy of an open system changes in response to heat flow (of either sign), mass transfer (of either sign), and irreversible processes (zero or positive).

2.2 Laws of Fluids

Two of the results of fluid mechanics, the continuity equation and the momentum equation, are needed throughout acoustics. Thermoacoustics also requires the fluid-mechanical expressions of the first and second laws. This section reviews these results, in anticipation of using simpler versions of them later. Some of these results are tedious to derive, especially with respect to keeping track of the viscous stress tensor, so these discussions will not be rigorous derivations, but only plausible explanations of the results. The rigorous derivations can be found in advanced fluid-mechanics textbooks [46, 53].

Each equation in this section expresses the time derivative of one variable in terms of the spatial derivatives of all the variables. The mass equation, usually called the continuity equation, gives $\partial \rho / \partial t$. The momentum equation gives $\partial \mathbf{v} / \partial t$. (The name is suitable because velocity $\mathbf{v}$ is the momentum per unit mass. As shown in Exercise 2.7, the momentum equation is sometimes rewritten in terms of the time derivative of the momentum per unit volume $\rho \mathbf{v}$.) The first law gives $\partial \left(\epsilon + |\mathbf{v}|^2 / 2 \right) / \partial t$, and the second law gives $\partial s / \partial t$. In all cases, the partial derivative $\partial / \partial t$ indicates the time derivative at a fixed location through which fluid may move.

Although the macroscopic control volumes introduced above are very useful, the derivation of the differential equations of fluid mechanics requires consideration of a microscopic control volume: a cube centered on location x, y, z with edge lengths dx, dy, dz.

2.2.1   Continuity (Mass)

The continuity equation expresses conservation of mass for a microscopic control volume in a fluid. The simple idea behind the equation is that the mass inside a cubical volume $dx \, dy \, dz$, as shown in Fig. 2.4, can only change as a result of net inflow or outflow through the six faces of the cube. The mass in the cube is $\rho \, dx \, dy \, dz$, so this conservation law can be written

$$d\rho \, dx \, dy \, dz = \sum dM, \qquad (2.21)$$

with the sum $\sum$ over the six faces of the cube. Dividing Eq. (2.21) by dt expresses the right side as the sum of six mass flow rates dM/dt. At the center of the cube, the mass-flow density is $\rho \mathbf{v}$. If $\rho \mathbf{v}$ is spatially uniform, then obviously there is no net mass deposited in the microscopic control volume, because whatever mass might go in the left side of the control volume must also go out the right side, whatever might go in the top must also go out the bottom, etc. Hence, gradients in $\rho \mathbf{v}$ must be responsible for changes $d\rho$ in the mass of fluid in the cube. To proceed, a three-dimensional Taylor-series expansion about the center of the cube is used to find the mass flow crossing each face. For example, the mass flow rate into the left face is

$$\left[\rho u - \frac{d}{dx} (\rho u) \frac{dx}{2} \right] dy \, dz, \qquad (2.22)$$

where u is the x component of $\mathbf{v}$. Similarly, v and w are the y and z components, respectively, of $\mathbf{v}$. When the sum of all six terms is formed, the naked ρu, ρv, and ρw

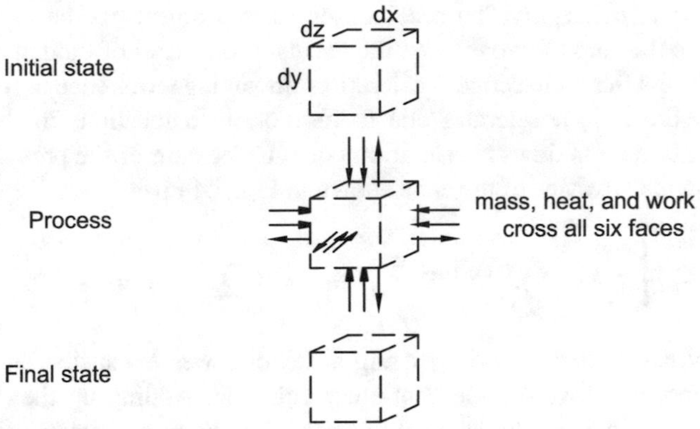

Initial state	
Process	mass, heat, and work cross all six faces
Final state	

Mass: $\quad \dfrac{\partial \rho}{\partial t} + \nabla \cdot (\rho v) = 0$

Momentum: $\quad \rho\left[\dfrac{\partial v}{\partial t} + (v \cdot \nabla)v \right] = -\nabla p + \nabla \cdot \sigma'$

Energy: $\quad \dfrac{\partial}{\partial t}\left(\rho \epsilon + \dfrac{1}{2}\rho |v|^2 \right) = -\nabla \cdot \left[\left(\rho h + \dfrac{1}{2}\rho |v|^2 \right)v - k\nabla T - v \cdot \sigma' \right]$

Entropy: $\quad \rho T\left[\dfrac{\partial s}{\partial t} + (v \cdot \nabla)s \right] = \nabla \cdot k\nabla T + (\sigma' \cdot \nabla)\cdot v$

Fig. 2.4 Illustration useful for the derivation of laws for microscopic fluid "systems." The microscopic control volume is a cubical volume $dx\,dy\,dz$ fixed in space through which fluid flows. The mass flux through each face of the cube is relevant to the continuity equation. The force on each face is relevant to the momentum equation. Mass, heat, and work can cross all six faces of the cube, changing the mass, momentum, energy, and entropy within the cube's boundary. The most useful equations resulting from these analyses are shown

terms cancel, leaving only the derivative terms. Dividing by $dx\,dy\,dz$ and expressing the result in the compact vector-calculus notation yields the well-known continuity equation,

$$\frac{\partial \rho}{\partial t} + \nabla \cdot (\rho \mathbf{v}) = 0, \qquad (2.23)$$

expressing conservation of mass in a fluid.

2.2.2 Momentum

The fluid momentum equation, also known as the Navier-Stokes equation, expresses one of Newton's laws: The mass of an object times its acceleration is the sum of the forces on the object. For an "object" consisting of a microscopic cubical element of fluid, the mass is $\rho\,dx\,dy\,dz$. The acceleration of this object can be expressed mathematically as the sum of two effects: the acceleration $\partial \mathbf{v}/\partial t$ of fluid at a given location in space, and the "convective acceleration" involving terms such as $u\,du/dx$, which tells how the fluid accelerates due to its motion to locations of different velocity. Writing this mass times acceleration equal to the sum of the pressure and viscous forces on the six faces of the cube shown in Fig. 2.4 gives

$$(\rho\,dx\,dy\,dz)\left[\frac{\partial \mathbf{v}}{\partial t} + (\mathbf{v}\cdot\nabla)\,\mathbf{v}\right] = \sum F_{\text{pressure}} + \sum F_{\text{viscous}}. \tag{2.24}$$

Adding up the pressure forces is easy, using a Taylor-series expansion about the center of the cube, as above for the continuity equation. Adding up the viscous forces involves tedium beyond the scope of this book, but the rigorous derivation can be found in advanced fluid-mechanics textbooks [46, 53]. The final result expresses Eq. (2.24) as

$$\rho\left[\frac{\partial \mathbf{v}}{\partial t} + (\mathbf{v}\cdot\nabla)\,\mathbf{v}\right] = -\nabla p + \nabla\cdot\boldsymbol{\sigma}', \tag{2.25}$$

where $\boldsymbol{\sigma}'$ is the nine-component viscous stress tensor. Two of its components are

$$\sigma'_{xx} = \mu\left(2\frac{\partial u}{\partial x} - \frac{2}{3}\nabla\cdot\mathbf{v}\right) + \zeta\nabla\cdot\mathbf{v}, \tag{2.26}$$

$$\sigma'_{xy} = \mu\left(\frac{\partial u}{\partial y} + \frac{\partial v}{\partial x}\right), \tag{2.27}$$

where μ is the viscosity and ζ is the bulk viscosity. The other seven components of $\boldsymbol{\sigma}'$ can be obtained from Eqs. (2.26) and (2.27) by switching among x, y, and z, and u, v, and w, in the obvious way.

Often gradients in the viscosities can be neglected, and often fluid motion can be regarded as essentially incompressible when considering momentum effects. With these approximations, Eq. (2.25) simplifies to

$$\rho\left[\frac{\partial \mathbf{v}}{\partial t} + (\mathbf{v}\cdot\nabla)\,\mathbf{v}\right] = -\nabla p + \mu\nabla^2\mathbf{v}. \tag{2.28}$$

This approximate momentum equation is more than sufficient for most thermoacoustics problems, with the notable exception of some streaming phenomena (see Chap. 7) for which gradients in viscosity cannot be neglected.

2.2.3   Energy

Next, another expression of the first law of thermodynamics is introduced, one for a microscopic stationary volume in a fluid—a microscopic control volume—into which or through which fluid can flow. Like the momentum equation above, this formulation is subtle enough that the discussion here will not be a rigorous derivation, but only a plausible explanation of the result. The rigorous derivation can be found in advanced fluid-mechanics textbooks [46, 53].

The derivation starts from Eq. (2.8) applied to the microscopic volume $dx\,dy\,dz$ shown in Fig. 2.4 through which the fluid flows. As discussed above near Eq. (2.4), the energy of the fluid in that volume is

$$\left(\rho\epsilon + \frac{1}{2}\rho\,|\mathbf{v}|^2\right) dx\,dy\,dz, \tag{2.29}$$

and according to Eq. (2.8) the change

$$d\left(\rho\epsilon + \frac{1}{2}\rho\,|\mathbf{v}|^2\right) dx\,dy\,dz \tag{2.30}$$

in that energy must be equal to

$$\sum dQ - \sum dW + \sum \left(h + \frac{|\mathbf{v}|^2}{2}\right) dM, \tag{2.31}$$

where each of the sums $\sum$ has six terms, one for each face of the cubic microscopic volume $dx\,dy\,dz$. The difficult part of the derivation, which is only outlined here, is the bookkeeping necessary to evaluate these sums.

For example, the first term (the sum of heats) is made up of six pieces, each a form of the heat-flux density $-k\nabla T$, which is the microscopic, vector version of Eq. (2.3). Gradients of heat-flux density are responsible for the net heat delivery to the control volume, in the same way that gradients in mass-flux density are responsible for net mass accumulation in the derivation of the continuity equation. With this idea in mind, the six terms in the heat sum can be written as a three-dimensional Taylor-series expansion about the midpoint of the cube, and added up, as was mass-flux density in the derivation of the continuity equation above. Keeping track of these details is tedious, but the result is simple: The net rate at which heat flows into the control volume is

$$\sum \frac{dQ}{dt} = -\nabla \cdot (k\nabla T)\,dx\,dy\,dz. \tag{2.32}$$

Similar bookkeeping can be done on the other two sums in Eq. (2.31). The mass-flow terms dM involve the mass-flux density $\rho\mathbf{v}$. The work terms dW are due only to viscous shear forces on the faces of the cube, because work associated with pressure forces normal to each cube face is already taken into account in the $h\,dM$ terms.

When all these details are properly accounted for, the result [46] of combining (2.30) and (2.31) is

$$\frac{\partial}{\partial t}\left(\rho\epsilon + \frac{1}{2}\rho\,|\mathbf{v}|^2\right) = -\nabla\cdot\left[-k\nabla T - \mathbf{v}\cdot\boldsymbol{\sigma}' + \left(\rho h + \frac{1}{2}\rho\,|\mathbf{v}|^2\right)\mathbf{v}\right]. \quad (2.33)$$

The quantity inside the square brackets on the right side is the energy flux density, because its divergence causes change in the energy per unit volume. Although the details of the rigorous derivation of Eq. (2.33) are much more complicated than necessary for an introductory course in thermoacoustics, the reader can at least appreciate that this expression comes directly from the first law: conservation of energy.

2.2.4   Entropy

In a microscopic control volume as shown in Fig. 2.4, the entropy is given by $\rho s\,dx\,dy\,dz$, where s is the entropy per unit mass. Then Eqs. (2.19) and (2.20) express the second law as

$$d\,(\rho s)\,dx\,dy\,dz = \sum\frac{dQ}{T} + \sum s\,dM + \rho\,(ds)_{\text{gen}}\,dx\,dy\,dz, \quad (2.34)$$

$$(ds)_{\text{gen}} \geq 0. \quad (2.35)$$

Dividing by $dt\,dx\,dy\,dz$ and keeping track of the sums as usual yields

$$\frac{\partial}{\partial t}\,(\rho s) = \nabla\cdot\left(\frac{k\nabla T}{T}\right) - \nabla\cdot(\rho\mathbf{v}s) + \rho\dot{s}_{\text{gen}}, \quad (2.36)$$

$$\dot{s}_{\text{gen}} \geq 0. \quad (2.37)$$

Using the product rule to expand the derivatives, subtracting s times the continuity equation from the result, and multiplying by T yields

$$\rho T\left(\frac{\partial s}{\partial t} + \mathbf{v}\cdot\nabla s\right) = \nabla\cdot k\nabla T - k\frac{|\nabla T|^2}{T} + \rho T\dot{s}_{\text{gen}}, \quad (2.38)$$

$$\dot{s}_{\text{gen}} \geq 0. \quad (2.39)$$

This final pair of equations can be taken as the expression of the second law for a fluid.

A similar-looking equation can be derived by combining the energy equation, momentum equation, and continuity equation, and using thermodynamic identities such as $h = \epsilon + p/\rho$ as needed. The derivation [46] is tedious but straightforward. The result, which is called the general equation of heat transfer for fluids, is

$$\rho T \left(\frac{\partial s}{\partial t} + \mathbf{v} \cdot \nabla s \right) = \nabla \cdot k \nabla T + \left(\boldsymbol{\sigma}' \cdot \nabla \right) \cdot \mathbf{v}. \tag{2.40}$$

Note that the derivation of Eq. (2.40) does not seem to depend on the second law. However, comparing this result to Eq. (2.38) shows that

$$\dot{s}_{\text{gen}} = \frac{k \, |\nabla T|^2}{\rho T^2} + \frac{1}{\rho T} \left(\boldsymbol{\sigma}' \cdot \nabla \right) \cdot \mathbf{v}. \tag{2.41}$$

Tedious manipulation [46] of the second term using the definition of $\boldsymbol{\sigma}'$ in Eqs.(2.26) and (2.27) leads to

$$\dot{s}_{\text{gen}} = \frac{k \, |\nabla T|^2}{\rho T^2} + \frac{\mu}{\rho T} \Phi + \frac{\zeta}{\rho T} \left(\nabla \cdot \mathbf{v} \right)^2, \tag{2.42}$$

where the viscous dissipation function

$$\Phi = \frac{1}{2} \sum_{i=x,y,z} \sum_{j=x,y,z} \left(\frac{\partial v_i}{\partial x_j} + \frac{\partial v_j}{\partial x_i} - \frac{2}{3} \delta_{ij} \nabla \cdot \mathbf{v} \right)^2, \tag{2.43}$$

and where $\delta_{ij} = 1$ when $i = j$ and $\delta_{ij} = 0$ when $i \neq j$. The fact that all terms on the right side of Eq. (2.42) are obviously ≥ 0, so that $\dot{s}_{\text{gen}}$ must be ≥ 0, can be regarded either as a verification that the continuity, momentum, and energy equations are consistent with the second law, or as a proof that $k \geq 0$, $\mu \geq 0$, and $\zeta \geq 0$ for all fluids!

2.3 Ideal Gases

This section summarizes ideal-gas properties and relations used throughout the text.

2.3.1 Thermodynamic Properties

An ideal gas is a fluid that obeys these two fundamental equations for pressure p and internal energy per unit mass ϵ:

$$p = \rho \Re T, \tag{2.44}$$

$$\epsilon = \frac{\Re T}{\gamma - 1}, \tag{2.45}$$

where T is temperature and ρ is density. For present purposes, these two equations can be considered as experimental facts, but they can also be derived from the statistics of molecular motion. The gas constant $\Re$, which is different for different gases, is related to the universal gas constant $\Re_{\text{univ}} = 8.314$ J/mol·K via the molar mass m:

$$\Re = \Re_{\text{univ}} / m. \tag{2.46}$$

The constant γ, usually called the ratio of specific heats (see below), is 5/3 for monatomic gases and is typically near 7/5 for common diatomic gases near ambient temperature. Except for monatomic gases, γ can be temperature dependent. The table at the end of this section gives values of m and γ for several common gases. All the other thermodynamic properties of gases (which do not include the transport properties—viscosity and thermal conductivity—discussed in the next subsection) can be obtained from Eqs. (2.44) and (2.45) and the values of m and γ:

The thermal expansion coefficient

$$\beta \equiv -\frac{1}{\rho} \left(\frac{\partial \rho}{\partial T} \right)_p = \frac{1}{T} \tag{2.47}$$

and the isothermal bulk modulus

$$B_T \equiv \rho \left(\frac{\partial p}{\partial \rho} \right)_T = p \tag{2.48}$$

are obtained by straightforward differentiation of Eq. (2.44). The isothermal compressibility $1/p$ is simply the inverse of the isothermal bulk modulus.

The isentropic[4] bulk modulus

$$B_s \equiv \rho \left(\frac{\partial p}{\partial \rho} \right)_s \tag{2.49}$$

is slightly more difficult to obtain, because an isentropic density change $d\rho$ is accompanied by both a pressure change dp and a temperature change dT. Differentiating Eq. (2.44) shows that these changes are interrelated by

$$dp = \rho \Re \, dT + \Re T \, d\rho. \tag{2.50}$$

---

[4]"Isentropic" means "at constant entropy." "Adiabatic" refers to conditions of zero heat transfer across a control surface. Hence, a reversible *and* adiabatic process is isentropic.

Neglecting kinetic energy, the first law says that the increase of internal energy per unit mass must equal the heat added per unit mass minus the work done by unit mass:

$$d\epsilon = dq - dw. \tag{2.51}$$

For an isentropic change, this becomes

$$(d\epsilon)_s = -dw, \tag{2.52}$$

with the change of internal energy obtained from Eq. (2.45),

$$d\epsilon = \frac{\Re}{\gamma - 1} dT, \tag{2.53}$$

and the work per unit mass given by

$$dw = p\, d\left(\frac{1}{\rho}\right) = -\frac{p}{\rho^2} d\rho, \tag{2.54}$$

with $(1/\rho)$ the volume per unit mass. Combining Eqs. (2.50) and (2.52)–(2.54) in order to eliminate dw, $d\epsilon$, and $d\rho$, while using Eq. (2.44) freely, quickly yields $dp = \gamma \Re T\, d\rho$ under these isentropic conditions, so that the isentropic bulk modulus is

$$B_s \equiv \rho\left(\frac{\partial p}{\partial \rho}\right)_s = \gamma p. \tag{2.55}$$

The isentropic compressibility $1/\gamma p$ is the inverse of the isentropic bulk modulus. The speed of sound, introduced more fully in the next chapter, is

$$a \equiv \sqrt{(\partial p/\partial \rho)_s} = \sqrt{\frac{\gamma p}{\rho}} = \sqrt{\gamma \Re T}, \tag{2.56}$$

and gives the speed with which compressional waves propagate through a gas under ordinary, isentropic conditions.

At constant volume (i.e., constant density), $dw = 0$ in Eq. (2.51), so the specific heat at constant volume (also called the isochoric specific heat) of the ideal gas is obtained by straightforward differentiation of Eq. (2.45):

$$c_v \equiv \left(\frac{dq}{dT}\right)_v = \frac{1}{\gamma - 1}\Re. \tag{2.57}$$

The constant-pressure (i.e., isobaric) specific heat is slightly more difficult to obtain, because $dw \neq 0$ in Eq. (2.51) at constant pressure. As the temperature is raised by dT at constant pressure, ϵ increases by $d\epsilon = \Re\, dT/(\gamma - 1)$ according to

Eq. (2.53), ρ changes by $d\rho = -(\rho/T)\,dT$ according to Eq. (2.44), and hence $dw = \Re\,dT$ according to Eq. (2.54). Combining these intermediate results by means of Eq. (2.51) yields the isobaric specific heat of an ideal gas:

$$c_p \equiv \left(\frac{dq}{dT}\right)_p = \left(\frac{d\epsilon}{dT}\right)_p + \left(\frac{dw}{dT}\right)_p = \frac{\gamma}{\gamma-1}\Re. \qquad (2.58)$$

Note that the ratio of specific heats $c_p/c_v = \gamma$, and this is also the ratio of the isentropic and isothermal bulk moduli. Thermodynamics shows that $c_p/c_v = B_s/B_T$ for all fluids, although the present derivation shows it only for an ideal gas.

The enthalpy h is defined in Eq. (2.7) as $h = \epsilon + p/\rho$. Combining Eqs.(2.44) and (2.45) easily yields the enthalpy of an ideal gas:

$$h = \frac{\gamma}{\gamma-1}\Re T = c_p T. \qquad (2.59)$$

The entropy s can be obtained through its fundamental definition

$$ds = \frac{dq}{T} \qquad (2.60)$$

and the first law of thermodynamics as expressed in Eq. (2.51). These can be combined to form

$$ds = \frac{d\epsilon}{T} + \frac{dw}{T} = \frac{\Re}{\gamma-1}\frac{dT}{T} - \Re\frac{d\rho}{\rho}, \qquad (2.61)$$

by using Eq. (2.45) to express $d\epsilon$ in terms of dT and Eqs. (2.54) and (2.44) to express dw in terms of $d\rho$. If γ is a constant, Eq. (2.61) can be integrated to obtain the entropy of the ideal gas:

$$s - s_0 = \frac{\Re}{\gamma-1}\ln\left(\frac{T}{T_0}\right) - \Re\ln\left(\frac{\rho}{\rho_0}\right), \qquad (2.62)$$

where the variables with subscript 0, together comprising only the constant of integration, indicate a reference state.

2.3.2 Transport Properties

Over a wide range in temperature, the viscosity and thermal conductivity of gases usually obey the power laws

$$\mu = \mu_0\,(T/T_0)^{b_\mu}, \qquad (2.63)$$

$$k = k_0\,(T/T_0)^{b_k}. \qquad (2.64)$$

For $T_0 = 300$ K, approximate values of μ_0, k_0, b_μ, and b_k are given in the table below for some gases of common interest in thermoacoustics.

	m (kg/mol)	γ	for $T_0 = 300$ K: μ_0 (kg/m·s)	b_μ	k_0 (W/m·K)	b_k
Air	0.02896	7/5	1.85×10^{-5}	0.76	0.026	0.89
Nitrogen	0.02801	7/5	1.82×10^{-5}	0.69	0.026	0.75
Helium	0.00400	5/3	1.99×10^{-5}	0.68	0.152	0.72
Neon	0.02018	5/3	3.2×10^{-5}	0.66	0.049	0.66
Argon	0.03995	5/3	2.3×10^{-5}	0.85	0.0180	0.84
Xenon	0.1313	5/3	2.4×10^{-5}	0.85	0.0058	0.84

2.3.3   Shortcuts

For quick and easy estimating of relevant thermoacoustic gas properties, it is useful to memorize a few ideal-gas identities,

$$p = \rho \Re T, \tag{2.65}$$

$$a^2 = \gamma \Re T, \tag{2.66}$$

$$c_p = \gamma \Re / (\gamma - 1), \tag{2.67}$$

$$\Re = \Re_{\text{univ}} / m, \tag{2.68}$$

and a few constants,

$$\Re_{\text{univ}} \simeq 8.3 \text{ J/mol·K}, \tag{2.69}$$

$$\mu \simeq \left(2 \times 10^{-5} \text{ kg/m·s}\right) \left[T/(300 \text{ K})\right]^{0.7} \text{ for most gases}, \tag{2.70}$$

$$\gamma \simeq 5/3 \text{ for monatomic gases or } 7/5 \text{ for air}, \tag{2.71}$$

$$\sigma \simeq 2/3 \text{ for pure gases, but lower for mixtures}, \tag{2.72}$$

in addition to the definitions $\sigma = \mu c_p / k$, $\lambda = a/f$, $\omega = 2\pi f$, $\delta_\nu = \sqrt{2\mu/\omega\rho}$, and $\delta_\kappa = \sqrt{2k/\omega\rho c_p}$ given in Eqs. (1.1)–(1.4). Remember that $\Re$ is the gas constant per unit mass (specific to each gas) and $\Re_{\text{univ}}$ is the universal gas constant.

2.3.4   Mixtures

If two gases a and b are mixed together, with mole fractions n_a and n_b respectively, the molar mass m of the mixture is

$$m = n_a m_a + n_b m_b \tag{2.73}$$

and the ratio of specific heats, γ, of the mixture is given by

$$\frac{1}{\gamma - 1} = \frac{n_a}{\gamma_a - 1} + \frac{n_b}{\gamma_b - 1}. \tag{2.74}$$

From these, all the other thermodynamic quantities can be calculated as described in Sect. 2.3.1 for pure gases. However, the transport properties of gas mixtures are more difficult to estimate [55], and both the thermodynamic and hydrodynamic equations become much more complicated [2, 46, 51] if diffusion of the components of the mixture relative to each other must be accounted for.

2.4   Some Consequences of the Laws

2.4.1   Carnot's Efficiency

A "thermal reservoir" is a source or sink of heat whose temperature changes only negligibly as heat is added or removed. Its essentially isothermal nature is usually imagined as being due to a large heat capacity, but it can also be due to some active agent that is external and irrelevant to the system under consideration.

The second law prohibits direct conversion of heat from a single thermal reservoir into work, as in the "impossible" illustration of Fig. 2.3b where the rapidly flowing water stream could be considered to be the single thermal reservoir. However, work can be created when two thermal reservoirs at different temperatures are involved, as illustrated schematically in Fig. 2.5a. Such a heat engine (sometimes known as a prime mover, or more simply as an engine) produces work W from heat Q_H absorbed from the high-temperature reservoir at T_H, rejecting waste heat Q_0 to the reservoir at the lower temperature T_0. Similarly, a refrigerator (see also "heat pump," Exercise 2.23) uses work to lift heat Q_C from a reservoir at one temperature T_C and reject waste heat Q_0 into another reservoir at temperature T_0.

The laws of thermodynamics place bounds on the efficiency of such devices. If the device is in steady state, the first law implies

$$Q_H = W + Q_0 \tag{2.75}$$

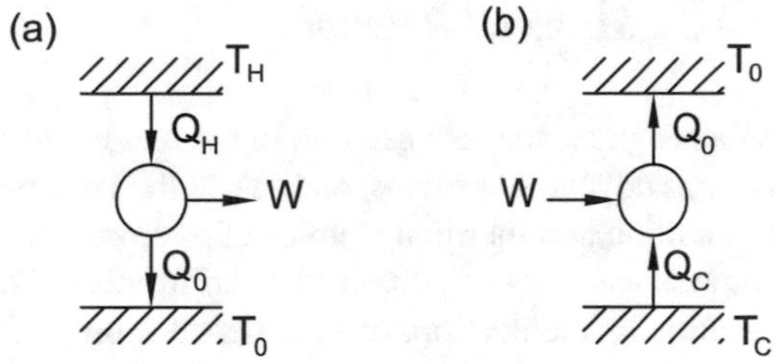

Fig. 2.5 A heat engine (**a**) and a refrigerator (**b**), showing energy flow directions and thermal reservoirs. The central circles represent the devices themselves

(a)

(b)

for the engine, and

$$Q_0 = W + Q_C \tag{2.76}$$

for the refrigerator. The second law is only slightly more subtle. If the device is in steady state, then its entropy doesn't change with time. Hence, for the engine Eqs. (2.15) and (2.16) can be combined to form

$$\frac{Q_0}{T_0} - \frac{Q_H}{T_H} \geq 0, \tag{2.77}$$

and similarly for the refrigerator

$$\frac{Q_0}{T_0} - \frac{Q_C}{T_C} \geq 0. \tag{2.78}$$

Combining Eq. (2.75) with Eq. (2.77) by eliminating what is usually the least interesting variable, Q_0, yields an inequality for the engine's efficiency. "Efficiency" is most often defined as "what you want, divided by what you have to spend to get it." For the engine, the efficiency η is the work produced by the engine divided by the heat supplied by the hot reservoir:

$$\eta = \frac{W}{Q_H} \leq \frac{T_H - T_0}{T_H}. \tag{2.79}$$

The temperature ratio on the right is called the Carnot efficiency, which bounds the actual efficiency of all engines. Similarly, for the refrigerator the coefficient of performance *COP* is the heat of refrigeration divided by the work consumed by the device:

$$COP = \frac{Q_C}{W} \leq \frac{T_C}{T_0 - T_C}, \tag{2.80}$$

with the inequality obtained by combining Eqs. (2.76) and (2.78), eliminating Q_0. The temperature ratio on the right is called the Carnot *COP*, which bounds the actual *COP* of all refrigerators.

2.4.2   *Maxwell Relations*

Most engines and refrigerators in the real world operate by carrying a fluid through
a thermodynamic process, and hence the laws presented in this chapter that govern
the motion and thermodynamics of microscopic elements of fluid are of paramount
importance. It is important to keep in mind that the thermodynamic "state" of a
small mass element dm of fluid depends on just two independent variables.[5] Often
the most natural independent variables to consider are p and T. Then all the other
thermodynamic state variables, such as ϵ, h, s, and ρ, are dependent functions of
those two independent variables, and are expressed as $\rho(p, T)$, $\epsilon(p, T)$, etc. However,
few problems are so simple that changing point of view and adopting a different pair
of variables as independent can be avoided; e.g., one might want to consider $h(\rho, T)$
instead of $h(p, T)$.

The Maxwell relations are one set of equations that are useful when changing
among thermodynamic variables. Maxwell relations arise because calculus shows
that, for a continuous function of two independent variables, the order of differentiation in mixed second derivatives is unimportant. For example,

$$\frac{\partial}{\partial S}\left(\left(\frac{\partial \mathcal{E}}{\partial V}\right)_S\right)_V = \frac{\partial}{\partial V}\left(\left(\frac{\partial \mathcal{E}}{\partial S}\right)_V\right)_S. \qquad (2.81)$$

The first law and the definitions of dQ and dW show that $d\mathcal{E} = T\,dS - p\,dV$, so that
$(\partial \mathcal{E}/\partial S)_V = T$ and $(\partial \mathcal{E}/\partial V)_S = -p$. Substituting these into Eq. (2.81) gives one of
the Maxwell relations:

$$-\left(\frac{\partial p}{\partial S}\right)_V = \left(\frac{\partial T}{\partial V}\right)_S. \qquad (2.82)$$

For a differential mass element of fluid, S becomes the entropy per unit mass s and
V becomes the volume per unit mass $1/\rho$, so the Maxwell relation (2.82) becomes

$$-\left(\frac{\partial p}{\partial s}\right)_V = \left(\frac{\partial T}{\partial (1/\rho)}\right)_S, \qquad (2.83)$$

which can be converted via calculus to

$$\left(\frac{\partial p}{\partial s}\right)_\rho = \rho^2 \left(\frac{\partial T}{\partial \rho}\right)_s. \qquad (2.84)$$

---

[5]With two independent variables here, we speak of a pure fluid. However, in a two-component
mixture one mole fraction must also be specified as an independent variable. Furthermore, the
three components of the velocity vector are additional independent variables, but they are typically
discussed separately because there is seldom reason to consider them anything but independent in
thermodynamics.

A second Maxwell relation is obtained by considering mixed second derivatives of h instead of ϵ. Since $h = \epsilon + p/\rho$ and $d\epsilon = dq - dw = T\,ds - p\,d(1/\rho)$,

$$dh = T\,ds + \frac{1}{\rho}\,dp, \tag{2.85}$$

so

$$\frac{\partial}{\partial s}\left(\left(\frac{\partial h}{\partial p}\right)_s\right)_p = \frac{\partial}{\partial p}\left(\left(\frac{\partial h}{\partial s}\right)_p\right)_s \tag{2.86}$$

yields the Maxwell relation

$$\left(\frac{\partial T}{\partial p}\right)_s = \left(\frac{\partial (1/\rho)}{\partial s}\right)_p = -\frac{1}{\rho^2}\left(\frac{\partial \rho}{\partial s}\right)_p. \tag{2.87}$$

Two other Maxwell relations, listed here without proof, are

$$\left(\frac{\partial s}{\partial p}\right)_T = \frac{1}{\rho^2}\left(\frac{\partial \rho}{\partial T}\right)_p, \tag{2.88}$$

$$\left(\frac{\partial p}{\partial T}\right)_\rho = -\rho^2\left(\frac{\partial s}{\partial \rho}\right)_T. \tag{2.89}$$

These relations are used to change from one pair of independent variables to another. For example, the transformation from $h(s,p)$ [as expressed in Eq. (2.85)] to $h(T,p)$ requires

$$ds = \left(\frac{\partial s}{\partial p}\right)_T dp + \left(\frac{\partial s}{\partial T}\right)_p dT \tag{2.90}$$

$$= -\frac{1}{\rho^2}\left(\frac{\partial \rho}{\partial T}\right)_p dp + \left(\frac{\partial s}{\partial T}\right)_p dT \tag{2.91}$$

$$= -\frac{\beta}{\rho}\,dp + \frac{c_p}{T}\,dT, \tag{2.92}$$

where a Maxwell relation is used to go from the first line to the second and the definitions of β and c_p are used to go from the second line to the third. Then combining Eqs. (2.85) and (2.92) gives

$$dh = c_p\,dT + \frac{1}{\rho}(1 - T\beta)\,dp, \tag{2.93}$$

the desired result.

All such relationships between the many variables of thermodynamics can of course be derived, but some of them [e.g., $(\partial h/\partial p)_\rho$] require a great deal of effort.[6] Fortunately, Bridgman's table [56], which requires only a single page in Bejan's book [45], displays all possible such partial derivatives in a compact, convenient format.

2.5 Exercises

2.1 Your mother's antique "grandfather" clock is powered by two slowly falling weights, which hang from chains that pass over sprockets that keep a steady, gentle torque on the gears as they turn. One of the weights powers the timekeeping mechanism, the other powers the chimes. In order to keep the clock operating continuously, mother must open the door to the clock, reach inside, and raise the clock's two 1-kg weights a distance of 0.5 m once every three days. About 10% of the energy stored in the weights is converted to acoustic energy that is radiated into the room (including "tick-tock" as well as chimes); the other 90% is used to overcome friction inside the clock. At what average rate does heat flow from the clock to the room? (The acceleration of gravity is $10\,\mathrm{m/s^2}$.) Formulate your answer by defining a "system" and discussing work delivered to the system, work delivered from the system, and energy stored in the system.

2.2 The discussion of Fig. 2.3 considered the system to be the dull drill plus the cannon barrel, which is a closed system because its mass does not change in time. Present a similar discussion in terms of the open system comprising the drill, the cannon barrel, and some water. Define the system boundary carefully, and discuss each term in the first and second laws for an open system.

2.3 Similar to Fig. 2.3, think of two other examples of irreversible processes, and sketch the impossible reverse processes that are allowed by the first law but prohibited by the second law. Define your systems clearly. Make one of your examples a closed system and the other an open system.

2.4 The elevator in a tall building is counterweighted and nearly frictionless, so that the empty elevator can be moved up and down with negligible expenditure of energy. Consider the whole building and its contents as a thermally insulated system. A 1000 kg cargo is placed in the elevator on the 10th floor, and a brake mechanism lowers the cargo and elevator to the ground floor 30 m below. How much does the energy of the system change? How much does the entropy of the system change? How much entropy is irreversibly created? Next, the cargo and elevator are raised

---

[6]Thermodynamics has a reputation as a difficult subject. This is mostly because there are so many variables, with different problems most easily done with different variables. The proper application of calculus to the transformation among all these variables is actually the principal challenge of thermodynamics.

back up to the 10th floor, using the elevator's electric motor whose power comes into the building via the electric power lines. The motor is 75% efficient; i.e., the work done lifting the cargo is 75% of the electric energy consumed by the motor. How much does the energy of the system change? How much does the entropy of the system change? How much entropy is irreversibly created? Do any of these answers depend on making a reasonable assumption about the temperature of the brake or motor? Can you avoid such an assumption by rewording the questions? After the lowering and raising are both complete, what must be removed from the building and its contents to return them to their initial state? How much of it must be removed?

2.5 In a particular turbulent flow of water down a particular pipe, the size of the smallest turbulent eddies is approximately 1 mm, so you might expect that a cube with edge lengths 0.01 mm is small enough to be considered a "microscopic element" with essentially uniform T, p, and $\mathbf{v}$. How many water molecules are in this cube? (The density of water is $\rho = 1$ gm/cm^3 and its molar mass is 18 gm. A mole is 6.02×10^{23}, just as a dozen is 12.)

2.6 Use Eq. (2.25) in the "laminar" limit, when $(\mathbf{v} \cdot \nabla) \mathbf{v}$ can be neglected, to show that flow of a viscous incompressible fluid through a long channel with uniform cross-sectional area obeys $\Delta p = ZU$, where Δp is the pressure difference driving the flow, U is the volume flow rate, and the flow impedance $Z = 128\mu \, \Delta x / \pi D^4$ for a circular channel of diameter D and length Δx, and $Z = 12\mu \, \Delta x / WD^3$ for a slit-shaped channel of length Δx, width W, and thickness $D \ll W$.

2.7 Add $\mathbf{v}$ times the continuity equation (2.23) to the momentum equation (2.25), in order to form an equation for the time derivative of the momentum per unit volume $\partial(\rho\mathbf{v})/\partial t$. Similarly, subtract the product of $(\epsilon + |\mathbf{v}|^2/2)$ and the continuity equation from the energy equation (2.33) to obtain $\partial \left(\epsilon + |\mathbf{v}|^2/2 \right)/\partial t$.

2.8 Water at room temperature flows through a 0.1-m-long converging nozzle at the end of a garden hose, so that the velocity in the nozzle is $u = (1 \text{ m/s}) \times [1 + (90 \text{ m}^{-1})x]$ where x is the distance from the entrance to the nozzle. Use Eq. (2.25) to predict the pressure necessary to force the water through the nozzle. Neglect viscosity.

2.9 Why doesn't the first law for a microscopic fluid element, Eq. (2.4), include rotational kinetic energy?

2.10 Show that the entropy equation for a fluid can be written as

$$\frac{\partial(\rho s)}{\partial t} = -\nabla \cdot \left(\rho s \mathbf{v} - \frac{k\nabla T}{T} \right) + \rho \dot{s}_{\text{gen}}. \tag{2.94}$$

What is the entropy-flux density in a fluid?

2.11 Calculate the Prandtl number for the gases in the table in Sect. 2.3.2 at 300 K.

2.12 For your favorite working gas, and your favorite temperature and pressure, look up γ, calculate the sound speed, density, and isobaric specific heat. Look up the viscosity and thermal conductivity. Pick your favorite frequency. Calculate λ, δ_κ, and δ_ν. Compare these lengths to various dimensions in your favorite piece of thermoacoustics hardware. Do you know the order of magnitude of $|\xi_1|$? (If you don't have a personal favorite gas, etc., then use air at 300 K and atmospheric pressure, 440 Hz, and dimensions in your office. Estimate $|\xi_1|$ using the fact that conversational acoustics has $|p_1| \simeq 10^{-6}$ bar.)

2.13 If the mole fraction of one component in a binary mixture is n_a, what is the mass fraction?

2.14 A mixture of helium and argon has a sound speed of 500 m/s at 300 K. What is the concentration of argon in the mixture?

2.15 How high must the frequency be in order to make $\delta_\nu \sim \lambda$ in a sound wave in air at ordinary atmospheric conditions?

2.16 What is the volume of 1 mole of an ideal gas at $p = 1$ bar and $T = 27\,°C$?

2.17 Thermodynamics review: Show that the ratio of isothermal compressibility to isentropic compressibility is the same as the ratio γ of isobaric to constant-volume specific heats.

2.18 Show that p/ρ^γ is constant for an isentropic change of state of an ideal gas. Derive the corresponding relationship between p and T.

2.19 Twelve grams of helium are at $p = 1$ bar and $T = 27\,°C$. The helium is isentropically expanded until its temperature falls to $-23\,°C$. What are the initial and final volumes? What is the final pressure? Compute the work done by the gas in two ways: by performing $\int p\,dV$ and by calculating the difference between the initial and final internal energies.

2.20 Show that Eq. (2.74) implies that the energies ϵ and h and the heat capacities c_p and c_v of the components of a mixture are additive, as if the energies of the two components of the mixture do not really affect one another.

2.21 The air from a glass tube was completely pumped out, and the tip of the tube was then fused shut. You break the tip of the tube, wait a second or two, and dip the broken end in a glass of water. Does the water rise into the tube, or do air bubbles come out of the tube, or do neither of these events occur? (after Bejan [45])

2.22 The initial state of a "system" consists of a mole of helium and a mole of argon, in separate tanks, each at $p = 1$ bar and $T = 27\,°C$. They are allowed to mix, and the mixture is brought to a final state of $p = 1$ bar and $T = 27\,°C$. Is the internal energy of the final state the same as that of the initial state?

2.23 The heat pump is similar to the refrigerator in Fig. 2.5, but its purpose is to deliver heat Q_H to a sink at high temperature T_H. It accomplishes this by using work W to pump heat from a source at ambient temperature T_0. The largest possible

Q_H/W is desired. Show that the first and second laws of thermodynamics require $Q_H/W \leq T_H/(T_H - T_0)$.

2.24 Describe a process that is neither reversible nor adiabatic, but is isentropic.

Chapter 3
Simple Oscillations

This chapter reviews simple oscillations and waves [5, 57–59], establishing the notation and background that will be used for thermoacoustic oscillations in all subsequent chapters. The chapter begins with the simple harmonic oscillator, and progresses to one-dimensional standing waves and traveling waves, which are the simplest building blocks for the more complicated waves of thermoacoustics.[1]

3.1 The Harmonic Oscillator and Complex Notation

The simplest oscillatory problem is the "simple harmonic oscillator." Consider the mass–spring system shown in Fig. 3.1a. One end of the spring is fixed to a rigid mount, and the other is attached to the mass M. The distance ξ describes the position of the mass as a function of time t, and $\xi = 0$ can be chosen as the mean position of the mass, i.e., the position of the mass when nothing is moving and the spring is relaxed.

The mass M times its acceleration $d^2\xi/dt^2$ must equal the sum of the forces on the mass. In Fig. 3.1a, the only force is $-K\xi$, the force exerted by the spring on the mass. Then

$$M\frac{d^2\xi}{dt^2} = -K\xi. \tag{3.1}$$

---

[1]Most of this chapter should seem very familiar to acousticians, who can probably read through it quickly, spending just enough effort to absorb the notation. Mechanical engineers and chemical engineers may be less familiar with oscillating systems, so they may need to invest more time here and work on the exercises at the end.

© Acoustical Society of America 2017
G.W. Swift, *Thermoacoustics*, DOI 10.1007/978-3-319-66933-5_3

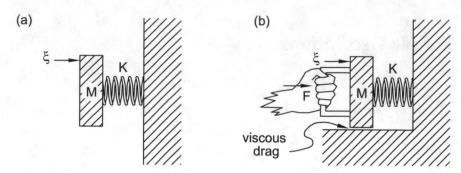

Fig. 3.1 A simple harmonic oscillator. (**a**) The mass M moves in response to the force exerted by a spring attached to it. The position of the mass is ξ, the spring constant is K, and the other end of the spring is attached to an immovable object. (**b**) In the damped, driven harmonic oscillator, three forces act on the mass: the spring exerts a force $-K\xi$, an external agent exerts a force $F(t)$, and laminar viscous drag exerts a force proportional to, and opposing, the velocity $d\xi/dt$

The solution to this simple differential equation can be written as

$$\xi(t) = \xi_a \cos(\omega t + \phi), \tag{3.2}$$

with $\omega = \sqrt{K/M}$ required by Eq. (3.1), and with two arbitrary constants ξ_a and ϕ.

This solution shows that the position of the mass oscillates, with a "sinusoidal" (i.e., sine or cosine) dependence on time. The trigonometric function has period 2π, so the temporal period of the oscillation is $2\pi/\omega$ and the frequency of the oscillation is

$$f = \omega/2\pi. \tag{3.3}$$

Because it occurs so frequently in oscillatory problems, ω is given a name of its own: the angular frequency of the oscillation. The amplitude of the oscillation ξ_a and the temporal phase of the oscillation ϕ depend on initial conditions (e.g., when and where the mass is initially released to start the oscillation).

3.1.1 Complex Notation

It is vital to become fluent with the notation to be used throughout the book for time-dependent variables such as the position of the oscillating mass in Fig. 3.1. When all time dependence is purely sinusoidal, at frequency f and angular frequency $\omega = 2\pi f$, variables such as position $\xi(t)$ *could* be written as in Eq. (3.2). However, it is much more convenient to rewrite Eq. (3.2) as

$$\xi(t) = \text{Re}\left[\xi_1 e^{i\omega t}\right], \tag{3.4}$$

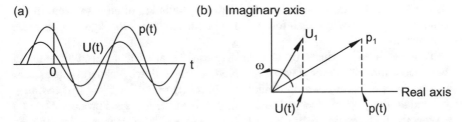

Fig. 3.2 Phasors show complex variables in the complex plane. (**a**) Two sinusoidally time-dependent variables $p(t)$ and $U(t)$, with U leading p by $30°$. (**b**) The phasor representation of these two variables can be imagined as two vectors whose magnitudes are the amplitudes of p and U, rotating counterclockwise at angular frequency ω. The actual, time-dependent values of the associated variables are the projections on the real axis of these rotating phasors

where ξ_1 is a complex number such that

$$|\xi_1| = \xi_a, \tag{3.5}$$

$$\text{phase}(\xi_1) = \phi. \tag{3.6}$$

This notation is convenient because a single symbol, with subscript 1, stands for both amplitude and phase, and because all the shortcuts of complex arithmetic can be used (with care). In this book, all variables with subscript 1 will be complex. The complex notation makes frequent use of $i = \sqrt{-1}$ and of the identity $e^{i\mathcal{F}} = \cos \mathcal{F} + i \sin \mathcal{F}$. Other routine manipulation of complex numbers is reviewed in the Exercises.[2]

In a graphical representation of this complex notation known as a phasor diagram, the complex variables are plotted on the complex plane with the real axis horizontal and the imaginary axis vertical. Conventionally, arrows from the origin "point" to each such complex variable, as shown in Fig. 3.2. If one imagines a phasor $\mathcal{F}_1$ in a phasor diagram rotating counterclockwise at angular frequency ω, then the actual time-dependent variable $\mathcal{F}(t) = \text{Re}\left[\mathcal{F}_1 e^{i\omega t}\right]$ represented by the phasor is its projection on the real axis.

Phasor diagrams are extremely useful for indicating the relationships in magnitude and phase among many variables in complicated thermoacoustic systems, especially traveling-wave systems. For example, if a second variable is proportional to i times a first variable, the phasor for the second variable will point $90°$ counterclockwise from the direction of the phasor for the first variable, and the second variable is said to "lead" the first variable by $90°$. Phasor diagrams are used so frequently that the axis labels "Real part" and "Imaginary part" are often omitted. Usually phasor diagrams are used only to illustrate the relative magnitudes and phases of variables—not their actual magnitudes—so a quantitative scale is

---

[2]Sometimes acousticians and physicists get lazy or forgetful, saying that ξ_1 is the oscillating position. But of course $\text{Re}[\xi_1 e^{i\omega t}]$ is actually the oscillating position.

also omitted. In fact, variables having different units are often displayed on the same plot, to convey phase information without cluttering the figure with scale information.

Time-averaged products of sinusoidal variables are encountered throughout thermoacoustics, whenever power is involved, and the following general expressions are often useful. For any two sinusoidal variables $F(t) = \text{Re}[F_1 e^{i\omega t}] = |F_1| \cos(\omega t + \phi_F)$ and $u(t) = \text{Re}[u_1 e^{i\omega t}] = |u_1| \cos(\omega t + \phi_u)$, the time-averaged product $\overline{F(t)\,u(t)}$ is obtained by averaging over one full period $1/f = 2\pi/\omega$. It is easy to show by substitution and integration that

$$\frac{\omega}{2\pi} \int_0^{2\pi/\omega} F(t)\,u(t)\,dt = \frac{1}{2}|F_1|\,|u_1|\cos(\phi_F - \phi_u) \tag{3.7}$$

$$= \frac{1}{2}\text{Re}[F_1 \widetilde{u_1}] = \frac{1}{2}\text{Re}[\widetilde{F_1} u_1], \tag{3.8}$$

where the overbar indicates time averaging, the tilde denotes complex conjugation (in which i is replaced by $-i$), and Re[ ] denotes the real part.

3.1.2 Damped, Driven Harmonic Oscillator

With this complex notation, the more complicated harmonic oscillator shown in Fig. 3.1b is easily analyzed. Three sinusoidal forces act on the mass: the spring's force $-K\xi$, an applied force F, and a viscous drag force $-R_{\text{mech}}\,d\xi/dt$ proportional to the mass's velocity $u = d\xi/dt$. Setting the mass times its acceleration equal to the sum of the forces yields

$$M\frac{d^2\xi}{dt^2} = -K\xi - R_{\text{mech}}\frac{d\xi}{dt} + F_a\cos(\omega t + \phi_F). \tag{3.9}$$

Here and elsewhere in this book, only "steady-state" oscillatory problems are considered, i.e., problems having amplitudes and phases of all variables independent of time. The question of how the system might evolve in time as it approaches such a steady state from some other initial state is ignored. So, in response to a driving force whose amplitude F_a and phase ϕ_F are independent of time, a solution $\xi(t)$ is sought that is sinusoidal in time, sharing the same frequency as the driving force, and with amplitude and phase independent of time. Hence, $\xi(t)$ must be of the form of Eq. (3.4).

One of the great benefits of the complex notation is that it converts differential equations in time into algebraic equations. For example, examination of Eq. (3.4) shows that the velocity $u = d\xi/dt$ of the moving mass can be expressed in complex notation as

$$u_1 = i\omega\xi_1. \tag{3.10}$$

Hence, in complex notation, Eq. (3.9) becomes simply

$$-\omega^2 M \xi_1 = -K\xi_1 - i\omega R_{mech}\xi_1 + F_1,$$ (3.11)

where $F_1 = F_a e^{i\phi_F}$. The rules of complex algebra provide the solution easily:

$$\xi_1 = \frac{F_1}{K - \omega^2 M + i\omega R_{mech}}.$$ (3.12)

This expression displays the complex notation to full advantage, and gives some of the useful intuition to be gained from this simple harmonic oscillator. Equation (3.12) shows that the amplitude of the motion $|\xi_1|$ is proportional to the amplitude of the force $|F_1|$. At low enough frequency, the mass and drag terms in the denominator are negligible compared to the spring term, so Eq. (3.12) reduces to $\xi_1 = F_1/K$. In this case the position and the force are in phase, with the mass being the farthest to the right at the instant of time when the force is greatest and to the right. The system "feels" like a spring. At the opposite extreme, for high enough frequency the spring and drag terms in the denominator are negligible compared to the mass term, so Eq. (3.12) reduces to $\xi_1 = -F_1/\omega^2 M$. In this case the position and the force are 180° out of phase: The position is farthest right when the force is leftward. The system "feels" inertial, like a mass. Between these two extremes, at the resonance frequency for which $K - \omega^2 M = 0$, Eq. (3.12) reduces to $\xi_1 = -iF_1/\omega R_{mech}$. The "$-i$" shows that the position lags the force by 90° in time. The system "feels" perfectly resistive.

The work done by a force F moving an object a distance $d\xi$ is $F\,d\xi$, and the power is $F\,d\xi/dt$. Hence, the time-averaged power $\dot{W}$ delivered by F in Fig. 3.1b is obtained by using Eqs. (3.12) and (3.8) with $u_1 = i\omega\xi_1$:

$$\dot{W} = \frac{1}{2}\text{Re}[F_1\widetilde{u_1}] = \frac{|F_1|^2}{2}\frac{R_{mech}}{(K/\omega - \omega M)^2 + R_{mech}^2}.$$ (3.13)

On resonance, $\dot{W}$ takes on its maximum value, $|F_1|^2/2R_{mech}$, a particularly simple result. Again, on resonance, the agent exerting the driving force cannot discern the presence of the spring or the mass—their effects on the motion completely "cancel" on resonance.

Often it is convenient to use a dimensionless version of the drag constant R_{mech}. Therefore, the quality factor Q of the damped harmonic oscillator is defined as

$$Q = \omega M/R_{mech}.$$ (3.14)

With this notation, Eq. (3.12) becomes

$$\xi_1 = \frac{F_1}{K - \omega^2 M + i\omega^2 M/Q}.$$ (3.15)

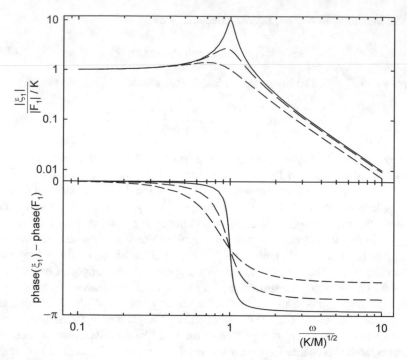

Fig. 3.3 Amplitude and phase of the position ξ of the damped, driven harmonic oscillator, as given in Eq. (3.15), for $Q = 10$ (*solid line*), $Q = 2.5$ (*longer-dash line*), and $Q = 1$ (*shorter-dash line*)

Fig. 3.4 For $Q = 2.5$, displacement phasors ξ_1 for six values of ω are shown, and the circular curve linking those phasors indicates the entire function $\xi_1(\omega)$ as ω ranges from 0 to infinity. The complex force F_1 is assumed to be real

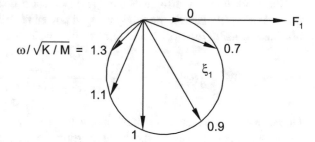

Figure 3.3 shows this solution as a function of ω, for three different values of Q. The Q controls the "sharpness" and height of the resonance. It is easiest to drive a damped harmonic oscillator to high amplitude if the Q is high and the driving frequency is near the resonance frequency. Figure 3.4 shows the solution in phasor representation for $Q = 2.5$.

3.2 Acoustic Approximations to the Laws of Gases

Next, the laws of Chap. 2 are revisited, with simplified forms presented here that are suitable for use in the next few chapters. These simplified forms will be called "Rott's acoustic approximation" in this book. They are slightly more restrictive than acousticians' usual acoustic approximation.[3]

Acousticians usually assume that the time-dependent parts of all variables are "small." Neglecting small variables would eliminate all of acoustics, but *products* of small variables can often be neglected in comparison to the small variables themselves. For example, in acoustics the $(\mathbf{v} \cdot \nabla)\,\mathbf{v}$ term in the momentum equation is usually neglected in comparison to the $\partial \mathbf{v}/\partial t$ term, because the former contains two factors of the small quantity $\mathbf{v}$ and hence is "small squared." Such "small squared" terms are called *second order,* while the small variables themselves are called first order.

Following Rott [12], in *this book's version* of the acoustic approximation additional assumptions are made. The acoustic propagation direction is x. The important physical length scales are assumed to obey

$$\delta_\kappa \ll \lambda, \tag{3.16}$$

$$\delta_\nu \ll \lambda, \tag{3.17}$$

inequalities that are easily satisfied under ordinary acoustic conditions. Only ideal gases are considered. For the gases of interest, bulk viscosity ζ is negligible. Steady-state sinusoidal oscillations of variables such as pressure, temperature, and density are considered,[4] and such variables are written in the complex notation that was introduced for force and displacement in the previous section. The density, temperature, and entropy are expressed as

$$\rho(x, y, z, t) = \rho_m(x) + \mathrm{Re}\left[\rho_1(x, y, z)e^{i\omega t}\right], \tag{3.18}$$

$$T(x, y, z, t) = T_m(x) + \mathrm{Re}\left[T_1(x, y, z)e^{i\omega t}\right], \tag{3.19}$$

$$s(x, y, z, t) = s_m(x) + \mathrm{Re}\left[s_1(x, y, z)e^{i\omega t}\right]. \tag{3.20}$$

The fact that the mean values (subscripts m) of these variables depend only on x, not on y or z, simply indicates the assumption that nothing but the oscillation can

---

[3]When we rely on these results later in the book, you will know where to find them and how to justify them. However, use these expressions with caution if you are working outside the realm of thermoacoustics as practiced at Los Alamos today, because some of the approximations applicable for us might not be applicable for you. For example, once we've introduced the ideal-gas assumption here, subsequent equations are not applicable for a liquid working fluid [60].

[4]Oscillations in some other variables, such as a or δ_ν, can be nonzero but are of no interest to us in this book, because careful analysis shows that only their mean values appear in the equations derived here. Such variables will appear without subscript m.

cause gradients perpendicular to the oscillation direction. The assumption that the time-dependent variables are "small" is expressed mathematically as $|\rho_1| \ll \rho_m$, $|T_1| \ll T_m$, etc. These expressions are only approximations to the truth—a periodic time-dependent variable can have many Fourier components[5] in addition to the fundamental component given here—but this approximation suffices for Chap. 4.

Most other variables have the same form as ρ, T, and s. The exceptions are p and $\mathbf{v}$. The mean pressure must be spatially uniform—independent of x as well as y and z—because *any* gradient in p_m would cause an acceleration of the gas, and the sinusoidal oscillations themselves are supposed to include all such time-dependent behavior. More subtly, the oscillating pressure can vary significantly only with x. The neglect of y and z gradients in p_1 is justified with an argument similar to that used by Prandtl in his historic analysis of steady-flow boundary layers, which is discussed in most fluid-mechanics textbooks (e.g., [46, 52, 53]). By definition, the velocity u along the propagation direction is much larger than the perpendicular components v and w. Carefully examining the individual components of the momentum equation (2.25),

$$\rho\left[\frac{\partial u}{\partial t} + (\mathbf{v}\cdot\nabla)\,u\right] = -\frac{\partial p}{\partial x} + (\mathbf{v}\cdot\boldsymbol{\sigma}')_x, \tag{3.21}$$

$$\rho\left[\frac{\partial v}{\partial t} + (\mathbf{v}\cdot\nabla)\,v\right] = -\frac{\partial p}{\partial y} + (\mathbf{v}\cdot\boldsymbol{\sigma}')_y, \tag{3.22}$$

and similarly for the z component, shows that each of the velocity-dependent terms in the y and z equations is much smaller than the corresponding term in the x equation. Therefore, $\partial p/\partial y$ and $\partial p/\partial z$ are similarly small. Hence, the pressure is always written

$$p(x, y, z, t) = p_m + \mathrm{Re}\left[p_1(x)e^{i\omega t}\right], \tag{3.23}$$

which is dramatically simpler than the similar expressions for ρ and most other variables.

In addition, the mean velocity is assumed to be zero (i.e., there is no steady gas motion in the absence of the acoustic oscillation), so the x component of the velocity is written

$$u(x, y, z, t) = \mathrm{Re}\left[u_1(x, y, z)e^{i\omega t}\right], \tag{3.24}$$

and similarly for the y and z components. The smallness of the time-dependent variables is expressed mathematically here as $|u_1| \ll a$.

Additional simplifications are sometimes possible with the realization that derivatives with respect to x tend to be of the order of $1/\lambda$ while derivatives

---

[5] Second-order terms will be considered in Chap. 5, and used in Chap. 7. To see how complicated it can get, see, e.g., Eq. (7.109).

perpendicular to x tend to be of the order of $1/\delta_\nu$ and $1/\delta_\kappa$, so x derivatives can often be neglected in comparison to y and z derivatives.

With these approximations, simplified versions of the equations of fluid mechanics can be derived for use in the next chapter. For example, the general continuity equation, Eq. (2.23),

$$\frac{\partial \rho}{\partial t} + \nabla \cdot (\rho \mathbf{v}) = 0, \tag{3.25}$$

can be simplified easily. Substituting the complex-notation expressions, Eqs. (3.18)–(3.20), (3.23), and (3.24), into Eq. (3.25), neglecting the second-order term involving the product of ρ_1 and $\mathbf{v}_1$, and noting that $\partial \rho_m/\partial t = 0$ yields a first-order equation:

$$\frac{\partial}{\partial t} \mathrm{Re}\left[\rho_1(x, y, z)e^{i\omega t}\right] + \nabla \cdot \left\{\rho_m(x)\,\mathrm{Re}\left[\mathbf{v}_1(x, y, z)e^{i\omega t}\right]\right\} = 0. \tag{3.26}$$

Recall that $\partial/\partial t$ is identical to $i\omega$, and note the freedom to interchange the order of derivatives and the "real" operator and the applicability of the more frequently encountered rules of complex arithmetic reviewed in Exercise 3.1. Hence, an equation like Eq. (3.26) can always be written in an abbreviated form:

$$i\omega\rho_1 + \nabla \cdot (\rho_m \mathbf{v}_1) = 0. \tag{3.27}$$

This expression is mathematically true, but in terms of physically realistic variables it should always be interpreted as an abbreviation of Eq. (3.26). Since ρ_m depends only on x, further simplification of the continuity equation leads to

$$i\omega\rho_1 + \frac{d\rho_m}{dx}u_1 + \rho_m \nabla \cdot \mathbf{v}_1 = 0, \tag{3.28}$$

the form to be taken as a starting point in later chapters.

Substituting Eqs. (3.18), (3.19), and (3.23) into the ideal-gas equation of state (2.44) yields

$$p_m + \mathrm{Re}\left[p_1(x)e^{i\omega t}\right]$$
$$= \{\rho_m(x) + \mathrm{Re}\left[\rho_1(x, y, z)e^{i\omega t}\right]\} \,\Re\,\{T_m(x) + \mathrm{Re}\left[T_1(x, y, z)e^{i\omega t}\right]\}. \tag{3.29}$$

The simple zeroth-order equation of state, $p_m = \rho_m \Re T_m$, is obtained by neglecting all small variables. Keeping first-order small terms in Eq. (3.29) but neglecting the even smaller second-order small term with the product of ρ_1 and T_1 gives the first-order equation of state, here written in abbreviated form:

$$p_1 = \rho_m \Re T_1 + \Re T_m \rho_1. \tag{3.30}$$

[This relationship can also be derived by differentiating Eq. (2.44) to yield

$$dp = \rho \mathfrak{R} \, dT + \mathfrak{R} T \, d\rho, \tag{3.31}$$

from which the acoustic version is obtained by associating the small variables such as p_1 with the small differentials such as dp.] The first-order equation of state is often more usefully written as

$$p_1/p_m = T_1/T_m + \rho_1/\rho_m, \tag{3.32}$$

which is typically used to express ρ_1 in terms of p_1 and T_1.

Similarly, Eqs. (2.92) and (2.59) become

$$s_1 = -\frac{p_1}{\rho_m T_m} + \frac{c_p T_1}{T_m}, \tag{3.33}$$

$$h_1 = c_p T_1 \tag{3.34}$$

for the ideal gas with $\beta = 1/T$.

Following a procedure similar to that for Eq. (3.28) above, the x component of the momentum equation (2.25) becomes

$$i\omega \rho_m u_1 = -\frac{dp_1}{dx} + \mu \left(\frac{\partial^2 u_1}{\partial y^2} + \frac{\partial^2 u_1}{\partial z^2} \right), \tag{3.35}$$

when x derivatives of u_1, being of the order of u_1/λ, can be neglected compared with its y and z derivatives, which are of the order of u_1/δ_ν. Similarly, the first-order approximation to the heat-transfer equation is obtained by substituting Eq. (2.92) for ds into Eq. (2.40), and neglecting the x derivative of T_1. The result is

$$i\omega \rho_m c_p T_1 + \rho_m c_p \frac{dT_m}{dx} u_1 = i\omega p_1 + k \left(\frac{\partial^2 T_1}{\partial y^2} + \frac{\partial^2 T_1}{\partial z^2} \right). \tag{3.36}$$

One of the reasons that thermoacoustics is at first confusing is that the acoustic variables can be wavelike functions of both position and time. Remember that the time dependence is always assumed to be exactly sinusoidal in Rott's acoustic approximation. The spatial dependence need not be sinusoidal but sometimes turns out to be nearly so over portions of an apparatus. When talking about the phase of such a wave, it is occasionally necessary to explicitly say *temporal* phase to identify interest in the phase of the *time* oscillation, not the x dependence.

3.3 Some Simple Oscillations in Gases

To gain familiarity with gas dynamics in Rott's acoustic approximation (both the physics and the complex notation), a review of some simple oscillations is presented in this section: a simple harmonic oscillator in which the spring is a gas spring, and standing and traveling sound waves.

3.3.1 The Gas Spring

As a first example of oscillating gas dynamics, consider the damped, driven harmonic oscillator shown in Fig. 3.5a, in which the "spring" is a sealed volume $V(t)$ of gas, at pressure $p(t)$, to the right of and sealed by the moving piston with mass M. If the face of the piston has area A on which the pressure acts, the equation of motion of the piston is similar to Eq. (3.9):

$$M\frac{d^2\xi(t)}{dt^2} = -Ap(t) - R_{\text{mech}}\frac{d\xi(t)}{dt} + F_a\cos(\omega t + \phi_F). \qquad (3.37)$$

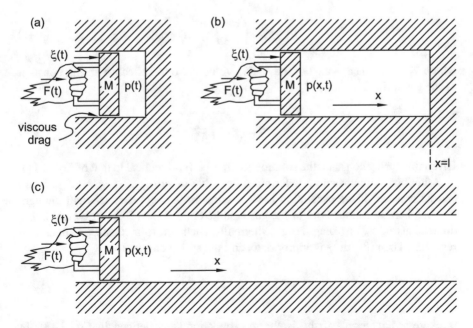

Fig. 3.5 Some simple oscillations in gases. (**a**) A damped, driven harmonic oscillator, in which the moving mass is a piston and the spring is a gas spring. (**b**) A column of gas with nonnegligible extent in the x direction, sealed at the *right end* $(x = l)$ and driven at the *left end*, in which a standing wave exists. (**c**) An infinitely long column of gas, driven at the *left end*, in which a traveling wave exists

The average pressure p_m can be assumed to be the same inside and out, so it exerts no net force on the piston—the only part of the pressure relevant to Eq. (3.37) is the oscillating part. Hence, converting to complex notation yields

$$-\omega^2 M \xi_1 = -Ap_1 - i\omega R_{\text{mech}} \xi_1 + F_1. \tag{3.38}$$

When the complex notation becomes familiar enough, one can write a complex equation such as Eq. (3.38) immediately, without the intermediate step of writing a differential equation such as Eq. (3.37).

The oscillating pressure p_1 is caused by volume changes. Assuming that p_1 is spatially uniform requires that the horizontal dimension of the gas spring be short enough and the frequency low enough: $V/A \ll \lambda$, where λ is the wavelength of sound defined in Eq. (1.1) and discussed below. For such a spatially uniform p,

$$dp = \left(\frac{dp}{d\rho}\right) d\rho, \tag{3.39}$$

$$\frac{d\rho}{\rho} = \frac{A\,d\xi}{V}. \tag{3.40}$$

Combining these, and switching to complex notation, yields

$$p_1 = \rho_m \left(\frac{dp}{d\rho}\right) \frac{A}{V_m} \xi_1, \tag{3.41}$$

where V_m is the mean value of the sealed volume. So the effective spring constant of the gas spring is

$$K = \rho_m \left(\frac{dp}{d\rho}\right) \frac{A^2}{V_m}. \tag{3.42}$$

Given this spring constant, the solution to Eq. (3.38) is identical to that of Eq. (3.11).

Additional circumstances determine the value of $dp/d\rho$. The gas temperature can be constant if the oscillation is slow enough, the geometry is small enough, and the thermal conductivity of the gas is large enough (i.e., if $\delta_\kappa \gg$ typical linear dimensions of V_m) to keep the gas thermally anchored to a constant-temperature reservoir. Then $(\partial p/\partial\rho)_T$ is appropriate, and the spring constant

$$K_T = B_T \frac{A^2}{V_m} = \frac{p_m A^2}{V_m} \tag{3.43}$$

is expressed in terms of the isothermal bulk modulus defined in Eq. (2.48). In another limiting case, the oscillations are adiabatic: There is negligible heat transfer to and from the gas during the oscillations. Adiabatic oscillations can occur if the thermal conductivity is low, the geometry is large, and the frequency is high (i.e., if

$\delta_\kappa \ll$ all linear dimensions of V_m); or if the piston and cylinder walls are perfectly insulating (difficult to achieve in real hardware). Then $(\partial p/\partial \rho)_s$ is appropriate, and the spring constant

$$K_s = B_s \frac{A^2}{V_m} = \frac{\gamma p_m A^2}{V_m} \tag{3.44}$$

is expressed in terms of the isentropic bulk modulus defined in Eq. (2.49). Comparing these two effective spring constants shows that adiabatic gas makes a stiffer spring than does isothermal gas.

3.3.2 Simple Sound Waves

If the gas spring is long, or the frequency is high, then the assumption that $p(t)$ in the gas spring is spatially uniform is not valid. This circumstance, illustrated in Fig. 3.5b, c, introduces the simplest one-dimensional sound waves. Viscosity is neglected, u_1 is assumed to be independent of y and z, and T_m is assumed to be independent of x. Hence, $\rho_m = p_m/RT_m$ is also independent of the horizontal coordinate x. Examination of the equations in Sect. 3.2 shows that two of them are relevant. The continuity equation becomes

$$i\omega \rho_1 + \rho_m \frac{du_1}{dx} = 0, \tag{3.45}$$

and the x component of the momentum equation becomes

$$i\omega \rho_m u_1 = -\frac{dp_1}{dx}. \tag{3.46}$$

These two equations involve three oscillating variables—u_1, p_1, and ρ_1—so a third equation is needed to define the solution. As with the gas spring, the third equation links p and ρ: the value of $\partial p/\partial \rho$ governing the oscillations. Most ordinary sound propagation takes place under adiabatic conditions: δ_κ defined in Eq. (1.2) is much smaller than distances to the nearest solid boundaries, and also $\delta_\kappa \ll \lambda$ so the gas in one part of the wave has no time for heat transfer with the gas in another part of the wave. Hence, for ordinary sound propagation, $(\partial p/\partial \rho)_s$ is the appropriate choice. For the gas spring it was most useful to express $(\partial p/\partial \rho)_s = B_s/\rho$, but here Eq. (2.56) will be used to write it in terms of the sound speed,

$$\left(\frac{\partial p}{\partial \rho} \right)_s = a^2, \tag{3.47}$$

i.e., $p_1 = a^2 \rho_1,$ \hfill (3.48)

for reasons that will be explained below. Then Eqs. (3.45) and (3.46) can be rewritten as

$$i\omega p_1 = -\rho_m a^2 \frac{du_1}{dx}, \tag{3.49}$$

$$i\omega u_1 = -\frac{1}{\rho_m}\frac{dp_1}{dx}, \tag{3.50}$$

which are coupled differential equations for $u_1(x)$ and $p_1(x)$. The general solution for either variable is easily shown by substitution to be sinusoidally periodic in x, with wavelength $\lambda = a/f = 2\pi a/\omega$. Depending on boundary conditions, such a solution can be most conveniently written in any of a number of ways, such as

$$C_c \cos kx + C_s \sin kx, \tag{3.51}$$

$$C \cos k(x - x_o), \tag{3.52}$$

$$C_+ e^{ikx} + C_- e^{-ikx}, \tag{3.53}$$

where $k = \omega/a = 2\pi/\lambda$ is called the wave vector, x_o is a real constant, and the C's are complex constants. The wave vector is determined by the constants in the differential equations, but the other constants in Eqs. (3.51)–(3.53) are determined by the boundary conditions.

Two special cases of boundary conditions give the simplest solutions. First, consider Fig. 3.5b, in which the right end of the gas space is sealed by an immovable solid wall at $x = l$. One boundary condition is therefore $u_1(l) = 0$. The most convenient form of the solution is then

$$p_1(x) = C \cos k(x - l), \tag{3.54}$$

with the amplitude and phase of the complex constant C related to the amplitude and phase of the applied force F at the left end. This type of wave is called *standing*. Examination of Ani. Wave /s shows why: The peaks and troughs of the wave appear to stand still horizontally as they oscillate vertically. This animation illustrates a wave oscillating horizontally in a duct of uniform cross section. The uppermost graph shows the pressure of the gas in the wave as a function of position and time. The middle graph shows the velocity in the gas as a function of position and time (with positive velocity to the right). The lower part of the display shows the motion, or displacement, of the gas in the duct, as the vertical lines move with the gas.

A second simple solution is obtained for the case of Fig. 3.5c, where the gas channel extends to $x = \infty$. Then the moving piston launches what is called a rightward *traveling* wave down the channel, described most conveniently by

$$p_1(x) = C e^{-ikx}, \tag{3.55}$$

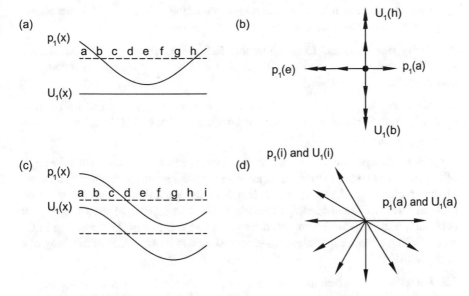

Fig. 3.6 Phasors at various locations in standing and traveling waves. The phasors for the pressure p_1 and the volume flow rate U_1 are shown. (**a**) The waves in Ani. Wave /s, corresponding to Eq. (3.54), as a function of position x, frozen at one particular time t. Eight equally spaced locations "a" through "h" are labeled. (**b**) Phasor diagram for Ani. Wave /s, corresponding to Eq. (3.54). The pressure phasors are along the real axis, and the volume-flow-rate phasors are along the vertical axis. (**c**) The waves in Ani. Wave /t, corresponding to Eq. (3.55), as a function of position x, frozen at one particular time t. Nine equally spaced locations "a" through "i" are labeled. (**d**) Phasor diagram for Ani. Wave /t, corresponding to Eq. (3.55). At each location, the pressure phasor and volume-flow-rate phasor point in the same direction

with, again, the amplitude and phase of the complex constant C related to the amplitude and phase of the applied force F. Examination of Ani. Wave /t shows why the solution is called traveling: The wave indeed appears to travel, to the right. Exercise 3.20 shows that this wave and waves of more complicated shape travel at speed a. The phasor representations for these standing and traveling waves are shown in Fig. 3.6.

These simple waves will be discussed more thoroughly early in the next chapter.

3.4 Exercises

3.1 Given four complex numbers $a_1 = ae^{i\phi_a}$, $b_1 = be^{i\phi_b}$, $c_1 = c_r + ic_i$, and $d_1 = d_r + id_i$, where $a, \phi_a, b, \phi_b, c_r, c_i, d_r,$ and d_i are real, write down the magnitudes of $a_1, b_1, c_1,$ and d_1; the phases of $a_1, b_1, c_1,$ and d_1; the magnitudes and phases of $a_1b_1, a_1/b_1,$ and c_1d_1; and the real parts of $a_1, 1/c_1,$ and $\widetilde{c}_1d_1$. Describe the latter results in words, e.g., "The phase of the product of two complex numbers is the sum of the phases of the numbers, and the magnitude of the product is ..." Draw phasor diagrams for a few numerical examples.

3.2 Verify Eqs. (3.7) and (3.8). Show that $\text{Re}[\int \mathcal{F}_1(x)\,dx] = \int \text{Re}[\mathcal{F}_1(x)]\,dx$, and similarly for a derivative. Show that $\text{Re}[ab_1] = a\text{Re}[b_1]$ if a is real and b_1 is complex. Show that $\text{Re}[ia_1] = -\text{Im}[a_1]$.

3.3 Show that $|a_1|^2 = a_1\widetilde{a_1}$. Show that $\text{Re}[1/a_1] = \text{Re}[a_1]/|a_1|^2$ and $\text{Im}[1/a_1] = -\text{Im}[a_1]/|a_1|^2$. Write down the real and imaginary parts of $1/(1-f_v)$ where $f_v = (1-i)\delta_v/2r_h$ (with r_h and δ_v real).

3.4 Show that a pendulum exhibits simple harmonic motion. Assume that the angle of oscillation is small. Derive an expression for the resonance frequency in terms of the acceleration of gravity g, the length of the pendulum l, and the mass M. (Are all three variables needed?) Crudely verify two aspects of the truth of your derivation (e.g., dependence on M, dependence on l, factor of 2π appearing in the result) with a simple experiment using your shoe and its shoe lace or anything else easily available.

3.5 For a damped, driven harmonic oscillator, $\sqrt{K/M}/2\pi$ is *defined* to be the resonance frequency. However, close examination of Fig. 3.3 shows that the maximum displacement amplitude does not occur exactly at $\omega = \sqrt{K/M}$, especially for lower Q's. Use Eq. (3.12) to show that the maximum velocity amplitude does occur at $\omega = \sqrt{K/M}$.

3.6 Show that the power supplied by F to the damped, driven oscillator is equal to the power dissipated by R_{mech}.

3.7 The drag force opposing the motion of the simple harmonic oscillator of Fig. 3.1 was assumed to be proportional to the velocity. However, frictional forces are sometimes modeled as having a magnitude that is independent of velocity. Explore the consequences of such a frictional force on the initial steps of the derivation in this chapter.

3.8 In a particular gas spring, the gas's thermal conductivity is high enough and the frequency is low enough that the gas has spatially uniform temperature T at each instant of time, but the volume's walls are not good conductors of heat, so the thermal contact to the infinite heat sink at T_m is imperfect, represented by $\dot{Q} = G(T - T_m)$ where G is a constant. Show that the magnitude of the effective spring constant is between the isothermal and adiabatic values, and that the spring not perfectly resilient, i.e., time-averaged work must be supplied by the force F to maintain the oscillation even if $R_{\text{mech}} = 0$.

3.9 Show that Eqs. (3.30) and (3.33) are consistent with Eq. (2.56).

3.10 Verify Eq. (3.36), starting from Eq. (2.40). Write out each step, and write a sentence explaining the justification for each step.

3.11 For a plane sound wave traveling in the x direction, $u_1 = u_o e^{-ikx}$, where $k = \omega/a$ and u_o is a real constant. Show that the ratio of $|(\mathbf{v} \cdot \nabla)\mathbf{v}|$ to $|\partial\mathbf{v}/\partial t|$ is u_o/a.

3.12 Verify that solutions of the form of Eqs. (3.51)–(3.53) satisfy Eqs. (3.49) and (3.50).

3.13 For the standing wave of Fig. 3.5b and Eq. (3.54), find C as a function of driving force F_1 and driving angular frequency ω. Is the solution simpler, or more distressing, if you assume $M = 0$? Interpret the solution in terms of how hard it would "feel" to maintain a wave of given, constant pressure amplitude at $x = l$ as a function of frequency.

3.14 The human ear can hear sound between about 20 Hz and 20,000 Hz, with "conversation" centered about 500 Hz. What are the wavelengths for each of these frequencies? (The speed of sound in air can be obtained from Eqs. (2.56) and (2.46), and the table near the end of Sect. 2.3.)

3.15 Show that $T_1/T_m = (\gamma - 1)p_1/\gamma p_m$ for an ideal gas experiencing isentropic pressure oscillations. What is the value of $(\gamma - 1)/\gamma$ for your favorite gas? What proportionality constant links T_1/T_m and p_1/p_m for an ideal gas experiencing isothermal pressure oscillations? How big is T_1 at a few locations in your favorite thermoacoustics hardware?

3.16 The pressure amplitude of ordinary "conversational" sound is about 10^{-6} bar. A typical frequency is 500 Hz. What is $|\xi_1|$? What is $|T_1|$?

3.17 Derive an expression for isothermal sound speed, instead of the usual adiabatic sound speed. Show that the isothermal sound speed is about 18% lower than the true, adiabatic speed in air, for which $\gamma = 7/5$. Can you think of circumstances in which sound propagates isothermally?

3.18 When one has an ear infection, swelling prevents the body from keeping p_m equal on the two sides of the eardrum. When one rides an elevator or airplane with this condition, pain occurs with changes in elevation of approximately 1000 ft. What is the change in p_m attending a change of altitude of 1000 ft.? How does that compare to the acoustic pressure amplitude that causes pain, which is $|p_1| \sim 200$ Pa?

3.19 You heard the thunder 5 s after you saw the lightning. How far away was the lightning?

3.20 Convert Eqs. (3.49) and (3.50) from complex notation back to linear, coupled partial differential equations for $p(x, t)$ and $u(x, t)$. Show that these equations are satisfied by

$$p(x, t) = p_m \mathcal{F}(x - at), \tag{3.56}$$

$$u(x, t) = (p_m/\rho_m a)\mathcal{F}(x - at), \tag{3.57}$$

where $\mathcal{F}$ is an arbitrary function. Interpret this result in terms of "propagation" at speed a of a disturbance whose spatial shape is described by $\mathcal{F}(x)$ and whose shape does not change with time.

3.21 Write the standard expression for *bulk* attenuation of sound in terms of ratios of the viscous and thermal penetration depths to the acoustic wavelength. Interpret the result in terms of the frequency dependences of these lengths. You can look up the expression for bulk attenuation in an acoustics textbook such as Kinsler et al. [5]. Justify the neglect of bulk attenuation of sound in your favorite thermoacoustics hardware or in one of the four examples introduced in Chap. 1.

3.22 Which of the equations in Sect. 3.2 depend on the assumption that $\lambda \gg \delta_\kappa, \delta_\nu$ so that some x derivatives can be neglected?

3.23 Does $p_1(x) = C \sin kx$ represent a standing wave or a traveling wave? Does $p_1(x) = Ce^{ikx}$ represent a standing wave or a traveling wave? Use the lossless first-order momentum equation to write down $u_1(x)$ in both cases. Use the lossless first-order continuity equation to check your expressions for $u_1(x)$.

3.24 How would you write $p_1(x)$ for a standing wave with a temporal phase shift of 90° relative to the standing wave of Eq. (3.54)? What about a quarter-wavelength spatial shift in the node locations?

Chapter 4
Waves

The focus of this chapter will be on pressure and velocity for wave propagation in the x direction in a channel or duct of arbitrary length whose y and z dimensions are much smaller than the acoustic wavelength. Adopting increasingly sophisticated perspectives on the continuity equation shows how oscillations in pressure are coupled to spatial gradients in velocity. Similarly, the momentum equation shows how oscillations in velocity are coupled to spatial gradients in pressure. Merging these two pictures yields wave propagation. The chapter begins with the lossless concepts of inertia and compressibility, later adding viscous loss to the momentum equation and thermal hysteresis and gain to the continuity equation. Overall, propagation along x depends primarily on inertia and compressibility, while viscous and thermal effects involve diffusion perpendicular to x. Throughout, the assumption of monofrequency, steady, small oscillations in an ideal gas is maintained—approximations that for brevity are called simply "Rott's acoustic approximation" in this book, as described in Sect. 3.2.

The principal variables are oscillating pressure p_1 and oscillating volume flow rate U_1, which is the integral of the x component of the oscillating velocity, u_1, over the cross-sectional area A of the channel. (Instead of "volume flow rate," many thermoacoustics publications say "volume velocity" or "volumetric velocity," a terminology commonly used in acoustics [58] but rarely in fluid mechanics.) At the transition between two channels, U_1 and p_1 are taken to be continuous (but see "Joining conditions" in Chap. 7). For example, at a transition between two ducts, U_1 out of the first duct has nowhere to go but into the second duct, so U_1 must be continuous. At such a junction, p_1 must also be continuous, because any pressure discontinuity represents a shock front, outside the realm of the acoustic approximation and today's thermoacoustics.

© Acoustical Society of America 2017
G.W. Swift, *Thermoacoustics*, DOI 10.1007/978-3-319-66933-5_4

4.1   Lossless Acoustics and Ideal Resonators

The simple lossless version of Rott's acoustic continuity equation, Eq. (3.45), can be rewritten

$$p_1 = -\frac{\gamma p_m}{i\omega A \, \Delta x} \Delta U_1 \qquad (4.1)$$

$$= -\frac{1}{i\omega C} \Delta U_1 \qquad (4.2)$$

for the short length Δx of channel of cross-sectional area A that is shown in Fig. 4.1a. Arriving at these equations requires use of Eq. (3.48) to express ρ_1 in terms of p_1 and use of the ideal-gas identity

$$\rho_m a^2 = \gamma p_m, \qquad (4.3)$$

which is a form of Eq. (2.56) that is worth memorizing. In Eq. (4.2), the compliance C of the channel is defined as

$$C = \frac{V}{\gamma p_m}. \qquad (4.4)$$

The compliance is the product of the volume $V = A \, \Delta x$ of the channel and the compressibility $1/\gamma p_m$. Larger volume or greater ("softer") compressibility means

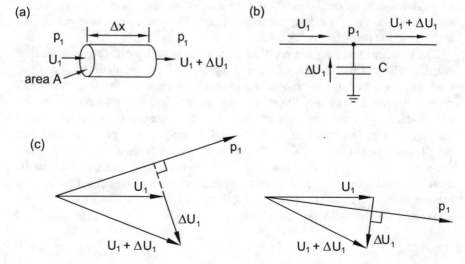

Fig. 4.1 (**a**) A short channel in which the compressibility of the gas is important. (**b**) Symbolic impedance diagram of the channel. The symbol C indicates compliance. (**c**) Two possible phasor diagrams for the channel

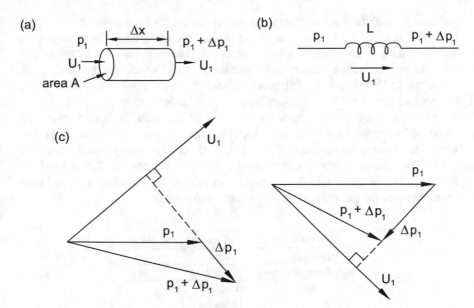

Fig. 4.2 (**a**) A short channel in which the inertia of the gas is important. (**b**) Symbolic impedance diagram of the channel. The symbol L signifies the inertance. (**c**) Two possible phasor diagrams for the channel

greater compliance. This combination of variables gives the springy, compressive properties of the gas in the channel.

Similarly, the lossless version of Rott's acoustic momentum equation, Eq. (3.46), can be rewritten

$$\Delta p_1 = -i\omega \frac{\rho_m \, \Delta x}{A} U_1 \qquad (4.5)$$

$$= -i\omega L U_1, \qquad (4.6)$$

for the short length of channel shown in Fig. 4.2a. Note that the channel illustrated is the same as in Fig. 4.1a; the figures differ only because attention is focused on different properties of this channel in the two situations. In Eq. (4.6),

$$L = \frac{\rho_m \, \Delta x}{A} \qquad (4.7)$$

is called the inertance of the channel. It is the product of the gas density ρ_m and the length Δx divided by the cross-sectional area A of the channel. This combination of variables describes the inertial properties of the gas in the channel.

Writing the continuity and momentum equations in this way allows an accurate analogy between acoustic systems and ac electric circuits, motivating the symbolic "circuit" impedance diagrams of Figs. 4.1b and 4.2b. The analogue of oscillating pressure p_1 is electric ac voltage; the analogue of oscillating volume flow rate U_1 is electric ac current. The analogue of compliance is electric capacitance to ground, and the analogue of inertance is series electric inductance. (Note that "inertance" sounds a little like the electrical word "inductance" and a little like the mechanical word "inertia.") This analogy [5], more extensively summarized in the table below, can provide useful guidance through the middle of the next chapter.[1]

When you look at the symbol for inertance in Fig. 4.2b, think of a long tube ("coiled" for compactness?) through which the dense gas must accelerate. When you look at the symbol for compliance in Fig. 4.1b, think of a spongy cushion between the two parallel lines, whose gap changes in response to pressure. When you look at the symbol for resistance in figures later in this chapter, think of a kinked tube through which the gas must overcome viscous resistance.

Acoustic networks	AC electric networks
Pressure p_1	Voltage V_1
Volume flow rate U_1	Current I_1
Compliance C	Capacitance C
Inertance L	Inductance L
Resistance R	Resistance R
Acoustic power $\dot{E}_2$	Electric power $\dot{W}_2$

Figures 4.1c and 4.2c show possible phasor diagrams for a compliance and an inertance, respectively. In Fig. 4.1c, note that $|U_1 + \Delta U_1|$ could be either larger or smaller than $|U_1|$; the only strict requirement imposed by Eq. (4.2) is that ΔU_1 must lag p_1 by 90°. Similarly, in Fig. 4.2c, the only strict requirement imposed by Eq. (4.6) is that Δp_1 must lag U_1 by 90°. Beyond these requirements, anything is possible, depending on the relative magnitudes and phases of p_1 and U_1.

Usually, both the inertance and the compliance of a channel simultaneously contribute to the behavior of the wave propagation in the channel, as illustrated in Fig. 4.3 for a channel of differential length dx. As before, note that dp_1 lags U_1 by 90°, and dU_1 lags p_1 by 90°.

The complex ratio of pressure to volume flow rate, $Z = -p_1/U_1$, is known as the acoustic impedance and is of great utility in discussing acoustic systems.[2] Sometimes it refers to the impedance $Z = -p_1/U_1$ at a location in a wave, but

---

[1]We will rely on this point of view extensively, because some readers (e.g., physicists and electrical engineers) have some prior experience with ac electric circuits. However, other engineers and scientists often gain no intuition from what is to them a totally unfamiliar analogy. Such readers, unfamiliar with ac electric circuits, can take the "circuit diagrams" to be no more than abstract representations of the acoustic impedance networks themselves, and we will use nothing but acoustic terminology in referring to these diagrams. See also Exercise 4.12.

[2]The minus sign in this definition of Z may surprise some readers. This choice of sign allows taking U_1 positive in the positive x direction *and* keeping the universally accepted sign conventions of differential calculus.

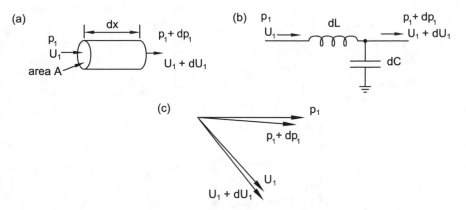

Fig. 4.3 (a) A very short channel in which both the compressibility and the inertia of the gas are important. (b) Symbolic impedance diagram of the channel. (c) A possible phasor diagram for the channel

sometimes it refers to the impedance of a component so that $Z = -\Delta p_1/U_1$ is related to the pressure difference across the component or $Z = -p_1/\Delta U_1$ describes the change in volume flow rate caused by the component. In the latter modes, Eq. (4.6) shows that the acoustic impedance of an inertance is $Z = i\omega L$, and Eq. (4.2) shows that the acoustic impedance of a compliance is $Z = 1/i\omega C$.

The double Helmholtz resonator, illustrated in Fig. 4.4, is a simple acoustic resonator consisting of two bulbs, each of volume V, connected by a short neck with length Δx and cross-sectional area A. All dimensions are shorter than the acoustic wavelength, so the lumped-impedance models of Figs. 4.1 and 4.2 are directly applicable: Each bulb is a compliance C, and the neck connecting the two bulbs is an inertance L. (In the idealized Helmholtz resonator, the compliance of the neck and inertial effects in the compliances are assumed negligible.) Imagine the inertial mass of gas in the neck bouncing back and forth sinusoidally against forces exerted by the gas springs in the two bulbs. Combining Eqs. (4.2) and (4.6) appropriately, it is easy to show that p_1 and U_1 are nonzero only if

$$i\omega L + 2/i\omega C = 0, \qquad (4.8)$$

and hence that the resonance frequency of the double Helmholtz resonator is given by

$$(2\pi f)^2 = \omega^2 = \frac{2}{LC} = 2\frac{\gamma p_m}{\rho_m}\frac{A}{V\,\Delta x} = a^2\frac{2A}{V\,\Delta x}. \qquad (4.9)$$

At this frequency, either Eq. (4.2) or (4.6) gives the relative magnitudes of p_1 and U_1, and their relative phases. The relative phases are illustrated in the phasor diagram in Fig. 4.4c, where the phase of U_1 in the inertance was arbitrarily picked to be zero.

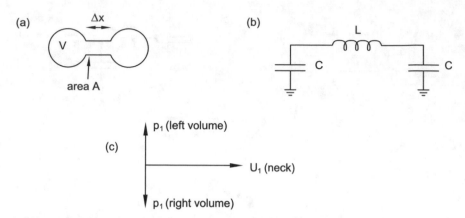

Fig. 4.4 The double Helmholtz resonator. (**a**) Schematic of the resonator, comprising two bulbs connected by a neck. (**b**) Impedance diagram of the resonator. (**c**) Phasor diagram

In lossless acoustic resonators more complicated than the Helmholtz resonator, the acoustic waves are described by the coupled, lossless continuity and momentum equations. Hence, understanding lossless acoustic waves in a duct or resonator can be based on simultaneous consideration of the concepts of compliance and inertance, as the thermophysical properties responsible for compliance, γp_m, and for inertance, ρ_m, combine to form the sound speed $a = \sqrt{\gamma p_m / \rho_m}$.

For example, reconsider the standing wave shown in Ani. Wave /s. First focus attention on the velocity node, where the gas never moves. This region of the duct is functionally similar to a bulb of the double Helmholtz resonator described above. With zero velocity, Eq. (4.6) indicates that there should be no spatial pressure gradient, and indeed the animation shows that the oscillating pressure has no gradient in the vicinity. Hence, the inertance per unit length in the duct is irrelevant in this vicinity. The nonzero oscillating p_1 in this vicinity, however, indicates that Eqs. (4.1) and (4.2) are important. This region of the duct has compliance per unit length, so that the oscillating pressure can only occur in the presence of a spatial gradient in velocity, which is needed to supply and remove the mass necessary to cause the local density to increase and decrease. The presence of i in Eq. (4.2) indicates a 90° temporal phase shift between the pressure oscillations and the oscillating velocity gradient. This temporal phase relationship is also evident in the animation.

Now focus attention on one of the pressure nodes in Ani. Wave /s. This region of the duct is functionally similar to the neck of the double Helmholtz resonator described above. Here, $p_1 = 0$, so Eq. (4.2) indicates that gradients in velocity must be zero and that the compliance per unit length in this portion of the duct is irrelevant. However, the nonzero oscillating velocity in this region and Eq. (4.6) show that here the oscillating pressure must have a nonzero gradient, as seen in the animation. Additionally, both Eq. (4.6) and the animation indicate a 90° temporal phase shift between pressure gradient and velocity.

Precisely at the pressure node in the standing wave, the compliance per unit length is irrelevant. Precisely at the velocity node, the inertance per unit length is irrelevant. Everywhere else in the standing wave, where neither p_1 nor U_1 are zero, both inertance and compliance per unit length contribute to the behavior of the wave. (This is, in fact, what makes it a wave instead of a lumped-impedance oscillator such as the double Helmholtz resonator.)

While separate regions of predominantly compliance and predominantly inertance are apparent in the standing wave, in the traveling wave illustrated in Ani. Wave /t each location is a time-shifted replica of any other location, with inertance and compliance equally important at each location.

Phasor diagrams for both wave animations were shown in Fig. 3.6.

Example: Standing-Wave Engine

The resonator of the standing-wave engine illustrated in Figs. 1.8 and 1.9 was particularly simple. Ignoring the branch to the refrigerator, it was essentially a half-wavelength-long resonator, closed at both ends so that the velocity nodes were at the ends and the pressure node was in the center. This overall wave shape is illustrated roughly in Ani. Standing /k.

However, note in Fig. 1.9 that the diameter of the central portion of the resonator was slightly smaller than the diameter of the two ends. This diameter variation served several functions, most of which will be discussed in Chap. 7. One of those functions is to lower the resonance frequency slightly below $f = a/2\Delta x$, where Δx is the overall length of the resonator. This frequency reduction is considered quantitatively in Exercise 4.1 at the end of this chapter. Nevertheless, using the concepts of inertance and compliance, it is easy to see qualitatively how this diameter variation leads to a lower resonance frequency: The half-wavelength resonator can be thought of crudely as consisting of two compliances on the two ends with an inertance the middle, so that the resonance crudely resembles that of the double Helmholtz resonator shown in Fig. 4.4. If the diameter of the inertance is reduced, L rises according to Eq. (4.7), and hence the resonance frequency of the CLC resonator drops as suggested by Eq. (4.9). From an alternative point of view, if the diameter of the compliances is increased, the value of C rises according to Eq. (4.4), and again the resonance frequency of the CLC circuit drops according to Eq. (4.9).

With the central diameter reduction drawn to scale in Fig. 1.9 (and with the branch to the refrigerator sealed off), the resonance frequency was 10% lower than the frequency for which the half wavelength equals the length of the apparatus. One often speaks loosely of such a resonator as a half-wavelength resonator, despite the fact that its length is significantly smaller than $\lambda/2$, because this phrase captures the essential nature of the standing wave: pressure maxima at the two ends and a velocity maximum in the center.

Example: Standing-Wave Refrigerator

The standing-wave refrigerator resonator, shown in Figs. 1.13 and 1.14, also employed variations in diameter to reduce the resonance frequency below that for which a full wavelength would equal the total path length around the resonator.

The apparatus had left–right symmetry, so driving the loudspeaker pairs 180° out of phase ensured that pressure nodes appeared at the top center and bottom center of the resonator. Hence, these were the locations of high velocity, where inertance was important.

The parts of the resonator near the loudspeakers had high pressure amplitude and hence were the regions where compliance is important. These compliance portions were enlarged in diameter relative to the inertance portions, so the resonance frequency was reduced.

Example: Lumped Approximation to Half-Wavelength Resonator

For the uniform-diameter half-wavelength resonator that is shown in Fig. 4.5a, the most accurate impedance diagram would have an infinite number of inertances dL and compliances dC, as suggested by the large number of them in Fig. 4.5a, which would yield the sinusoidal x dependences of p_1 and U_1 shown. Numerical integrations for accurate results might have 10 or 20 such segments, with each being integrated by a Runge-Kutta method so that each is effectively subdivided even further.

However, the essence of the resonance is simply what is shown in the phasor diagram of Fig. 4.5b, with pressure oscillations at the two ends having equal magnitudes and opposite phases, and the velocity in the center phased 90° from the pressures. Hence, if numerically accurate results are not needed, the essence can be captured with the simpler impedance diagram shown in Fig. 4.5c, in which the central third of the resonator is modeled as an inertance and the two ends as

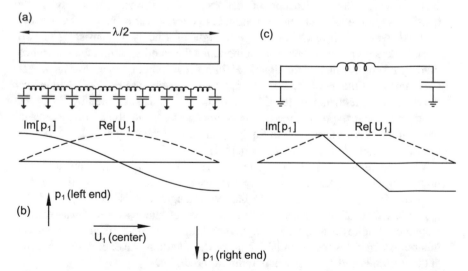

Fig. 4.5 (**a**) The half-wavelength resonator, with sinusoidal spatial distributions of pressure $p_1(x)$ and volume flow rate $U_1(x)$, is properly regarded as an infinite series of infinitesimal inertances and compliances. (**b**) The phasors at the ends and center are particularly simple. (**c**) Two compliances and one inertance capture these simple end behaviors, with less accuracy in values of p_1 and U_1 at intermediate x

compliances. The accuracy of this crude approximation is suggested by the close resemblance of the trapezoidal distributions for p_1 and U_1 shown in Fig. 4.5c to the sinusoids shown in Fig. 4.5a. (To consider the accuracy of this crude approximation, see Exercise 4.2.)

4.2 Viscous and Thermal Effects in Large Channels

To introduce the key concepts of dissipation and gain into the lossless acoustics picture developed so far, this section presents two simple "complete" thermoacoustics problems: the ordinary viscous and thermal attenuation of sound propagating in a large channel, due to the sound wave's interaction with the solid channel wall.

Consider sound propagating in the x direction in an ideal gas within a channel with cross-sectional area A and perimeter Π that are both independent of x. The hydraulic radius[3]

$$r_h = A/\Pi \qquad (4.10)$$

is conventionally defined as the ratio of the channel's cross-sectional area to its perimeter. (The hydraulic radius can also be thought of as the ratio of gas volume to gas–solid contact area, or as the distance from a typical parcel of gas to the nearest solid surface. Note that the hydraulic radius of a circular channel is *half* of the circle's actual radius!) This section treats only channels for which $\delta_\nu \ll r_h$ and $\delta_\kappa \ll r_h$, allowing what is called the boundary-layer approximation to be used. The usual complex notation for the time-oscillating quantities (pressure p, temperature T, velocity component u in the x direction, density ρ) is

$$p = p_m + \mathrm{Re}\left[p_1(x)e^{i\omega t}\right] + \cdots , \qquad (4.11)$$

$$u = \mathrm{Re}\left[u_1(x,y)e^{i\omega t}\right] + \cdots , \qquad (4.12)$$

$$T = T_m + \mathrm{Re}\left[T_1(x,y)e^{i\omega t}\right] + \cdots , \qquad (4.13)$$

$$\rho, \text{ etc.} = \text{similar to } T, \qquad (4.14)$$

with μ, k, etc. constant. In this section, the channel is assumed to be spatially isothermal, so T_m is independent of x in Eq. (4.13). The coordinate y measures the perpendicular distance from the wall of the channel, with $y = 0$ at the wall, as shown in Fig. 4.6.

---

[3] Alternative dimensions sometimes used to characterize pore size are the hydraulic diameter $= 4r_h$ and the Kozeny radius $= 2r_h$.

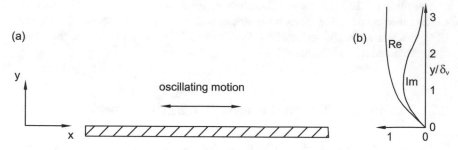

Fig. 4.6 (a) The coordinate system used in this section. In the boundary-layer approximation, the opposite wall of the channel is at such large y that it does not appear in this figure. (b) The real and imaginary parts of the y-dependent factor in square brackets in Eq. (4.16)

4.2.1 Viscous Resistance

To develop a quantitative understanding of viscous effects, consider the simplified situation illustrated in Fig. 4.6. In order to find the y dependence of the x component of the gas velocity, use the x component of the momentum equation, Eq. (3.35), for which Rott's acoustic approximation is

$$i\omega\rho_m\, u_1(x,y) = -\frac{dp_1(x)}{dx} + \mu\frac{\partial^2 u_1(x,y)}{\partial y^2}\,. \tag{4.15}$$

This equation is the appropriate approximation to Newton's law ($m\,du/dt = F$) for a differential volume of gas: The left side is mass times acceleration, and the right side is the sum of the forces—the pressure force and the viscous force.

Equation (4.15) is an ordinary differential equation for $u_1(x,y)$ as a function of y. With two boundary conditions, $u_1(y = 0) = 0$ at the solid surface and u_1 finite as $y \to \infty$, its solution is

$$u_1 = \frac{i}{\omega\rho_m}\left[1 - e^{-(1+i)y/\delta_v}\right]\frac{dp_1}{dx}\,. \tag{4.16}$$

This function shows how viscosity reduces the magnitude of the oscillating velocity and shifts its phase.

The complex notation is compact and easy to manipulate, but difficult to interpret intuitively. To gain a visual appreciation for the boundary-layer velocity expression, Eq. (4.16), study Ani. Viscous /m, which displays the particle displacement $\xi(y,t) = \mathrm{Re}\left[u_1(y)e^{i\omega t}/i\omega\right] \propto \mathrm{Re}\left[\left(1 - e^{-(1+i)y/\delta_v}\right)e^{i\omega t}\right]$ with y vertical and the acoustic-oscillation direction x horizontal. The moving line can be imagined to be a very thin cloud of smoke moving with the oscillating gas. The tic-mark spacing on the vertical axis is the viscous penetration depth δ_v. Gas that is much closer than δ_v to the solid boundary is nearly at rest. Gas that is much farther than δ_v from the solid boundary experiences essentially no viscous shear: It moves with a velocity and displacement that are independent of y. This inviscid motion is purely inertial: The acceleration is in phase with the force, which is $-dp_1/dx$, so the displacement is in phase with $+dp_1/dx$. Gas that is roughly δ_v from the nearest solid surface moves with a modified, y-dependent velocity and a significant, y-dependent phase shift. You should be able to visualize the time evolution shown in the animation being a smooth progression in time between the real and imaginary parts shown in Fig. 4.6, from real part to minus imaginary part to minus real part to imaginary part and back to real part.

Acoustic boundary-layer problems often need the spatial average, over the cross-sectional area A, of a function $\mathcal{F}(y)$ that goes to zero rapidly outside the boundary layer, such as the exponential term in Eq. (4.16). Using Eq. (4.10) defining the hydraulic radius, such an average can be written as

$$\langle \mathcal{F} \rangle = \frac{1}{A} \int \mathcal{F} \, dA = \frac{1}{\Pi r_h} \int_0^{r_h} \mathcal{F}(y) \, \Pi \, dy \simeq \frac{1}{r_h} \int_0^{\infty} \mathcal{F}(y) \, dy, \qquad (4.17)$$

where the final step is permissible because the boundary-layer function $\mathcal{F}$ is already so small at $y = r_h$ that extending the integration to infinity adds nothing. Thus, the spatial average of Eq. (4.16) is

$$\langle u_1 \rangle = \frac{i}{\omega \rho_m} \left[1 - (1 - i) \frac{\delta_v}{2r_h} \right] \frac{dp_1}{dx}. \qquad (4.18)$$

In the previous section, the lossless momentum equation led to Eq. (4.6) and to the concept of inertance for a short channel with length Δx and area A. Now, including viscous losses, Eq. (4.18) shows that the corresponding expression is

$$\Delta p_1 = -\frac{i\omega \rho_m \, \Delta x / A}{1 - (1 - i)\delta_v / 2r_h} U_1 \qquad (4.19)$$

$$= -Z U_1 \qquad (4.20)$$

$$= -(i\omega L + R_v) U_1, \qquad (4.21)$$

so viscosity adds viscous resistance R_v in series with inertance L to comprise total series acoustic impedance $Z = R_v + i\omega L$ as shown in Fig. 4.7b. In this boundary-layer approximation, the inertance and resistance are

$$L \simeq \frac{\rho_m \, \Delta x}{A} \qquad (4.22)$$

$$R_v \simeq \frac{\mu \Pi \, \Delta x}{A^2 \delta_v} = \frac{\mu S}{A^2 \delta_v} \qquad (4.23)$$

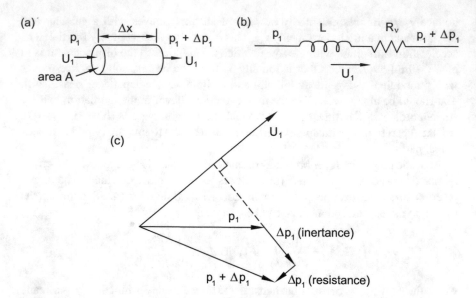

Fig. 4.7 The momentum equation describes inertial and resistive effects in a channel. (**a**) A short channel, of length Δx and cross-sectional area A. (**b**) Impedance diagram showing inertance L and resistance R_v. (**c**) A typical phasor diagram for this situation, showing that the pressure difference Δp_1 is the sum of inertial and resistive components

to lowest order in δ_v/r_h, where $S = \Pi \Delta x$ is the surface area of the channel.[4] Note that R_v and L are real; hence, the "i" in Eq. (4.21) implies that the inertial contribution to the pressure difference leads the resistive contribution by 90°, as shown in Fig. 4.7c.

So any length of channel can be thought of as having resistance caused by viscosity and inertance caused by mass. A fat channel, with $\delta_v \ll r_h$, has $R_v \ll \omega L$. In fact, in that limit the ratio $\omega L/R_v$ is proportional to $A/\Pi\delta_v$, so it is useful to think of the core of the channel, with volume approximately $A \Delta x$, as being responsible for the inertance, while the lossy inner skin of the channel, with volume approximately $\Pi \Delta x\delta_v$ equal to the channel's surface area times the viscous penetration depth, as being responsible for the resistance.

Example: Orifice Pulse-Tube Refrigerator
Consider the lumped-element components at the top of the orifice pulse-tube refrigerator shown in Fig. 1.18: a compliance, an inertance, and two valves. The purpose of these components was to provide an adjustable acoustic impedance at the top of the pulse-tube refrigerator, with the adjustments made by the valve settings. A simple example occurs through neglect of the resistances in the inertance and compliance, so that the network can be regarded as shown in

---

[4]If the algebra required to go from Eq. (4.19) to Eq. (4.23) seems mysterious, see Exercise 3.3.

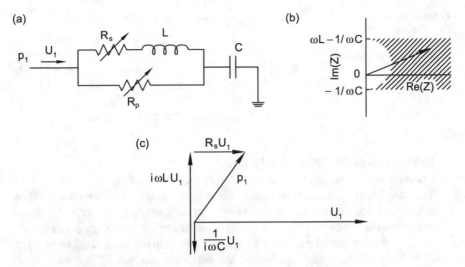

Fig. 4.8 (a) Impedance diagram for the acoustic network at the *top* of the orifice pulse-tube refrigerator. (b) The *shaded zone* of the plot shows values of Z accessible by adjusting the two valves. (c) Phasor diagram for one case in which the valve labeled "R_p" is closed

Fig. 4.8a. One operating point had 40 Hz oscillations in 3.1-MPa helium, with the network at 300 K. The compliance had a volume of 9.8 L, so according to Eq. (4.4) $C = 1.9 \times 10^{-9}\,\mathrm{m^3/Pa}$ and the magnitude of its acoustic impedance is $1/\omega C = 2.1\,\mathrm{MPa\cdot s/m^3}$. The inertance had a length of 2.5 m and a diameter of 2.4 cm, so $L = 26{,}800\,\mathrm{Pa\cdot s^2/m^3}$ according to Eq. (4.7), and the magnitude of its acoustic impedance is $\omega L = 6.7\,\mathrm{MPa\cdot s/m^3}$. To keep this example simple, the two valves can be regarded as resistances R_s and R_p, adjustable from zero to infinity. Then the total acoustic impedance (see Exercise 4.13) provided by the network is

$$Z = \frac{1}{1/(i\omega L + R_s) + 1/R_p} + \frac{1}{i\omega C}. \qquad (4.24)$$

Plotting Eq. (4.24) for all values of R_s and R_p shows that adjustment of the two valves allowed access to the region of complex Z that is shaded in Fig. 4.8b.

Suppose the valves were set so that $R_p = \infty$ (i.e., closed) and $R_s = 3.35\,\mathrm{MPa\cdot s/m^3}$. Then Eq. (4.24) shows that $Z = (3.35 + i4.6)\,\mathrm{MPa\cdot s/m^3}$, so the phase of Z was 54°. The phasor diagram for this case is shown in Fig. 4.8c, where the phase of U_1 was arbitrarily chosen to be zero.

A numerical integration (using DELTAE [61]) of a turbulent version of Eq. (4.19) along the length of the inertance showed that $\omega L = 6.9\,\mathrm{MPa\cdot s/m^3}$ and that the inertance contributed a resistance $R_v \simeq 0.8\,\mathrm{MPa\cdot s/m^3}$ to the network. Hence, the use of Eq. (4.22) to estimate L was accurate, but neglect of R_v relative to ωL was not very good.

Fig. 4.9 Crudest
representation of the
resonator of the
standing-wave refrigerator

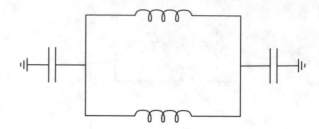

Example: Standing-Wave Refrigerator

Consider one of the central tees in the standing-wave refrigerator of Figs. 1.13
and 1.14. These locations were pressure nodes, so compliance (though it could be
calculated) was irrelevant here; these were locations of high velocity, so inertance
and series resistance were important. Because of the way the black cones fit into
the tees, the horizontal passage through each tee actually comprised three ducts
in series: a 12.7-cm-long, 11.2-cm-diam duct between two 5.7-cm-long, 10.2-
cm-diam ducts. L and R_v can be computed for each of these, and summed, at
a typical operating point: 300 K, 92 Hz, 8% argon and 92% helium at a mean
pressure of 324 kPa. Since δ_v is only 0.3 mm, the boundary-layer Eqs. (4.22)
and (4.23) are appropriate. The resulting sums are $L = 24.1$ Pa·s^2/m^3, $\omega L =
13{,}900$ Pa·s/m^3, and $R_v = 75$ Pa·s/m^3. Indeed, $R_v \ll \omega L$, as expected for a large-
diameter component. These two tees, with parts of the cones nearby, contributed
the inertance necessary for the full-wave resonance. Although the resonator was
not really lumped-element, it could be represented crudely as such, as shown in
Fig. 4.9.

4.2.2 Thermal-Relaxation Conductance

Viscosity is only one of the two principal causes of acoustic attenuation at a solid
boundary. Thermal relaxation can be equally important wherever oscillating pressure exists, as oscillating pressure causes oscillating temperature. Rott's acoustic
approximation to the general equation of heat transfer [46], Eq. (3.36), expresses
the fact that the temperature changes with time because of the sum of three effects:
pressure changes, thermal conduction, and velocity with a temperature gradient. An
example of the pressure-induced temperature changes is shown in Ani. Thermal /w.
The animation shows temperature, pressure, and gas motion in a half-wavelength
resonance. The temperature oscillations are both proportional to and in temporal
phase with the pressure oscillations. Such adiabatic temperature oscillations are
present in all sound waves in free space, although for typical "audio" amplitudes
they are too small to be readily noticed.

To start with a simple case in which temperature changes due to both pressure
changes and thermal conduction are important, consider Fig. 4.10, showing oscillat-

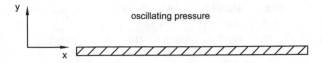

y

oscillating pressure

x

Fig. 4.10 The coordinate system used in this section. In the boundary-layer approximation, the opposite wall of the channel is at such large y that it has no effect here

ing pressure near a solid boundary. With the usual complex notation for the relevant variables, as expressed above in Eqs. (4.11)–(4.14), Eq. (3.36) reduces to

$$i\omega\rho_m c_p\, T_1 - i\omega\, p_1 = k\frac{\partial^2 T_1}{\partial y^2} \tag{4.25}$$

in the channel if $dT_m/dx = 0$ so that the $u_1\, dT_m/dx$ term can be neglected. A solid usually has sufficient heat capacity and thermal conductivity to enforce $T_1 = 0$ on the gas at the solid surface $y = 0$. The other necessary boundary condition is that $T_1 (\infty)$ is finite. Then Eq. (4.25) is an ordinary differential equation for $T_1(y)$, identical in form and boundary condition to Eq. (4.15) for $u_1(y)$. When the similarity is exploited, the solution can be written

$$T_1 = \frac{1}{\rho_m c_p}\left[1 - e^{-(1+i)y/\delta_\kappa}\right] p_1, \tag{4.26}$$

and its spatial average can be written

$$\langle T_1 \rangle = \frac{1}{\rho_m c_p}\left[1 - (1-i)\frac{\delta_\kappa}{2r_h}\right] p_1. \tag{4.27}$$

Equation (4.26) shows how thermal contact with the solid surface reduces the magnitude and shifts the phase of the oscillating temperature, similar to the effect of viscosity on oscillating velocity discussed above. Gas that is much farther than δ_κ from the nearest solid surface is essentially adiabatic, experiencing adiabatic temperature oscillations $T_1 = (1/\rho_m c_p) p_1$ in phase with the pressure oscillations (as shown in Ani. Thermal /w). The temperature of gas that is much closer than δ_κ to the nearest solid surface oscillates little: It is locked to the time-independent wall temperature T_m by the heat capacity of the solid wall. At approximately δ_κ from the nearest solid surface, these oscillations are reduced in magnitude and shifted in phase. In this boundary-layer limit, the functional dependence of Eq. (4.26) is shown in Ani. Thermal /y, in which the tics on the y axis are separated by δ_κ. This animation shows the oscillating temperature as a function of time and distance y from the wall,

$$T(y,t) = T_m + \frac{1}{\rho_m c_p}\mathrm{Re}\left[\left(1 - e^{-(1+i)y/\delta_\kappa}\right) p_1 e^{i\omega t}\right], \tag{4.28}$$

with the motion of the gas toward and away from the wall illustrated by the vertical straight lines moving left and right.

Using Eq. (3.32), the spatially averaged density can be expressed as

$$\langle \rho_1 \rangle = -\frac{\rho_m}{T_m} \langle T_1 \rangle + \frac{\rho_m}{p_m} p_1. \tag{4.29}$$

Averaging Eq. (3.28) over the cross-sectional area of the channel gives

$$i\omega \langle \rho_1 \rangle + \rho_m \frac{d \langle u_1 \rangle}{dx} = 0. \tag{4.30}$$

Using Eqs. (4.29) and (4.27), and eliminating c_p by using Eq. (2.58) in the form

$$c_p = \frac{\gamma}{\gamma - 1} \frac{p_m}{\rho_m T_m}, \tag{4.31}$$

yields

$$i\omega \left[1 + (\gamma - 1)(1 - i) \frac{\delta_\kappa}{2r_h} \right] \frac{p_1}{\gamma p_m} + \frac{d \langle u_1 \rangle}{dx} = 0 \tag{4.32}$$

as an acoustic expression of the continuity equation in the presence of thermal relaxation effects.

In the previous section, the lossless continuity equation led to Eq. (4.2) and the concept of compliance. Here, the heat-transfer equation modifies the continuity equation, leading to Eq. (4.32), which includes both compliance and thermal relaxation. Equation (4.32) can be rewritten for a short channel shown in Fig. 4.11a with volume $V = A \Delta x$ as

$$p_1 = -Z \Delta U_1, \tag{4.33}$$

where Z is the parallel combination of a compliance and a resistance,

$$\frac{1}{Z} = i\omega C + \frac{1}{R_\kappa}, \tag{4.34}$$

as shown in Fig. 4.11b. The effective compliance is given by

$$C \simeq \frac{V}{\gamma p_m} \tag{4.35}$$

to lowest order in δ_κ / r_h, just as in Eq. (4.4). The effective resistance R_κ that is in parallel with the compliance is

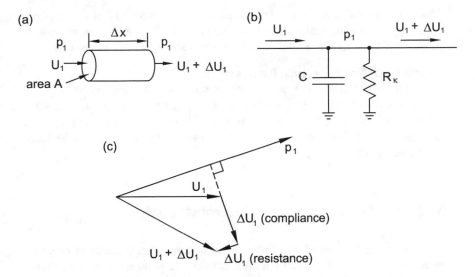

Fig. 4.11 The continuity equation describes compliant and resistive effects in a channel. (**a**) A short channel, of length Δx and cross-sectional area A. (**b**) Impedance diagram showing compliance C and parallel resistance R_κ. (**c**) A typical phasor diagram for this situation, showing that the volume-flow-rate change ΔU_1 is the sum of compliant and resistive components

$$R_\kappa = \frac{2\gamma p_m}{\omega\,(\gamma - 1)\,S\delta_\kappa}\,, \tag{4.36}$$

inversely proportional to the volume $S\delta_\kappa$ that experiences thermal relaxation. Note that Eq. (4.34) shows that a thermal-relaxation resistance is negligible when it is infinite, while the viscous resistance of Eq. (4.21) is negligible when it is zero.

So any channel can be thought of as having both compliance and thermal-relaxation resistance, which cause ΔU_1 to change from one end of the channel to the other. A wide-open volume V, with $V \gg S\delta_\kappa$, has $\omega C \gg 1/R_\kappa$, so the compliance dominates the net impedance. It is useful to think of the core of the volume V, with isentropic bulk modulus $B_s = \gamma p_m$, as being responsible for the compliance, while space near the inner surface, with effective volume $S\delta_\kappa$ equal to the channel's surface area times the thermal penetration depth, has a bulk modulus intermediate between the isentropic and isothermal bulk moduli and is responsible for the thermal-relaxation resistance.

Example: Standing-Wave Engine

Consider one of the outer "hot ducts" (outboard of the hot heat exchangers) in the standing-wave engine of Fig. 1.9. This location was near a velocity node, so inertance and series resistance were unimportant. The compliance and its parallel thermal-relaxation resistance can be computed at a typical operating point: 3-MPa helium, 390 Hz, 800 K. Each hot duct had a volume of 120 cm^3 and a surface area of 130 cm^2 (accounting properly for the surface area and blocked volume due to electric feedthroughs passing through this space). Since δ_κ is only 0.2 mm, boundary-layer Eqs. (4.35) and (4.36) are applicable, yielding $C = 2.4 \times 10^{-11}$ m^3/Pa, $1/\omega C = 17$ MPa·s/m^3, and $R_\kappa = 2350$ MPa· s/m^3.

Example: Orifice Pulse-Tube Refrigerator
The compliance at the top of the orifice pulse-tube refrigerator of Fig. 1.18 had a volume of 9.8 L and a surface area of approximately 0.3 m^2. At 40 Hz with 3.1-MPa helium, $1/\omega C = 2.1$ MPa·s/m^3 was calculated a few pages ago. Equation (4.36) shows that $R_\kappa = 950$ MPa·s/m^3. Hence, $1/R_k$ can be safely neglected in comparison to ωC.

4.3 Inviscid Boundary-Layer Thermoacoustics

Now we will finally proceed to a thermoacoustics problem at the heart of a thermoacoustic engine or refrigerator—a problem with a significant mean temperature gradient along the direction of acoustic oscillation. The simplest such problem neglects viscosity, which is unrealistic because viscous and thermal penetration depths in gases are typically about the same size. Nevertheless, to gain intuition about thermoacoustics, we will begin with a brief consideration of this unrealistic inviscid problem.

Consider a nonzero temperature gradient dT_m/dx along the direction x of acoustic oscillations, with a plane solid boundary at $y = 0$, as shown in Fig. 4.12. Start from Rott's acoustic approximation to the general equation of heat transfer, Eq. (3.36), substitute the appropriate form of Rott's acoustic approximation for the variables

$$p = p_m + \mathrm{Re}\left[p_1(x)e^{i\omega t}\right] + \cdots , \tag{4.37}$$

$$u = \mathrm{Re}\left[u_1(x, y)e^{i\omega t}\right] + \cdots , \tag{4.38}$$

$$T = T_m(x) + \mathrm{Re}\left[T_1(x, y)e^{i\omega t}\right] + \cdots , \tag{4.39}$$

$$\rho, \text{ etc.} = \text{similar to } T, \tag{4.40}$$

and keep first-order terms to obtain the appropriate equation of heat transfer,

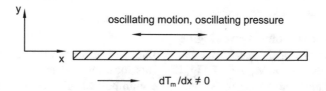

Fig. 4.12 The coordinate system used in Sect. 4.3. In the boundary-layer approximation, the opposite wall of the channel is at such large $y \gg \delta_\kappa$ that it does not appear in this figure

Fig. 4.13 The real and imaginary parts of the y-dependent factor in Eq. (4.42)

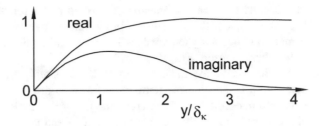

$$\rho_m c_p \left(i\omega T_1 + u_1 \frac{dT_m}{dx} \right) - i\omega p_1 = k \frac{\partial^2 T_1}{\partial y^2}. \qquad (4.41)$$

This differential equation for $T_1(y)$ is similar to Eq. (4.25) but with an additional nonzero dT_m/dx term. With the assumption that the gas viscosity is zero, u_1 is independent of y in this section. The necessary boundary conditions are $T_1(0) = 0$ and $T_1(\infty)$ is finite. The solution is

$$T_1 = \left(\frac{p_1}{\rho_m c_p} - \frac{u_1}{i\omega} \frac{dT_m}{dx} \right) \left[1 - e^{-(1+i)y/\delta_\kappa} \right], \qquad (4.42)$$

which resembles Eq. (4.26), but with more complexity arising from the $u_1 \, dT_m/dx$ term. This solution is the product of two factors. The first factor gives the overall magnitude of the oscillating temperature at large distances from the solid boundary, and the second factor describes the y dependence of that oscillating temperature close to the solid boundary.

The y-dependent factor is the same complex boundary-layer function discussed above, in Eqs. (4.16) and (4.26), and in Anis. Viscous /m and Thermal /y. View Ani. Thermal /y again, and consider Fig. 4.13, which shows the real and imaginary parts of the second factor in Eq. (4.42). The time evolution shown in the animation can be visualized as a smooth progression in time between the real and imaginary parts shown in Fig. 4.13, from real part to minus imaginary part to minus real part to imaginary part and back to real part.

The magnitude factor in Eq. (4.42) is itself the sum of two terms, each of which is easy to understand. The first term is simply the adiabatic temperature oscillation: When the pressure goes up the temperature goes up, and when the pressure goes down the temperature goes down. The second term is due to gas motion along x, along the temperature gradient, and is due to the Eulerian, fixed-in-the-laboratory point of view used in the conventional formulation of the equations of fluid mechanics. In all of these equations, such as Eqs. (4.37)–(4.42) in the present section, (x, y) refers to a fixed location in space, past which gas moves. The $u_1 \, dT_m/dx$ term in Eq. (4.42) simply reflects this reference frame: When a gas with a temperature gradient moves adiabatically past that point in space, the temperature at that point in space changes. (The alternative, Lagrangian point of view focuses greater attention on a particular parcel of gas as it moves, not on a fixed location in space. The Lagrangian viewpoint is often better for developing intuition, so it is used in most of the animations. In the Lagrangian viewpoint, a parcel of gas far from a solid boundary experiences $T_1 = p_1/\rho_m c_p$, quite independent of its velocity or whether a nonzero dT_m/dx exists.)

So the magnitude factor in Eq. (4.42) is simply the linear superposition of adiabatic pressure-induced temperature oscillations and adiabatic motion-induced temperature oscillations. The magnitude factor is complex: It can have any sign and phase, depending on the relative phases of p_1 and u_1 and the relative magnitude of the pressure term relative to the motion term.

For standing-wave phasing, the entire magnitude factor in Eq. (4.42) can be zero, if

$$\frac{p_1}{\rho_m c_p} = \frac{u_1}{i\omega} \frac{dT_m}{dx}. \tag{4.43}$$

In this special circumstance, the gas properties, standing-wave impedance, and temperature gradient conspire so that the pressure-induced temperature oscillation and motion-induced temperature oscillation are equal in magnitude but opposite in sign. This circumstance is illustrated in Ani. Standing /c, where the upper part of the display shows a standing wave in a resonator with a stack near the left end, with blue marker lines showing the moving gas. One particular parcel of gas is highlighted with a moving blue dot in the stack. The yellow oval marks the region that is shown magnified at the left center of the display, which shows that same parcel of moving gas and short fragments of the two stack plates adjacent to it. The volume of that parcel of gas changes in response to pressure and temperature. At the bottom left of the display is a plot of temperature vs position, in which the temperature of the parcel is the blue trace, and that of the nearby plate in the stack is the white line. Here, $r_h \sim \delta_\kappa$, so the thermal contact is poor. The parcel's temperature oscillation is due entirely to adiabatic pressure oscillation, and its temperature oscillation and motion just happen to match the local temperature gradient dT_m/dx exactly, so that the temperature *at a fixed location* is independent of time.

This circumstance is sufficiently important for standing-wave engines and refrigerators that it has a specific name, the critical temperature gradient:

$$\nabla T_{\text{crit}} = \frac{\omega A |p_1|}{\rho_m c_p |U_1|}. \tag{4.44}$$

[The notation $(\nabla T_m)_{\text{crit}}$, though more precise, is too awkward.] Inviscid standing-wave engines have $|dT_m/dx| > \nabla T_{\text{crit}}$, and inviscid standing-wave refrigerators have $|dT_m/dx| < \nabla T_{\text{crit}}$, as can be seen by reviewing Anis. Standing /e and Standing /r introduced in Chap. 1. Although the reality of nonzero viscosity blurs the boundary considerably, ∇T_{crit} as defined in Eq. (4.44) still provides a useful benchmark.

Equation (4.42) is complex enough—in the literal sense of "complex" with real and imaginary parts, and in the common English meaning of "complex"—that the combination of the complex y-dependent factor with the complex magnitude factor can yield essentially any sign, phase, and overall magnitude. Much of the rich variety encountered in thermoacoustics—standing wave behavior, traveling wave behavior, small-pore behavior and well-spaced behavior—arises from the inherent complexity of Eq. (4.42), expressing the simple thermal contact between an acoustic wave and an adjacent solid boundary parallel to the wave propagation direction.

4.4 General Thermoacoustics

Having gained some intuition with the inviscid, boundary-layer thermoacoustics problem in the previous section and the boundary-layer dissipation problems earlier, we now proceed with a fully general derivation of the dynamic equations of thermoacoustics, including viscosity and arbitrary shape and size of channels.

4.4.1 The Math

Naturally, the equations are based on Rott's acoustic approximation, with the relevant variables now written as

$$p = p_m + \text{Re}\left[p_1(x)e^{i\omega t}\right], \tag{4.45}$$

$$U = \text{Re}\left[U_1(x)e^{i\omega t}\right], \tag{4.46}$$

$$u = \text{Re}\left[u_1(x, y, z)e^{i\omega t}\right], \tag{4.47}$$

$$T = T_m(x) + \text{Re}\left[T_1(x, y, z)e^{i\omega t}\right], \tag{4.48}$$

$$\rho = \text{similar to } T, \tag{4.49}$$

$$\mu = \mu(x), \tag{4.50}$$

$$a, k, \text{ etc.} = \text{similar to } \mu. \tag{4.51}$$

This section will focus most on p_1 and U_1, as described by the momentum and continuity equations.

For clarity, the assumption of large solid heat capacity is maintained, so that the temperature of the solid material in the stack is simply $T_m(x)$, independent of time, y, and z. Additionally, only ideal gases are considered. (For finite solid heat capacity or non-ideal gas effects, see [11].)

The presence of nonzero dT_m/dx has no effect on the momentum equation in Rott's acoustic approximation. Hence, the x component of the momentum equation becomes

$$i\omega \rho_m u_1 = -\frac{dp_1}{dx} + \mu \left[\frac{\partial^2 u_1}{\partial y^2} + \frac{\partial^2 u_1}{\partial z^2}\right]. \tag{4.52}$$

Equation (4.52) is a differential equation for $u_1(y, z)$, with boundary condition $u_1 = 0$ at the solid surface. The solution is

$$u_1 = \frac{i}{\omega \rho_m} [1 - h_\nu(y, z)] \frac{dp_1}{dx}. \tag{4.53}$$

This defines the complex function $h_\nu(y, z)$, which depends on the specific channel geometry under consideration [40].

Integrating both sides of Eq. (4.53) with respect to y and z over the cross-sectional area A of the channel gives the volume flow rate U_1 on the left side and converts h_ν on the right side into its spatial average f_ν. Solving for dp_1 yields

$$dp_1 = -\frac{i \omega \rho_m \, dx / A}{1 - f_\nu} U_1. \tag{4.54}$$

In effect, this approximation to the momentum equation is the origin of pressure gradient in thermoacoustics: The motion U_1 of the gas causes the pressure gradient. If $f_\nu = 0$ or has some other purely real value, the pressure gradient is entirely "inertial," as indicated by the "i" in Eq. (4.54), but when $\mathrm{Im}[f_\nu] \neq 0$ the presence of viscosity and the stationary boundaries adds a resistive component to the pressure gradient and also effectively changes the magnitude of the inertial contribution.

The function h_ν and its spatial average f_ν are known for many geometries. The algebra can be intimidating, but Fig. 4.14 shows that the solutions for most of these geometries are actually very similar. As was discussed above, for wide-open channels for which the boundary-layer approximation is appropriate, and with $y = 0$ at the wall,

$$h = e^{-(1+i)y/\delta}, \tag{4.55}$$

$$f = \frac{(1-i)\,\delta}{2 r_h}. \tag{4.56}$$

If $y = 0$ is the center between parallel plates of separation $2y_0 = 2r_h$,

$$h = \frac{\cosh\left[(1+i)\,y/\delta\right]}{\cosh\left[(1+i)\,y_0/\delta\right]}, \tag{4.57}$$

$$f = \frac{\tanh\left[(1+i)\,y_0/\delta\right]}{(1+i)\,y_0/\delta}. \tag{4.58}$$

For circular pores of radius $\mathcal{R} = 2r_h$,

$$h = \frac{J_0\left[(i-1)\,r/\delta\right]}{J_0\left[(i-1)\,\mathcal{R}/\delta\right]}, \tag{4.59}$$

$$f = \frac{2 J_1\left[(i-1)\,\mathcal{R}/\delta\right]}{J_0\left[(i-1)\,\mathcal{R}/\delta\right]\,(i-1)\,\mathcal{R}/\delta}, \tag{4.60}$$

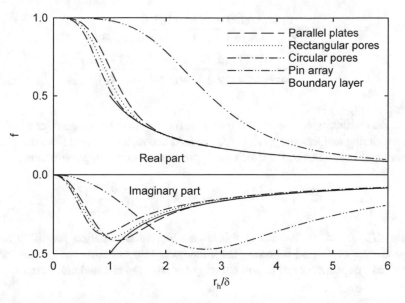

Fig. 4.14 Spatial-average function f for several geometries. Here, the rectangular pores have 6:1 aspect ratio, and the pin array has $r_o/r_i = 6$. The boundary-layer limit is approached at large r_h in all geometries. Using r_h/δ_ν on the horizontal axis yields f_ν on the vertical axis, and using r_h/δ_κ yields f_κ

where the coordinate $r = \sqrt{y^2 + z^2}$. These functions[5] are also known for rectangular channels [40] of dimensions $2y_0 \times 2z_0$:

$$h = 1 - \frac{16}{\pi^2} \sum_{m,n \text{ odd}} \frac{\sin(m\pi y/2y_0)\sin(n\pi z/2z_0)}{mnC_{mn}}, \tag{4.61}$$

$$f = 1 - \frac{64}{\pi^4} \sum_{m,n \text{ odd}} \frac{1}{m^2 n^2 C_{mn}}, \tag{4.62}$$

where

$$C_{mn} = 1 - i\frac{\pi^2 \delta^2}{8y_0^2 z_0^2} \left(m^2 z_0^2 + n^2 y_0^2\right), \tag{4.63}$$

and for the spaces between pins oriented along the direction of acoustic oscillations, each of radius r_i and arranged in a triangular array [62] with center-to-center spacing $\sqrt{2\pi}\, r_o / \sqrt[4]{3} \simeq 1.905 r_o$:

---

[5]Derivation of Eqs. (4.59) and (4.60) is tedious, and derivations of Eqs. (4.61)–(4.65) are *extremely* tedious.

$$h \simeq \frac{Y_1(\alpha_o) J_0(\alpha) - J_1(\alpha_o) Y_0(\alpha)}{Y_1(\alpha_o) J_0(\alpha_i) - J_1(\alpha_o) Y_0(\alpha_i)}, \tag{4.64}$$

$$f \simeq -\frac{2\alpha_i}{\alpha_o^2 - \alpha_i^2} \frac{Y_1(\alpha_o) J_1(\alpha_i) - J_1(\alpha_o) Y_1(\alpha_i)}{Y_1(\alpha_o) J_0(\alpha_i) - J_1(\alpha_o) Y_0(\alpha_i)}, \tag{4.65}$$

where $\alpha = (i - 1) r/\delta$.

In the continuity equation, an expression for the spatial average over y and z of the oscillating temperature, $\langle T_1 \rangle$, is required. As above, it is derived from the general equation of heat transfer, for which the appropriate acoustic approximation is

$$\rho_m c_p \left(i\omega T_1 + u_1 \frac{dT_m}{dx} \right) - i\omega p_1 = k \left[\frac{\partial^2 T_1}{\partial y^2} + \frac{\partial^2 T_1}{\partial z^2} \right] \tag{4.66}$$

when $dT_m/dx \neq 0$. Regarding this as a differential equation for the y and z dependences of T_1, and following the same procedure as before but now including the y and z dependences of u_1 and allowing for arbitrary channel cross section [40], gives[6]

$$T_1 = \frac{1}{\rho_m c_p} (1 - h_\kappa) p_1 - \frac{1}{i\omega A} \frac{dT_m}{dx} \frac{(1 - h_\kappa) - \sigma (1 - h_\nu)}{(1 - f_\nu)(1 - \sigma)} U_1 \tag{4.67}$$

and

$$\langle T_1 \rangle = \frac{1}{\rho_m c_p} (1 - f_\kappa) p_1 - \frac{1}{i\omega A} \frac{dT_m}{dx} \frac{(1 - f_\kappa) - \sigma (1 - f_\nu)}{(1 - f_\nu)(1 - \sigma)} U_1, \tag{4.68}$$

which are comparable to Eqs. (4.26), (4.27), and especially (4.42).

In the continuity equation, the x dependence of ρ_m must now be included:

$$i\omega \langle \rho_1 \rangle + \frac{d}{dx} (\rho_m \langle u_1 \rangle) = 0. \tag{4.69}$$

Substituting Eqs. (4.68) and (4.29) into this yields

$$dU_1 = -\frac{i\omega A \, dx}{\gamma p_m} [1 + (\gamma - 1) f_\kappa] p_1 + \frac{(f_\kappa - f_\nu)}{(1 - f_\nu)(1 - \sigma)} \frac{dT_m}{T_m} U_1. \tag{4.70}$$

This expression finds easy physical interpretation as a complete thermoacoustic approximation to the continuity equation. The terms on the right-hand side in Eq. (4.70) show that a gradient in U_1 can be caused either by pressure or by velocity along the temperature gradient. Consider the pressure term first. If $f_\kappa = 0$, there is

---

[6]This is most easily verified by direct substitution into Eq. (4.66), using the general property that h_ν and h_κ satisfy $\partial^2 h/\partial y^2 + \partial^2 h/\partial z^2 = 2ih/\delta^2$, and using Eqs. (4.53) and (4.54) as needed.

no thermal contact between gas and solid, so the density oscillations are adiabatic. In this case, $1/\gamma p_m$ is the correct compressibility, and the compliance of the segment of channel of length dx must be $A\,dx/\gamma p_m = dV/\gamma p_m$, as was shown in Eq. (4.4). At the other extreme, if $f_\kappa = 1$, the thermal contact between gas and solid is perfect, so the gas is anchored at the local solid temperature. In this case, the isothermal compressibility $1/p_m$ is appropriate. For intermediate thermal contact, an effective compressibility $[1 + (\gamma - 1)f_\kappa]/\gamma p_m$, intermediate in magnitude and with nontrivial phase, describes the spatial average of the density oscillations in response to pressure oscillations. Next, consider the velocity term, in the easily interpreted inviscid limit (with $f_\nu = 0$ and $\sigma = 0$ so that it is simply $f_\kappa U_1\,dT_m/T_m$. If $f_\kappa = 0$, there is no thermal contact between gas and solid, so the velocity term is zero because the temperature of a parcel of gas does not change as it moves along x. In the other extreme, if $f_\kappa = 1$, the gas is always at the local solid temperature, so that as a parcel of gas flows toward higher T_m its density decreases and its velocity increases. In the more general, and more interesting, intermediate regime, oscillating motion of the gas along the temperature gradient leads to complex density oscillations.

Equations (4.54) and (4.70) are applicable to a wide variety of thermoacoustic circumstances, and may be considered two of the principal tools of thermoacoustic analysis. Many earlier expressions in this chapter are simplified limits of these equations, and the next subsection will closely examine the effective compliance, inertance, and resistances indicated by these two equations.

[One more detail: The graphs and equations for f vs r_h/δ presented so far have listed several different geometries of interest mostly for standing-wave thermoacoustics. It is possible to follow through the same kind of analysis for the stacked-screen regenerators that are usually used in Stirling systems. Here is an outline of the derivation [63]. Assuming a sinusoidal volume flow rate through a stacked-screen regenerator, and assuming that published steady-flow data [64] for stacked-screen viscous pressure drop and heat transfer are valid at each instant of time during this oscillation, a Fourier transform of the appropriate continuity and momentum equations can be taken, with those very complicated published curves of pressure drop and heat transfer built in. The results can be rearranged to yield f_κ and f_ν as functions of r_h/δ_κ and r_h/δ_ν, the volumetric porosity of the screens, and the peak Reynolds number of the flow. The results are only valid for small r_h/δ_κ, which is the regime of interest for traveling-wave devices. The results look qualitatively like the low-r_h/δ parts of the curves in Fig. 4.14, with slightly different results for different porosities in the screen bed and for different Reynolds numbers (i.e., different acoustic amplitudes). Another minor detail that is discussed in Ref. [63]: The f_κ that shows up in the p_1 term of the continuity equation is actually a little different than the f_κ that shows up in the U_1 term. See Chap. 7 for further discussion.]

Finally, at this point Eqs. (4.54) and (4.70) can be combined, eliminating U_1, to obtain a second-order differential equation in p_1:

$$[1 + (\gamma - 1)f_\kappa]p_1 + \frac{\gamma p_m}{\omega^2}\frac{d}{dx}\left(\frac{1 - f_\nu}{\rho_m}\frac{dp_1}{dx}\right) - \frac{a^2}{\omega^2}\frac{f_\kappa - f_\nu}{1 - \sigma}\frac{1}{T_m}\frac{dT_m}{dx}\frac{dp_1}{dx} = 0\,.$$

$$(4.71)$$

This is Rott's "wave" equation [12], a milestone in the development of thermoacoustics. (Strictly, it is called a Helmholtz equation [58], because the time derivatives have been replaced by $i\omega$'s.) For numerical computations, it is easiest to use Eqs. (4.54) and (4.70) separately, and greater intuition can usually be gained by considering Eqs. (4.54) and (4.70) separately.

4.4.2 The Ideas

The principal results of the last subsection were the thermoacoustic versions of the momentum and continuity equations, Eqs. (4.54) and (4.70):

$$dp_1 = -\frac{i\omega\rho_m\,dx/A}{1-f_\nu}U_1,$$

$$dU_1 = -\frac{i\omega A\,dx}{\gamma p_m}[1 + (\gamma - 1)f_\kappa]p_1 + \frac{(f_\kappa - f_\nu)}{(1-f_\nu)\,(1-\sigma)}\frac{dT_m}{T_m}\,U_1.$$

The thermoviscous function f allows the description of the three-dimensional phenomena in the channel with these two one-dimensional equations. In this subsection, we will try to gain an intuitive appreciation of these two equations, following the outline indicated in Fig. 4.15. In the figure, the channel of length Δx is considered in two ways: in terms of the momentum equation to obtain its inertance and viscous resistance, and in terms of the continuity equation to obtain its compliance, thermal-relaxation resistance, and thermally induced volume-flow-rate source. Combining these two points of view yields a complete "five-parameter" impedance model for thermoacoustics, just as the momentum and continuity equations provide a complete description of the dynamics linking p_1 and U_1.

Begin with the momentum equation. Rewriting Eq. (4.54) in the form

$$dp_1 = -(i\omega l\,dx + r_\nu\,dx)\,U_1, \tag{4.72}$$

as shown schematically in the left part of Fig. 4.15, shows that the inertance and viscous resistance per unit length of channel can be written

$$l = \frac{\rho_m}{A}\frac{1 - \mathrm{Re}\,[f_\nu]}{|1-f_\nu|^2} \tag{4.73}$$

and

$$r_\nu = \frac{\omega\rho_m}{A}\frac{\mathrm{Im}\,[-f_\nu]}{|1-f_\nu|^2}. \tag{4.74}$$

The boundary-layer expressions earlier in this chapter, Eqs. (4.22) and (4.23), are simply limiting forms of Eqs. (4.73) and (4.74) for large r_h/δ_ν.

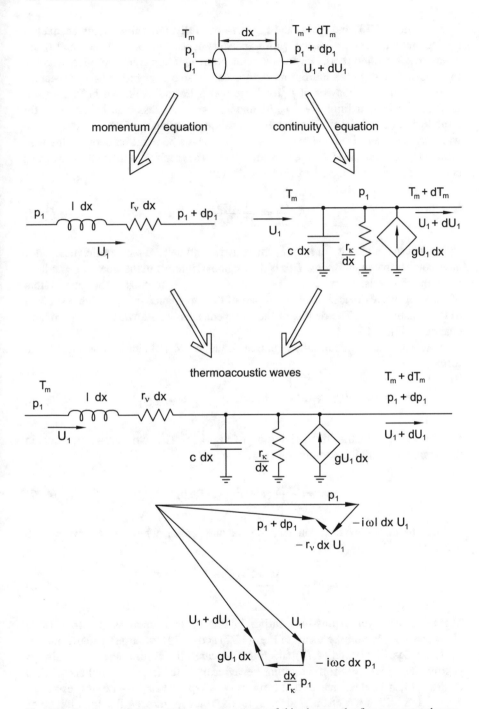

Fig. 4.15 Summary of the most important concepts of this chapter: the five-parameter thermo-acoustic impedance model

Study Eqs. (4.73) and (4.74) and Fig. 4.14 carefully, and think of any channel as having inertance and resistance, with the details depending on f_ν. Both l and r_ν are always positive (and real). Hence, the momentum equation can never give behavior that looks like negative inertial mass, nor can it give a negative flow resistance. From the limiting behavior of f_ν for large hydraulic radius shown in Fig. 4.14, it is clear that in that limit $l \rightarrow \rho_m/A$ and $r_\nu \rightarrow 0$. At less-than-infinite but still large hydraulic radius, r_ν rises above zero as the importance of viscous drag at the wall increases, and l rises above ρ_m/A as the viscous penetration depth at the wall effectively reduces the available flow area. This is the regime of interest in resonator components and in pulse tubes. Forming the ratio

$$\frac{r_\nu}{\omega l} = \frac{\mathrm{Im}\,[-f_\nu]}{1 - \mathrm{Re}\,[f_\nu]}, \tag{4.75}$$

shows that in the vicinity of $r_h \simeq \delta_\nu$ the resistive and inertial parts of the impedance have comparable magnitudes. This is the regime of interest in the stacks of standing-wave thermoacoustic devices. In the smallest passages, such as in the regenerators of traveling-wave devices, the inertial part of the impedance is negligible compared to the resistive part. The details of these impedances are shown for one particular channel in Fig. 4.16.

Now examine the continuity equation. Equation (4.70) can be rewritten in the form

$$dU_1 = -\left(i\omega c\,dx + \frac{1}{r_\kappa}dx\right)p_1 + g\,dx\,U_1, \tag{4.76}$$

as shown schematically in the right part of Fig. 4.15. The two familiar symbols in the figure represent the compliance per unit length

$$c = \frac{A}{\gamma p_m}\,(1 + [\gamma - 1]\,\mathrm{Re}\,[f_\kappa]) \tag{4.77}$$

and the thermal-relaxation conductance per unit length, which is the inverse of a resistance

$$\frac{1}{r_\kappa} = \frac{\gamma - 1}{\gamma}\frac{\omega A}{p_m}\,\mathrm{Im}\,[-f_\kappa]. \tag{4.78}$$

The boundary-layer expressions earlier in this chapter, such as Eqs. (4.35) and (4.36), are simply limiting forms of Eqs. (4.77) and (4.78) for large hydraulic radius.

Study Eqs. (4.77) and (4.78) and Fig. 4.14 carefully. Both c and r_κ are always positive (and real). From the limiting behavior of f_κ for large channel size shown in Fig. 4.14, it is clear that in that limit $c \rightarrow A/\gamma p_m$ and $r_\kappa \rightarrow \infty$. At less-than-infinite but still large channel size, $1/r_\kappa$ rises above zero as the importance of thermal relaxation at the wall increases, and c rises above $A/\gamma p_m$ as the thermal penetration depth at the wall contributes a greater compressibility per unit volume

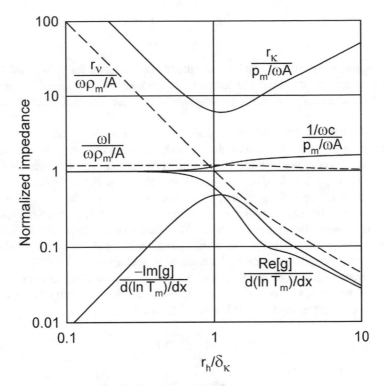

Fig. 4.16 Normalized impedances per unit length, ωl, r_ν, $1/\omega c$, r_κ, and g, for a gas with $\gamma = 5/3$ and $\sigma = 2/3$ in a parallel-plate channel. On this log-log plot, the slopes approach 0, $+2$, or -2 at small r_h/δ_κ and 0, $+1$, or -1 at large r_h/δ_κ

than the compressibility far from the wall. This is the regime of interest in resonator components and in pulse tubes. Forming the ratio

$$\frac{1}{\omega r_\kappa c} = \frac{(\gamma - 1)\,\mathrm{Im}\,[-f_\kappa]}{1 + (\gamma - 1)\,\mathrm{Re}\,[f_\kappa]}, \tag{4.79}$$

shows that in the vicinity of $r_h \simeq \delta_\kappa$ the thermal-relaxation resistance and the compliance parts of the impedance are of the same order of magnitude. This is the regime of interest in the stacks of standing-wave thermoacoustic devices. In the smallest passages, such as in the regenerators of traveling-wave devices, the thermal-relaxation-resistance part of the impedance is negligible compared to the compliance part. The details of these impedances are shown for one particular channel in Fig. 4.16.

The third, new symbol in the impedance diagram for the continuity equation represents a controlled source $g\,dx\,U_1$ (or sink, depending on sign) of volume flow rate [23, 42], proportional to the local volume flow rate U_1 itself, with proportionality constant

$$g = \frac{(f_\kappa - f_\nu)}{(1-f_\nu)\,(1-\sigma)} \frac{1}{T_m} \frac{dT_m}{dx}, \tag{4.80}$$

which represents a sort of complex gain/attenuation constant for volume flow rate, and which arises only when the temperature gradient dT_m/dx along the channel is nonzero. The details of this impedance are shown for one particular channel in Fig. 4.16.

The gU_1 term in the continuity equation has no dependence on p_1, so it must not represent any sort of compressibility. To understand this term, let $p_1 = 0$ so that c and r_κ do not confuse the issues in this paragraph. The dependence of the gU_1 term on dT_m/dx is key. In a pulse tube or other large-diameter thermal buffer tube, f_κ and f_ν are very small, so that $g \simeq 0$ even though $dT_m/dx \neq 0$. In this case, the gU_1 term in the continuity equation simply says that whatever volume flow rate goes in one end comes out the other end. The behavior is essentially the same as the displacement of a solid piston with a temperature gradient and mass-density gradient. At the opposite extreme, if a nonzero temperature gradient dT_m/dx exists along a channel with very small pore size, such as in a regenerator, the volume-flow-rate source term gU_1 is very important [23]. The small-channel limit of Eq. (4.80) is important and easy to appreciate[7]: For $r_h \ll \delta_\kappa$ and $r_h \ll \delta_\nu$,

$$gU_1 \simeq \frac{1}{T_m} \frac{dT_m}{dx} U_1. \tag{4.81}$$

In this case, the gU_1 term in the continuity equation says that $dU_1/U_1 = dT_m/T_m$: The volume flow rate is amplified in proportion to the temperature rise (or attenuated in proportion to a temperature drop). This is easy to understand as constancy of first-order mass flow $\rho_m U_1$, which for an ideal gas is equivalent to constancy of U_1/T_m: Whatever mass flow goes in one end must come out the other.

The even greater complications of the intermediate regime, $r_h/\delta_\kappa \sim 1$, encountered in the stacks of standing-wave systems, are suggested by the boundary-layer limit of Eq. (4.80):

$$gU_1 \simeq \frac{1-i}{2} \frac{1}{1+\sqrt{\sigma}} \frac{\delta_\kappa}{r_h} \frac{1}{T_m} \frac{dT_m}{dx} U_1. \tag{4.82}$$

In this case, the volume-flow-rate source is proportional to the volume flow rate itself, but with a phase shift of $-45°$.

If now the momentum and continuity pictures are combined, as shown at the bottom of Fig. 4.15, a complete, general five-parameter impedance picture of thermoacoustics in any channel is created. The principal variables p_1, U_1, T_m, dT_m, and the geometry can be regarded as given, so the impedance diagram serves as a reminder of how the continuity and momentum equations yield dp_1 and dU_1.

---

[7]But not easy to obtain. Use l'Hôpital's rule twice.

In most circumstances in a given location in a thermoacoustic system, many or most of the five components in this general impedance model can be neglected. With $\delta_\kappa \sim \delta_\nu$ for ideal gases, this table summarizes relative sizes and importance:

Resonator	$\delta \ll r_h$	$r_\nu \ll \omega l$	$1/r_\kappa \ll \omega c$	$g = 0$
Pulse tube	$\delta \ll r_h$	$r_\nu \ll \omega l$	$1/r_\kappa \ll \omega c$	$g \sim 0$
Stack	$\delta \sim r_h$	$r_\nu \sim \omega l$	$1/r_\kappa \sim \omega c$	g complex
Regenerator	$\delta \gg r_h$	$r_\nu \gg \omega l$	$1/r_\kappa \ll \omega c$	$g \simeq \nabla T_m/T_m$

Further insight can be achieved by considering the relative sizes of p_1 and U_1 in a given component, to determine whether either l or c can be neglected.

Example: Standing-Wave Engine
Each stack in the engine of Figs. 1.8 and 1.9 was of parallel-plate construction, with gaps between plates of 0.010 in., so $r_h = 0.13$ mm. At the stack center, where the temperature was about 550 K, the thermal and viscous penetration depths were $\delta_\kappa = 0.12$ mm and $\delta_\nu = 0.10$ mm at a typical operating point with 3-MPa helium at 390 Hz. Hence, with $r_h/\delta_\kappa = 1.1$ and $r_h/\delta_\nu = 1.3$, all of the impedance components—inertance, viscous resistance, compliance, thermal-relaxation resistance, and volume-flow-rate source—could be important. However, the stacks were close to the velocity nodes of the standing wave, so it is likely that the inertance and viscous resistance were less important than the other three components. The impedance diagram shown in Fig. 4.17 probably has adequate detail to convey the most important features of the apparatus. The overall resonance is portrayed as Helmholtz-like, with an inertance between two composite compliances. Although this level of detail does not show the wave nature of the x dependences of p_1 and U_1 in the apparatus, it does successfully show that the pressures on the left and right halves of the apparatus were 180° out of phase, with the pressure on the left half of the apparatus leading the volume flow rate through the inertance by 90° and the pressure on the right half of the

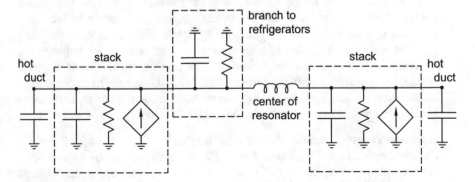

Fig. 4.17 Schematic impedance diagram for the standing-wave engine example

apparatus lagging the volume flow rate through the inertance by 90° (positive volume flow rate being to the right). Judging by eye in Fig. 1.9 that the volumes of gas in each stack and hot duct were comparable, the stack compliance must have been as important as the hot-duct compliance; i.e., the stack compliance caused ΔU_1 across the stack to be nearly as large as U_1 itself at the hot end of the stack. It turns out that the thermal-relaxation resistance and volume-flow-rate source in the stack caused a smaller change in $|U_1|$ and only about a 5° phase shift between U_1 at the left and right ends of each stack. The effects of the branch to the refrigerators were comparable.

A numerical integration of Eqs. (4.54) and (4.70) gives more detailed information about p_1 and U_1 in this system. I gave DELTAE all the geometry of half of the apparatus, the known acoustic impedance of the branch to the refrigerator, the helium pressure, temperatures, etc., picked one particular operating amplitude, and asked DELTAE to tabulate p_1 and U_1 as a function of position in the apparatus. (See Appendix B for details.) The result is shown in Fig. 4.18. I chose the phase of p_1 to be zero at $x = 0$. Then for a nearly standing wave, the pressure should remain nearly real and should look largely like a cosine of x, and the volume flow rate should be nearly imaginary and should look largely like a sine of x, as shown in the figure. The dependence of A on x, apparent in the scale drawing, causes $|p_1|$ and $|U_1|$ to deviate from perfect trigonometric functions, but these small deviations are not obvious in the figure. For the most part, dp_1/dx is due to inertance and U_1, and dU_1/dx is due to compliance and p_1.

However, there are small out-of-phase components to p_1 and U_1, which have been multiplied by 20 and 10 respectively in Fig. 4.18 in order to display them clearly. As with many standing-wave systems, the phase difference between p_1 and U_1 remains so close to 90° that phasor diagrams are of limited usefulness. The step in $\mathrm{Re}[U_1]$ occurs at the branch to the refrigerators, and accounts for the U_1 flowing into the refrigerators. The upward slope of $\mathrm{Re}[U_1]$ through the stack is due to the volume-flow-rate source term throughout the stack, and the upward slope of $\mathrm{Im}[p_1]$ is due to r_v in the stack. The weaker slopes of $\mathrm{Re}[U_1]$ and $\mathrm{Im}[p_1]$ elsewhere are partly due to r_κ and r_v elsewhere, but also from c and l interacting with $\mathrm{Im}[p_1]$ and $\mathrm{Re}[U_1]$, respectively.

To put more of the reality of the numerical integration into the impedance diagram shown in Fig. 4.18a, more impedance components might be included. Since $\mathrm{Im}[p_1]$ changes dramatically through the stack, addition of R_v in the stack might be wise; and inclusion of a compliance for the gentle cone between the ambient heat exchanger and the central, straight portion of the resonator would account for the change of $\mathrm{Im}[U_1]$ from -0.058 to $-0.098\,\mathrm{m^3/s}$ from one end of the cone to the other. There is never a unique "best" crude impedance diagram— it is always a matter of judgment what "essential" features should be represented.

Example: Thermoacoustic-Stirling Heat Engine

The regenerator in the engine of Figs. 1.22 and 1.23 was a pile of stainless-steel screen with a hydraulic radius of 42 μm. At a typical operating point, at 80 Hz and with 3-MPa helium gas, and with the center of the regenerator at

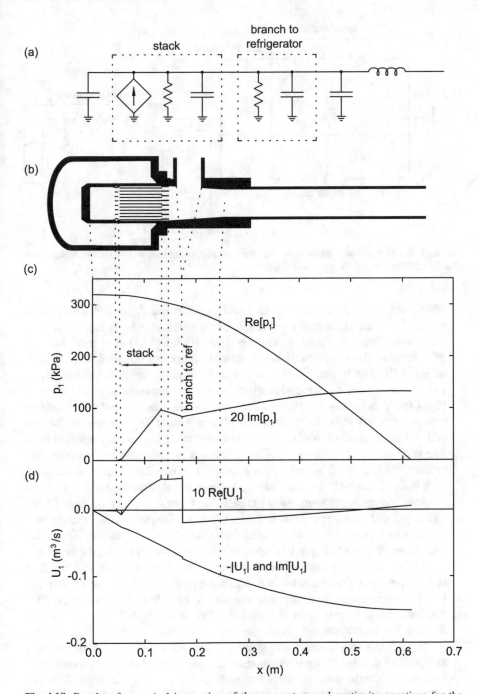

Fig. 4.18 Results of numerical integration of the momentum and continuity equations for the standing-wave engine example first introduced in Figs. 1.8 and 1.9. (**a**) Relevant portions of crude impedance diagram. (**b**) Scale drawing of apparatus. (**c**) p_1. (**d**) U_1. The numerical integration unrealistically assigned the entire impedance of the "branch to refrigerator" at the point $x = 0.18$ m, causing unrealistic discontinuities there

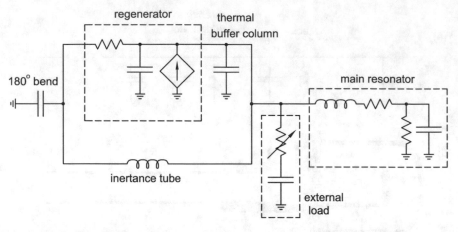

Fig. 4.19 Crude impedance diagram for thermoacoustic-Stirling heat engine example. See Exercise 4.14 for the corresponding phasor diagram

650 K, the penetration depths were $\delta_\kappa = 300\,\mu m$ and $\delta_\nu = 250\,\mu m$. Hence, $r_h \ll \delta$, so inertance and thermal-relaxation resistance were negligible, and the volume-flow-rate source is well described by Eq. (4.81). The regenerator is adequately modeled with the three impedance components labeled "regenerator" in Fig. 4.19. The thermal buffer tube was 9 cm diam, so it had $\delta \ll r_h$, so its most important dynamic characteristic was its compliance, also shown in Fig. 4.19. The feedback path is most roughly modeled as an inertance (the straight section) in series with a compliance (the 180° U bend at the left end). The resistances associated with these two components are neglected here. The resonator to the right of the junction was approximately an inertance (the uniform-diam section) in series with a compliance (the big volume on the end), with both of these components having associated resistances. Neither of these resonator components were really lumped, but those details are neglected here. The adjustable load on the system, comprising an adjustable valve in series with a tank, can be modeled as a resistance in series with a compliance. Hence, the impedance diagram of Fig. 4.19 shows the most important features of this system.

Animation Tashe /t shows these features for the thermoacoustic-Stirling heat engine, omitting the resonator and adjustable load. Study the animation and Fig. 4.19 together. In the animation, you should be able to "see" many features that have been discussed in this chapter. The expected behavior of each of the two large compliances should be apparent, with U_1 "in" differing from U_1 "out." Close examination will show the 90° phase difference between ΔU_1 and p_1: The density of the blue gas-marker lines rises while the pressure rises. The joining conditions—conservation of p_1 and U_1 at the transitions between components—should be apparent at the ends of the inertance and easy to imagine at the three-way junction. It should be apparent that the velocity through the regenerator, which is resistive, is in phase with Δp_1, while the velocity through

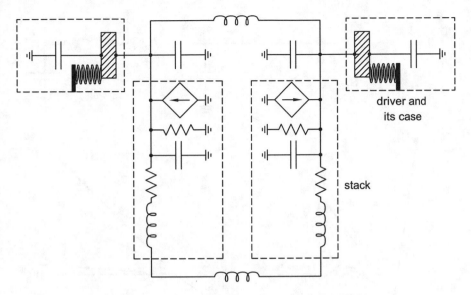

Fig. 4.20 Crude impedance diagram for the standing-wave refrigerator example

the inertance lags Δp_1 by 90°. I know that the compliance and volume-flow-rate source associated with the regenerator are correctly programmed in this animation too, but I confess they are too subtle for me to see here.

Example: Standing-Wave Refrigerator

Figure 4.20 shows a lumped-impedance diagram for the standing-wave refrigerator of Figs. 1.13 and 1.14. The stacks had $r_h/\delta \sim 1$, so all of the impedance components—inertance, viscous resistance, compliance, thermal-relaxation resistance, and volume-flow-rate source—could be important in them. For the standing-wave engine above, the stacks were close to the velocity nodes of the standing wave, so their inertance and viscous resistance could be neglected to some extent. In this refrigerator, the stacks were not so close to the velocity nodes, so we cannot neglect these momentum-equation contributions. The loudspeakers contributed significantly to the dynamics, so at least the moving mass and spring constant of each should be included. I've made up new symbols for these—a rectangular brick for mass, and something that reminds me of an automotive coil spring for the spring constant.

Example: Orifice Pulse-Tube Refrigerator

The impedance diagram of Fig. 4.21a symbolically represents the most important dynamic features of the orifice pulse-tube refrigerator of Figs. 1.17 and 1.18. The rightmost compliance C is the so-called compliance itself. The adjacent inertance L and resistance R_s represent the refrigerator's inertance tube and two valves, with valve settings the same as in the related example near Fig. 4.8 above: The parallel valve R_p in Fig. 4.8 is closed, and the series valve R_s in Fig. 4.8 is adjusted

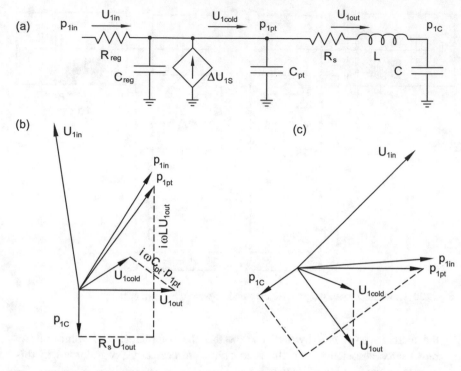

Fig. 4.21 (**a**) Crude impedance diagram of the orifice pulse-tube refrigerator. (**b**) Phasor diagram, with the phase of $U_{1\text{out}}$ set to zero. (**c**) Same phasor diagram, but rotated so that the phase of p_{1pt} is set to zero

so that $R_s = \omega L/2$. The compliance of the pulse tube is significant, and is shown as C_{pt}. With $r_h/\delta \sim 0.1$ in the regenerator, its compliance C_{reg}, viscous resistance R_{reg}, and volume-flow-rate source $\Delta U_{1S} = -U_{1\text{in}}(T_{\text{in}} - T_{\text{cold}})/T_{\text{in}}$ represent the regenerator well.

I used the impedance diagram of Fig. 4.21a and quantitative information about the hardware geometry to construct the phasor diagram of Fig. 4.21b, starting from the right end of the impedance network and working to the left. Part of the work was already done in Fig. 4.8c above; p_{1C}, $U_{1\text{out}}$, and p_{1pt} are copied from that figure. The compliance of the pulse tube then determines $U_{1\text{cold}}$; next C_{reg} and ΔU_{1S} determine $U_{1\text{in}}$; finally R_{reg} determines $p_{1\text{in}}$.

Example: Imperfect Inertance

Sometimes it is impossible to provide a desired phase shift in Z using an inertance, because the unavoidable compliance and viscous resistance can cause phase shifts of the opposite sign, as illustrated in Fig. 4.22. These variables are not completely independent, because the cross-sectional area, volume, and

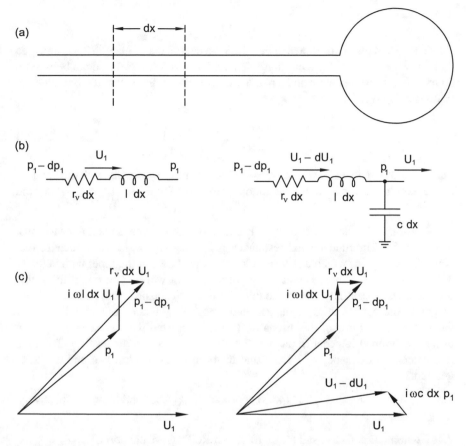

Fig. 4.22 (a) A lossy inertance in series with a compliance, such as might be used at the end of a small pulse-tube refrigerator. A short length dx of the lossy inertance is highlighted. (**b**) Impedance diagrams for the short length dx of the lossy inertance; *on the left*, the compliance per unit length is negligible, while *on the right* it is not negligible. (**c**) Phasor diagrams for the short length dx of the lossy inertance; *on the left*, the compliance per unit length is negligible, while *on the right* it is not negligible. The phase shift in Z can be either positive or negative, depending on the magnitudes of the phasors that are proportional to dx

surface area of a tube are not independent. Suppose that the viscous resistance r_v and the compliance c per unit length of this inertance are not negligible, as shown in the figure, and suppose it is desired that Z_{in} should lead Z_{out}. Figure 4.22b shows that, even though l works in the correct direction, causing p_{1in} to lead p_{1out}, the compliance c works to shift U_{1in} ahead of U_{1out}, so that the impact on the phase of Z could have either sign, depending on the quantitative details of l, r_v, and c.

4.5 Exercises

4.1 The sealed resonator shown in Fig. 4.23—a central duct of length $2\Delta x_c$ and area A_c between two other ducts of length Δx_b and area A_b—is intermediate between a double Helmholtz resonator and a plane-wave resonator. Show that the fundamental frequency is given by

$$\frac{A_b}{A_c} \tan \frac{\omega \Delta x_c}{a} \tan \frac{\omega \Delta x_b}{a} = 1. \tag{4.83}$$

Show that this reduces to something sensible when $A_b = A_c$. Show that it reduces to the expression for a double Helmholtz resonator for $A_c < A_b$, $\Delta x_c \ll \lambda$, and $\Delta x_b \ll \lambda$.

4.2 Consider a lossless duct of uniform cross-sectional area, sealed at both ends, of length Δx. The fundamental resonance frequency is $f = a/2\,\Delta x$, because that is the frequency for which a half-wavelength wave "fits" in the pipe, with velocity nodes at the sealed ends. If you knew nothing about waves, you might think this resonator was essentially like the double Helmholtz resonator, with an inertial mass of gas in the central third of the resonator bouncing against the compliances of gas in the outer thirds of the resonator. If you calculated the resonance frequency using this lumped-impedance picture, how different would your result be from the true resonance frequency $a/2\,\Delta x$? What if you used the central half and the outer quarters?

4.3 Verify that Eqs. (4.57) and (4.58) approach Eqs. (4.55) and (4.56), respectively, as $r_h/\delta \to \infty$.

4.4 Combine Eqs. (4.54) and (4.70) to obtain Rott's wave equation.

4.5 Check the derivations of all equations in this chapter, except Eqs. (4.57)–(4.65).

4.6 Look closely at Anis. Viscous and Oscwall. Use "pause" on your keyboard to study "Viscous" very carefully. It looks like the gas motion about a penetration depth

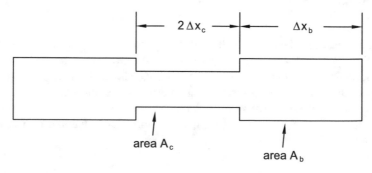

Fig. 4.23 Geometry for Exercise 4.1

away from the wall *leads* the motion farther from the wall. Does this make sense, or did the author make a minus-sign mistake when he made the animation? Interpret Ani. Viscous in terms of the superposition of the "acoustic mode" and "vorticity mode" discussed in advanced acoustics texts [58, 59].

4.7 The specific acoustic impedance of a wave in a duct is defined as

$$z = \frac{A p_1}{U_1}. \tag{4.84}$$

Show that the specific acoustic impedance of a lossless plane traveling wave is $\pm \rho a$. Show that the specific acoustic impedance of a lossless plane standing wave is $\pm \rho a \tan [\omega (x - x_0) / a]$. (Ignore minus signs.) As the wave crosses from a first duct of area A_f to a second duct of area A_s, which impedance is continuous, z or Z?

4.8 Sketch $|p_1|$ and either $|U_1|$ or $|\langle u_1 \rangle|$ vs position for your favorite piece of hardware. If useful, sketch the phasors.

4.9 Estimate the inertance, compliance, and resistances of some components in your favorite piece of thermoacoustics hardware. Compare R_ν, ωL, $1/\omega C$, and R_κ. Do the relative magnitudes make sense?

4.10 A Helmholtz resonator consists of a 4-L spherical volume V and a cylindrical neck having a diameter of $2r = 2$ cm and a length $\Delta x = 5$ cm. It is filled with air at 300 K and 1 bar. What is the inertance of the neck? How does it compare with the mass of gas in the neck, and what physics must you invoke to compare these two things that have different units of measure? What is the compliance of the volume? What is the resonance frequency? At resonance, how does ωL compare with $1/\omega C$? How does λ compare with the various dimensions in the problem? How do δ_ν and δ_κ compare with the other dimensions?

4.11 Draw a reasonably detailed impedance diagram for your favorite piece of thermoacoustics hardware.

4.12 In this chapter, my choice of U_1 and p_1 as the variables of greatest interest led to the forms of the momentum and continuity equations that we used, and from there to the forms of the impedance diagrams that we used. However, other choices can be made, with equal success if employed self-consistently. Your task: reconstruct some of the important results and figures of this chapter from a different point of view, with the variables of greatest interest being the average gas velocity $\langle u_1 \rangle$ and the force $F_1 = A p_1$ exerted by the gas at x on the gas at $x + dx$. To construct impedance diagrams, invent symbols for moving mass m, spring K, dashpots R_ν and R_κ, and thermally induced velocity source $d \langle u_{1S} \rangle$. Draw your version of the impedance diagram for the most general thermoacoustic element of length dx, thinking carefully about which components should be drawn in parallel (sharing the same $\langle u_1 \rangle$) and in series (sharing the same F_1). What are the joining conditions from one segment to the next, corresponding to our use of continuity of p_1 and U_1

in the text? Do you think your version is more intuitively understandable than the version used in the text? Why or why not?

4.13 Use Rott's acoustic momentum equation to show that, if any two acoustic components having complex acoustic impedances Z_a and Z_b are connected in series, the net complex acoustic impedance Z is given by $Z = Z_a + Z_b$. Similarly, for $dT_m/dx = 0$, use Rott's acoustic continuity equation to show that, if any two acoustic components having complex acoustic impedances Z_a and Z_b are connected in parallel, the net complex acoustic impedance Z is given by $1/Z = 1/Z_a + 1/Z_b$. Work out a few examples for complicated impedance diagrams in this chapter, such as the four-component network "main resonator" in Fig. 4.19.

4.14 Assign names to the pressures and volume flow rates at key locations in Fig. 4.19. (Remember to define a positive direction for volume flow rate.) Construct a phasor diagram, by examining the figure and Ani. Tashe /t. Don't worry about details—just try to get relative phase angles correct to within ±45°. Repeat this exercise for Fig. 4.20.

4.15 Add neck viscous resistance and bulb thermal-relaxation resistance to the impedance diagram of the double Helmholtz resonator of Fig. 4.4b. Write expressions for these resistances in terms of the dimensions shown in Fig. 4.4a and the gas properties. Add a piston driver to one bulb of Fig. 4.4a, so that the resonator can be driven on resonance. Which way do the pressure phasors shift in Fig. 4.4c, if the U_1 phasor is kept along the real axis? Under what circumstances is the effect of the thermal-relaxation resistance negligible compared to that of the viscous resistance? [65]

4.16 Draw phasors for Ani. Viscous /m: Take dp_1/dx to be real, and draw phasors for $u_1(y)$ for $y = n\delta_v/4$, n an integer from 0 to 20. You can decide whether to omit the arrow heads to avoid clutter, or whether it is clearer to displace the phasors for successive n vertically instead of having all phasors come from the same origin as in Fig. 3.6.

4.17 Show that Eq. (4.68) simplifies to become

$$\langle T_1 \rangle = \frac{p_1}{\rho_m c_p}(1 - f_\kappa) - \frac{U_1}{i\omega A}(1 - f_\kappa)\frac{dT_m}{dx} \qquad (4.85)$$

when $\sigma = 0$. Explain the simplification of the last term by considering moving parcels of gas.

4.18 In the discussion near Eqs. (4.43) and (4.44), $T_1(y) = 0$ for all y for standing-wave phasing when $\mu = 0$ and $dT_m/dx = \nabla T_{crit}$. (a) Investigate this issue for standing-wave phasing when $\mu \neq 0$, in boundary-layer approximation. Show that there is no single value of dT_m/dx that makes $T_1 = 0$ for all y. Show that $T_1 \simeq 0$ for $y \ll \delta_\kappa$ and $y \ll \delta_v$ when $dT_m/dx = (1 + \sqrt{\sigma})\nabla T_{crit}$ [66]. (b) Making $r_h \ll \delta_\kappa$ is another way to ensure $T_1 \simeq 0$. For traveling-wave phasing, is there any other way?

Chapter 5
Power

Products of first-order variables (such as p_1 and U_1) represent power, which is of central importance in thermoacoustic engines and refrigerators. Introductory acoustics textbooks teach that

$$\frac{\omega}{2\pi} \oint \text{Re}\left[p_1 e^{i\omega t}\right] \text{Re}\left[\mathbf{v}_1 e^{i\omega t}\right] dt = \frac{1}{2}\text{Re}\left[p_1 \widetilde{\mathbf{v}}_1\right], \qquad (5.1)$$

which is called the acoustic intensity, is the time-averaged "power per unit area" in a sound wave. It's not so simple in thermoacoustics. Great care must be taken to identify which type of power we are talking about, because there are so many important types of energy and power in thermodynamics—enthalpy, heat, Gibbs free energy, etc.

The concept of acoustic intensity is so familiar to acousticians that we are reluctant to abandon it. Hence, this book will make as much use as possible of the integral of the acoustic intensity across the cross-sectional area of the channel:

$$\dot{E}_2(x) = \frac{\omega}{2\pi} \oint \text{Re}\left[p_1(x)e^{i\omega t}\right] \text{Re}\left[U_1(x)e^{i\omega t}\right] dt \qquad (5.2)$$

$$= \frac{1}{2}\text{Re}\left[p_1 \widetilde{U_1}\right] = \frac{1}{2}\text{Re}\left[\widetilde{p_1} U_1\right] \qquad (5.3)$$

$$= \frac{1}{2}|p_1|\,|U_1|\cos\phi_{pU}, \qquad (5.4)$$

where ϕ_{pU} is the phase angle between p_1 and U_1 and the tilde denotes complex conjugation. This is the *acoustic power* flowing in the x direction, with the subscript 2 showing that it is second order—the product of two first-order quantities. Note also that it is a time average: The *instantaneous* power delivered along x is of no interest; $\dot{E}_2$ is the power averaged over an integral number of cycles of the wave.[1]

---

[1] Whispering generates about 10^{-9} W of acoustic power. Shouting generates about 10^{-3} W.

© Acoustical Society of America 2017
G.W. Swift, *Thermoacoustics*, DOI 10.1007/978-3-319-66933-5_5

Similarly, enthalpy is very familiar to mechanical and chemical engineers, because it is the central energy of fluid dynamics, of great utility when considering the first law of thermodynamics. Rott's acoustic approximation to the time-averaged *total power* flowing in the positive x direction is derived later in this chapter:

$$\dot{H}_2(x) = \frac{1}{2}\rho_m \text{Re}\left[h_1 \widetilde{U_1}\right] - (Ak + A_{\text{solid}}k_{\text{solid}}) \frac{dT_m}{dx}, \tag{5.5}$$

where h is the enthalpy per unit mass.

This chapter and Chap. 6 present the use and meaning of $\dot{E}_2$ and $\dot{H}_2$. These have the subscript "2" because they are of second order in smallness (e.g., $|p_1| |u_1| \lll p_m a$). In Chap. 3 we called such variables "small squared" and neglected them compared to the merely "small" first-order variables. However, the expressions for power derived below have no first-order power terms, so the second-order terms, though formally small, are the largest terms under consideration and cannot be neglected.[2]

5.1   Acoustic Power

Acoustic power depends strongly on the phase angle between p_1 and U_1. Whether using Eq. (5.2), (5.3), or (5.4) for acoustic power $\dot{E}_2$, it is immediately apparent that $\dot{E}_2 = 0$ when the phase between pressure and volume flow rate is 90°, i.e., for standing-wave phasing. Taking the positive x direction to the right, a pure rightward traveling wave has $\dot{E}_2 > 0$ and a pure leftward traveling wave has $\dot{E}_2 < 0$. Whenever $|\phi_{pU}| < 90°$, acoustic power flows in the positive x direction; when this angle is between 90° and 180°, acoustic power flows in the negative x direction, as illustrated in Fig. 5.1.

The utility of $\dot{E}_2(x)$ in acoustics is due largely to its intuitive appeal as describing a sort of flux of mechanical power past the location x. This interpretation is possible because the volume flow rate $U_1(x)$ is equal (to first order only!) to the volume flow rate of the particular slab of gas whose average position is at x. Letting V be the oscillating part of the volume of gas to the left of this slab of gas, so $U = dV/dt$,

$$\dot{E}_2(x) = \frac{\omega}{2\pi} \oint pU \, dt = \frac{\omega}{2\pi} \oint p \, dV, \tag{5.6}$$

which is the standard thermodynamic expression [Eq. (2.2)] for the time-averaged rate at which work is done by a piston. Hence, $\dot{E}_2(x)$ can be interpreted as giving the

---

[2]It is extremely fortunate that $\dot{E}_2$ and $\dot{H}_2$ depend only on products of first-order variables such as p_1 and T_1, as shown in this chapter, because otherwise Chaps. 3 and 4 would have required the great additional effort necessary to obtain expressions for second-order pressure, temperature, etc., as will be introduced briefly in Chap. 7.

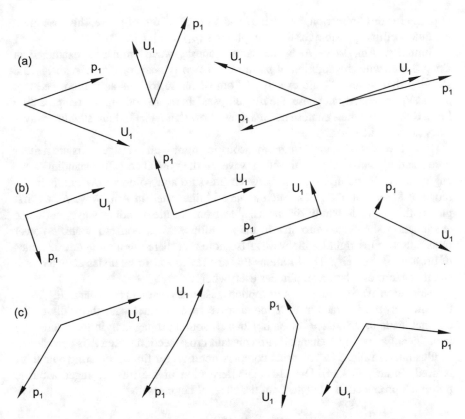

Fig. 5.1 (a) Typical phasor diagrams for positive acoustic power flow in the positive x direction. The angle between p_1 and U_1 is acute. (b) Typical phasor diagrams for zero acoustic power flow. The angle between p_1 and U_1 is 90°. (c) Typical phasor diagrams for acoustic power flow in the negative x direction. The angle between p_1 and U_1 is obtuse

work done by a slab of gas, whose average position is x, on the gas in front of it, as if the slab of gas whose average position is x were a solid piston.[3]

These points are illustrated in Anis. Wave /v and Wave /u. Animation Wave /v is the same rightward-traveling wave examined in Chap. 3, but with additional purple ellipses representing $\dot{E}_2$ at two typical locations. The horizontal coordinate tracing the ellipse is equal to the location of the slab of gas in question; the vertical coordinate is p. Hence, the purple area of the ellipse is proportional to $\oint p\,dV$, and so this area can be taken as a representation of $\dot{E}_2$. The clockwise rotation tracing the ellipse indicates that $\dot{E}_2 > 0$. With Fig. 3.6 showing that p_1 and U_1 are exactly

---

[3]Be wary of pushing this interpretation too far. In a stack or regenerator, this imaginary solid piston would have to absorb or reject nonzero *net* heat! The next chapter will show rigorously that $\dot{E}_2$ is the ability to do work, but only in some circumstances. See discussion below Eq. (6.26).

in phase for this animation, i.e., that p_1 and V_1 are 90° out of phase, this case gives the fattest ellipse possible for these amplitudes $|p_1|$ and $|U_1|$.

Similarly, Ani. Wave /u is the same standing-wave animation examined in Chap. 3, but with the addition of pV traces at two typical locations. With Fig. 3.6 showing that p_1 and U_1 are exactly 90° out of phase for this animation, i.e., that p_1 and V_1 are exactly in phase, the pV "ellipses" here are nothing but reciprocating lines. Since these lines enclose no area, they show that $\dot{E}_2 = 0$. Pure standing waves carry no acoustic power.

Typical waves of interest in thermoacoustic engines and refrigerators are neither pure standing wave nor pure traveling wave. In the typical so-called standing-wave engine or refrigerator, the phase between pressure and volume flow rate is in the range of 85–95° in the stack. Such a wave is illustrated in Ani. Wave /k. At first glance this wave is barely distinguishable from a pure standing wave, but close examination shows nonzero area in the pV ellipses, i.e., nonzero acoustic power transmitted to the right by the wave. Also note that there are no longer true nodes of pressure or velocity. The locations that at first appear to be nodes actually have small nonzero oscillation amplitudes everywhere.

Animation Wave, with any of its /options, assumes that $\dot{E}_2$ is constant; $d\dot{E}_2/dx = 0$. However, the interaction of a sound wave in a channel with the walls of the channel leads to $d\dot{E}_2/dx \neq 0$. As in Chap. 4, sound propagating in the x direction in an ideal gas within a channel with constant cross-sectional area A is considered, as illustrated in Fig. 5.2. The usual complex notation for time-oscillating quantities is used, namely Eqs. (4.11)–(4.14). The derivation of the time-averaged acoustic power $d\dot{E}$ produced in a length dx of the channel begins with

$$\frac{d\dot{E}}{dx} = \int \frac{d\,(\overline{pu})}{dx}\,dy\,dz, \tag{5.7}$$

where the overbar denotes time averaging and the integral is over the cross-sectional area A of the channel. Rewriting Eq. (5.7) in complex notation and expanding the x derivative gives

$$\frac{d\dot{E}_2}{dx} = \frac{1}{2}\,\mathrm{Re}\left[\widetilde{U}_1\frac{dp_1}{dx} + \widetilde{p}_1\frac{dU_1}{dx}\right]. \tag{5.8}$$

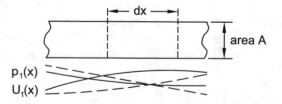

Fig. 5.2 Schematic of acoustic power flowing down a channel of area A. The absorption or production of acoustic power within the length dx is of interest. *Solid and dashed curves* represent real and imaginary parts

The gradients dp_1/dx and dU_1/dx are obtained from the momentum and continuity equations in Chap. 4, where Eqs. (4.72) and (4.76) provide a convenient form. Making these substitutions gives

$$\frac{d\dot{E}_2}{dx} = -\frac{r_v}{2}|U_1|^2 - \frac{1}{2r_\kappa}|p_1|^2 + \frac{1}{2}\text{Re}\left[g\widetilde{p_1}U_1\right]. \tag{5.9}$$

In the impedance diagram for a thermoacoustic component, such as the general impedance diagram of Fig. 4.15, Eq. (5.9) shows that only three of the five impedance components affect acoustic power. Inertance and compliance have no direct effect on acoustic power; they only cause $p_1(x)$ and $U_1(x)$ to evolve in ways that keep $\frac{1}{2}\text{Re}[\widetilde{p_1}U_1]$ independent of x.

The first two terms in Eq. (5.9) are always negative. The first term gives the viscous dissipation of sound and the second term gives the less intuitively obvious thermal-relaxation dissipation. The third term, which can have either sign, is of the greatest interest in thermoacoustic engines and refrigerators. The rest of this section will be devoted to interpreting these three terms.

5.1.1 Acoustic Power Dissipation with $dT_m/dx = 0$

With $dT_m/dx = 0$, Eq. (5.9) reduces to

$$\frac{d\dot{E}_2}{dx} = -\frac{r_v}{2}|U_1|^2 - \frac{1}{2r_\kappa}|p_1|^2 . \tag{5.10}$$

This expression gives the dissipation of acoustic power per unit length in a channel, due to viscous and thermal processes at the channel walls. Fortunately, this expression is easy to interpret because it is so simple. Both terms are negative, so both always represent dissipation: If power flows in the positive x direction, then $|\dot{E}_2(x + dx)| < |\dot{E}_2(x)|$, while if power flows in the negative x direction, then $|\dot{E}_2(x)| < |\dot{E}_2(x + dx)|$. Both terms in $d\dot{E}_2/dx$ are independent of $\dot{E}_2$ itself; i.e., the local dissipation of acoustic power is independent of the local transmission of acoustic power. There is a clean separation between viscous and thermal-relaxation effects: The first term, describing viscous dissipation, is proportional to $|U_1|^2$ and is independent of thermal conductivity, while the second term, describing thermal-relaxation dissipation, is proportional to $|p_1|^2$ and is independent of viscosity.

In the boundary-layer approximation, in which all dimensions of the channel are much larger than the penetration depths, Eq. (5.10) can be written

$$\frac{d\dot{E}_2}{dS} = -\frac{1}{4}\rho_m\left|\frac{U_1}{A}\right|^2 \delta_v \,\omega - \frac{1}{4}\frac{|p_1|^2}{\gamma p_m}(\gamma - 1)\,\delta_\kappa\,\omega, \tag{5.11}$$

where S is the surface area of the channel. This expression gives the ordinary dissipation of acoustic power in a large channel, and is easily remembered in terms of the energy density. The average kinetic energy per unit volume $\rho_m |U_1/A|^2/4$ times δ_v is roughly the kinetic energy in the gas within a viscous penetration depth of the solid surface, per unit area of the surface. The viscous dissipation term shows that this energy is dissipated by viscous shear at an average rate ω. Similarly, the thermal-relaxation term shows that the average adiabatic compressive energy density $|p_1|^2/4\gamma p_m$ stored in unit volume of the gas is dissipated at an average rate $(\gamma - 1)\omega$ within a region of thickness δ_κ near the solid surface. The extra factor $(\gamma - 1)$ appears because this thermal dissipation is proportional to the difference between the isothermal and isentropic compressibilities, which is proportional to $(\gamma - 1)$.

The viscous term arose from the

$$\mathrm{Re}\left[\widetilde{U}_1 \frac{dp_1}{dx}\right]$$

term in Eq. (5.8), and can be better appreciated by considering Ani. Viscous /m. Close examination of the animation shows that the gas motion far from the solid boundary is in phase with dp_1/dx, so the velocity $u_1(\infty)$ far from the solid boundary is 90° out of phase from dp_1/dx. Hence, the gas far from the solid boundary, where the dynamics is purely inertial, has $\mathrm{Re}[\widetilde{u_1}\, dp_1/dx] = 0$ and contributes nothing to $d\dot{E}_2/dx$. The dissipation arises close to the solid boundary, where the phase of u_1 is shifted by viscous interaction with the wall. Naively, when I watch Ani. Viscous I see layers of gas sliding relative to one another, with viscous "friction" between the layers turning "mechanical" energy into heat.

Example: Standing-Wave Refrigerator

In Chap. 4, the viscous resistance of the lower central tee of the standing-wave refrigerator was found to be $R_v = 75$ Pa·s/m^3. At a typical operating point, $|U_1| = 0.16$ m^3/s. Hence this component dissipated 1.0 W of acoustic power, due to viscosity. (There was negligible p_1 in this component, and hence negligible thermal-relaxation dissipation.) This is less than 2% of the total acoustic power supplied to the resonator by the loudspeakers. The center of this component was a center of symmetry of the apparatus, so it must be that $\dot{E}_2 = 0$ at the center. Hence, 0.5 W must have flown into the tee from each side. Taking the positive x direction toward the right, $\dot{E}_2 = +0.5$ W at the left end of the tee, and $\dot{E}_2 = -0.5$ W at the right end of the tee.

Example: Orifice Pulse-Tube Refrigerator

Near Fig. 4.8, the impedance network at the end of the orifice pulse-tube refrigerator was introduced, with an example in which $R_s = 3.35$ MPa·s/m^3. At a typical operating point, with $|U_1| = 0.04$ m^3/s, the flow resistance R_s dissipated 2700 W of acoustic power. This impedance network consumed a significant fraction of the 9000 W of acoustic power supplied to the entire refrigerator.

Compared to viscous dissipation, the dissipation of acoustic power by thermal relaxation is more difficult to appreciate. Since thermal-relaxation dissipation is independent of U_1, the issues can be clarified by considering a location in the wave where $u_1 = 0$. Consider Ani. Thermal /e, showing such a wave near a wall in boundary-layer approximation. (Note that the y direction is horizontal in this animation, so everything fits on the display.) As in Chap. 4, the temperature oscillates in phase with p_1 far from the wall. Close to the wall, the wall's heat capacity provides a thermal anchor that reduces the amplitude of the temperature oscillation and shifts its phase.

To consider dissipation of acoustic power in this region, imagine a piston moving with the gas, and compute the work done by that piston on the gas to the left of it. The work done by a piston is $\oint p \, dV$. In the animation, the small dot traces out the intersection of the pressure line and the volume between the wall and the imaginary piston (whose average position is $4\delta_\kappa$ from the wall), so the area of the little ellipse drawn by the moving dot is proportional to $\oint p \, dV$. That slim elliptical area represents the work that the imaginary piston does on the gas between it and the wall. It's not zero, because of thermal relaxation of the gas to the wall. Gas immediately adjacent to the surface experiences isothermal density and pressure oscillations, which are perfectly springy; gas far from the surface experiences isentropic density and pressure oscillations, which are also perfectly springy. In between, the gas approximately δ_κ from the surface experiences a complex, hysteretic cycle of density changes in response to the pressure oscillations: first an increase in density due to quasi-adiabatic compression by the sound wave, then a further increase in density as thermal relaxation to the surface removes heat from the gas, then a decrease in density due to quasi-adiabatic expansion by the sound wave, and finally a further decrease in density as thermal relaxation to the surface delivers heat to the gas. Since this gas experiences thermal expansion at low pressure and thermal contraction at high pressure, it absorbs work from the sound wave. This gas, approximately δ_κ *from the surface*, is the most effective at absorbing work from the sound wave, whereas in the case of viscous dissipation the gas *at the surface* is most effective.

The lowest plot in Ani. Thermal /e shows the area of such a work ellipse as a function of the distance of the imaginary piston from the wall. This curve is steepest at $y/\delta_\kappa \simeq 1$, which indicates again that the gas approximately δ_κ from the surface dissipates the most acoustic power.

Example: Standing-Wave Engine
In Chap. 4 the thermal-relaxation resistance of one of the hot ducts in the standing-wave engine was calculated as $R_\kappa = 2350 \, \mathrm{MPa \cdot s/m^3}$. At a typical operating point of $|p_1| = 300 \, \mathrm{kPa}$, the thermal relaxation dissipation is $|p_1|^2 / 2R_\kappa = 20 \, \mathrm{W}$. Hence, if this loss could be eliminated in both hot ducts, the 1 kW of acoustic power delivered to the refrigerator would rise by 4%.

Example: Orifice Pulse-Tube Refrigerator
The compliance at the top of the orifice pulse-tube refrigerator had only $|p_1| = 90$ kPa. Hence, even though it had 0.3 m^2 of surface area, the thermal-relaxation dissipation of acoustic power, obtained from the second term of Eq. (5.11), was only 5 W.

5.1.2 Acoustic Power with Zero Viscosity

Equation (5.9), two sections above, was

$$\frac{d\dot{E}_2}{dx} = -\frac{r_v}{2}|U_1|^2 - \frac{1}{2r_\kappa}|p_1|^2 + \frac{1}{2}\mathrm{Re}\,[g\widetilde{p_1}U_1].$$

The previous subsection examined the first two terms, which always consume acoustic power, exist independent of any temperature gradient along x, and represent viscous and thermal-relaxation dissipation, respectively. This subsection considers the third term, which is called the source/sink term because it can either produce or consume acoustic power and which exists only if $dT_m/dx \neq 0$. This term is difficult to fully understand, especially because g involves both f_κ and f_v. To build understanding, the qualitative discussion in this subsection will neglect viscosity in g, setting $f_v = 0$ and $\sigma = 0$. In this limit, the third term of Eq. (5.9) can be written

$$\frac{1}{2}\mathrm{Re}\,[g\widetilde{p_1}U_1] = \frac{1}{2}\frac{1}{T_m}\frac{dT_m}{dx}\mathrm{Re}\,[\widetilde{p_1}U_1]\,\mathrm{Re}\,[f_\kappa] + \frac{1}{2}\frac{1}{T_m}\frac{dT_m}{dx}\mathrm{Im}\,[\widetilde{p_1}U_1]\,\mathrm{Im}\,[-f_\kappa]$$

(5.12)

$$= \frac{1}{T_m}\frac{dT_m}{dx}\dot{E}_2\mathrm{Re}\,[f_\kappa] + \frac{1}{2}\frac{1}{T_m}\frac{dT_m}{dx}\mathrm{Im}\,[\widetilde{p_1}U_1]\,\mathrm{Im}\,[-f_\kappa].$$ (5.13)

Expressing it this way shows that $\mathrm{Re}\,[f_\kappa]$ is important for acoustic power in traveling-wave engines and refrigerators, in which $\mathrm{Re}\,[\widetilde{p_1}U_1]$ is large, while $\mathrm{Im}\,[-f_\kappa]$ is important for acoustic power in standing-wave engines and refrigerators, in which $\mathrm{Im}\,[\widetilde{p_1}U_1]$ is large.

5.1.2.1 Traveling Waves

Begin by considering the regenerator of a traveling-wave engine. To make the most of the first term on the right side of Eq. (5.12) or (5.13), $\mathrm{Re}\,[f_\kappa]$ should be as large as possible. Examination of Fig. 4.14 shows that this is accomplished at $r_h \ll \delta_\kappa$, where $\mathrm{Re}\,[f_\kappa] \simeq 1$. Under these circumstances, the thermal-relaxation resistance r_κ is negligibly large, as shown by Eq. (4.78). To generate (not dissipate) acoustic power requires that dT_m/dx and $\dot{E}_2$ share the same sign; i.e., the temperature must increase through the regenerator in the direction of acoustic power flow. This situation is illustrated in Anis. Tashe /s and Tashe /r. Animation Tashe /s shows an overview of a traveling-wave engine, namely a traditional Stirling engine. The sign of the circulation of the pV ellipses indicates that acoustic power flows from left to right. The temperature rises in the direction of this acoustic power flow, so the first term in Eq. (5.12) or (5.13) shows that the acoustic power should increase from left to right. Indeed, in the animation the right pV ellipse has greater area than the left pV ellipse, showing this increase in acoustic power. If these two pistons were connected to a common crankshaft, that crankshaft would supply acoustic power to the gas at the left and remove it from the gas at the right. The extra acoustic power, represented by the difference between the right and left ellipse areas, is the net acoustic power generated by the engine, available to do external work such as generating electricity.

Animation Tashe /r shows a close-up view inside the regenerator of Ani. Tashe /s. One small parcel of gas is highlighted, oscillating left and right while experiencing oscillating pressure. The phasing between pressure and motion is predominantly traveling wave: The gas moves to the right while the pressure is high, and moves to the left while the pressure is low, so the acoustic power flows from left to right. Temperatures are shown in the lower-left plot, where the temperature of the highlighted parcel of gas, shown as the blue trace, is always locked to the local solid temperature, represented by the white line. This indicates that $r_h \ll \delta_\kappa$; equivalently, $f_\kappa = 1$. The volume of the highlighted parcel of gas is shown on the horizontal axis of the lower-right plot, with the pressure plotted vertically. The gas expands while it moves to the right, because its temperature rises; it contracts while it moves to the left, because its temperature falls. This is the physical effect responsible for the volume-flow-rate source gU_1 in a traveling-wave engine. The net effect—clockwise circulation on the pV diagram—is the production of acoustic power, because the expansion takes place while the gas is at high pressure and the contraction takes place while the gas is at low pressure. The difference between the right piston's work and the left piston's work in Ani. Tashe /s is the total acoustic power produced by all the parcels of gas in the regenerator, each behaving as shown in Ani. Tashe /r.

Example: Thermoacoustic-Stirling Heat Engine

The thermoacoustic-Stirling heat engine of Figs. 1.22 and 1.23 followed the same thermodynamic cycle as discussed in the previous two paragraphs, but without pistons. The acoustic network that served the function of the two pistons and their crankshaft is shown schematically in Ani. Tashe /u. The acoustic power at several key locations is shown as purple ellipses. The size and signs of these pV ellipses show how acoustic power is produced in the regenerator, with some of the produced power flowing out of the display to the right and the remainder fed back to the left end of the regenerator through the inertance at the bottom. This animation is based on the assumption that the regenerator is the only location where acoustic power is produced or dissipated. Hence, the areas of the lower two pV ellipses and the pV ellipse to the left of the regenerator are equal. The shape change at equal ellipse areas is due to C and L. Similarly, the areas of the two pV ellipses at the ends of the thermal buffer tube are equal, but with a shape change due to the intervening C. Continuity of p_1 and U_1 at the three-way junction on the right guarantees that the area of the pV ellipse representing power flowing out of the display to the right is the difference between the areas of the pV ellipses at the right end of the thermal buffer tube and the right end of the inertance.

For a typical operating point of the actual engine of Figs. 1.22 and 1.23, $\dot{E}_{2,0} = 1250\,\text{W}$ of acoustic power flowed into the ambient end of the regenerator and $\dot{E}_{2,H} = 3150\,\text{W}$ of acoustic power flowed out of the hot end of the regenerator, a ratio of $\dot{E}_{2,H}/\dot{E}_{2,0} = 2.5$. The $\dot{E}_2\,dT_m/dx$ term in Eq. (5.13) shows that, ideally, this ratio could be higher: Setting $r_v = 0$ and $r_\kappa = \infty$ in Eq. (5.9), and letting $f_\kappa \to 1$ in Eq. (5.13), yields

$$\frac{d\dot{E}_2}{dx} = \frac{1}{T_m}\frac{dT_m}{dx}\dot{E}_2, \tag{5.14}$$

which is easily integrated to obtain $\dot{E}_{2,H}/\dot{E}_{2,0} = T_H/T_0 = (1000\,\text{K})/(300\,\text{K}) = 3.3$. The other two terms in Eq. (5.9), especially the viscous term, account for this lost acoustic power in the regenerator. (In addition, some 400 W of acoustic power was dissipated in viscous and thermal-relaxation effects in the inertance and compliance that form the acoustic network at the bottom and left end of the display in Ani. Tashe. These losses are not shown in the ideal case shown in the animation.)

The regenerator of a traveling-wave refrigerator is very similar to that of a traveling-wave engine. The primary difference is that T_m decreases in the direction of positive acoustic power flow in the refrigerator, so the third term in Eq. (5.9) is negative, representing the consumption of acoustic power. As with the engine, $\text{Re}\,[f_\kappa]$ should be as large as possible, so $r_h \ll \delta_\kappa$ is desirable. This situation is illustrated in Anis. Ptr /s and Ptr /r. Animation Ptr /s shows an overview of a Stirling refrigerator. The sign of the circulation of the pV ellipses indicates that acoustic power flows from left to right. The temperature decreases in the direction of this power flow, so the first term in Eq. (5.12) shows that the acoustic power should decrease from left to right. Indeed, the right pV ellipse has smaller area than the left pV ellipse, showing this decrease in acoustic power. If these two pistons were connected to a common crankshaft, that crankshaft would supply acoustic power to the gas at the left and remove it from the gas at the right. The difference between these two powers, represented by the difference between the left and right ellipse areas, is the net mechanical power required by the refrigerator, supplied by external means such as an electric motor.

Animation Ptr /r shows a close-up view inside the regenerator of Ani. Ptr /s. One small parcel of gas is highlighted, oscillating left and right while experiencing oscillating pressure. The phasing between pressure and motion is predominantly traveling wave: The gas moves to the right while the pressure is high, and moves to the left while the pressure is low, so the acoustic power flows from left to right. Temperatures are shown in the lower-left plot, where the temperature of the highlighted parcel of gas, shown as the blue trace, is always locked to the local solid temperature, represented by the white line. This indicates that $r_h \ll \delta_\kappa$; equivalently, $f_\kappa = 1$. The volume of the highlighted parcel of gas is shown on the horizontal axis of the lower-right plot, with the pressure plotted vertically. The gas expands while it moves to the left, because its temperature rises; it contracts while it moves to the right, because its temperature falls. The net effect—counterclockwise circulation on the pV diagram—is the consumption of acoustic power, because the expansion takes place while the gas is at low pressure and the contraction takes place while the gas is at high pressure. The difference between the right piston's work and the left piston's work in Ani. Ptr /s is the total acoustic power consumed by all the parcels of gas in the regenerator, each behaving as shown in Ani. Ptr /r.

Example: Orifice Pulse-Tube Refrigerator
The orifice pulse-tube refrigerator of Figs. 1.17 and 1.18 followed the same thermodynamic cycle as discussed in the previous two paragraphs, but without pistons. The acoustic network that allowed the apparatus to follow this cycle is shown schematically in Ani. Ptr /p. The acoustic power at several key locations is indicated by purple ellipses. The size and signs of these pV ellipses show how acoustic power flows into the system from the left (supplied by external means, such as a motored piston or a thermoacoustic engine) and is absorbed in the regenerator and the *RLC* impedance network at the right end of the display. The areas of the two pV ellipses at the ends of the pulse tube are equal.

For a typical operating point of the actual refrigerator of Figs. 1.17 and 1.18, 8800 W of acoustic power flowed into the ambient end of the regenerator and 3000 W of acoustic power flowed out of the cold end of the regenerator. The dT_m/T_m factor in Eq. (5.12) shows that, ideally, the ratio of these two powers, 2.9, could be the ratio of the engine's hot and ambient temperatures, $(300\,\text{K})/(120\,\text{K}) = 2.5$. The other two terms in Eq. (5.9), especially the viscous term, account for the additional absorption of acoustic power in the regenerator.

5.1.2.2 Standing Waves

Next consider the stacks of standing-wave engines and refrigerators, continuing to imagine zero viscosity so that the source/sink term in $d\dot{E}_2/dx$ is simple. The first term in Eq. (5.12) or (5.13) is zero for standing-wave phasing. To make the most of the second term in Eq. (5.12) or (5.13), $\text{Im}\,[-f_\kappa]$ should be as large as possible. Examination of Fig. 4.14 shows that this is accomplished at $r_h \sim \delta_\kappa$, where $\text{Im}\,[-f_\kappa] \simeq 0.4$. Equation (4.78) shows that r_κ cannot be neglected in this situation, so the source/sink term and the thermal-relaxation term in Eq. (5.9) should be considered together. Neglecting viscosity, Eqs. (5.12), (5.9), and (4.78) generate

$$\frac{d\dot{E}_2}{dx} = -\frac{1}{2}\,|p_1|^2\,\frac{(\gamma-1)\,\omega A\,\text{Im}\,[-f_\kappa]}{\gamma p_m} + \frac{1}{2}\text{Im}\,[\widetilde{p_1}\,U_1]\,\frac{1}{T_m}\frac{dT_m}{dx}\text{Im}\,[-f_\kappa]. \qquad (5.15)$$

Using Eqs. (4.44) and (4.31) to rearrange this expression, and assuming standing-wave phasing with U_1 leading p_1 by 90°, yields[4]

$$\frac{d\dot{E}_2}{dx} = \frac{1}{2} |p_1|^2 \omega A \frac{\gamma - 1}{\gamma p_m} \text{Im}\,[-f_\kappa] \left(\frac{dT_m/dx}{\nabla T_{\text{crit}}} - 1\right), \qquad (5.16)$$

where

$$\nabla T_{\text{crit}} = \frac{|p_1|/\rho_m c_p}{|U_1|/A\omega}$$

is the critical temperature gradient, which was first encountered in Chap. 4 and defined by Eq. (4.44).

The situation where $dT_m/dx = \nabla T_{\text{crit}}$, for which Eq. (5.16) suggests that acoustic power is neither produced nor absorbed, is illustrated in Ani. Standing /c. In the upper left of the display is a stack, with blue marker lines showing the moving gas. One particular parcel of gas is highlighted in blue in the stack. The yellow oval marks the region that is shown magnified at the left center of the display, which shows that same parcel of moving gas and short fragments of the two stack plates adjacent to it. The volume of that parcel of gas changes in response to pressure and temperature. At the bottom of the display are plots of pressure vs volume of the parcel of gas, and temperature vs position. In the temperature plot, the temperature of the parcel is the blue trace, and that of the nearby plate in the stack is the white line.

In the previous, traveling-wave, animations, the blue trace and white line were superimposed because small r_h ensured good thermal contact between gas and solid. Here, $r_h \sim \delta_\kappa$, so the thermal contact is not so good. The alignment of the gas and solid temperatures is due only to the fact that $dT_m/dx = \nabla T_{\text{crit}}$ for this animation. The parcel's temperature oscillation is due entirely to adiabatic pressure oscillation, and its temperature oscillation and motion just happen to match the local temperature gradient dT_m/dx exactly. Since the gas oscillations are merely adiabatic, the pV trace for this parcel is simply a reciprocating adiabat, so this parcel of gas neither produces nor absorbs acoustic power. This operating condition does not appear in the stacks of useful thermoacoustic engines and refrigerators; $dT_m/dx = \nabla T_{\text{crit}}$ is only of academic interest, because it separates the regimes of standing-wave engine and standing-wave refrigerator.

Animation Standing /e shows an overview of a standing-wave engine. The display is basically the same as for Ani. Standing /c discussed above, but now the temperature gradient is steeper, with $dT_m/dx > \nabla T_{\text{crit}}$, so that Eq. (5.16) indicates that acoustic power is produced. Because of the steep temperature gradient,

---

[4]The choice here and in Sect. 5.2.2 to assume that U_1 leads p_1 makes $dT_m/dx > 0$ the interesting case. The opposite choice would require defining ∇T_{crit} to be negative. For the least confusing sign convention, I usually start by choosing the $+x$ direction so that $dT_m/dx > 0$. Hence, in Anis. Standing, the x axis is directed leftward.

the gas parcel's adiabatic temperature oscillation and motion do not match the local solid temperature gradient, so oscillating heat transfer between the gas and the solid occurs, as indicated by the time-dependent red arrows in the magnified view of the stack. The phasing is such that thermal expansion occurs at high pressure and thermal contraction at low pressure, so the pV ellipse circulates clockwise, indicating that the parcel produces acoustic power. The acoustic power thus generated by all the parcels of gas in the stack is the power generated by the engine, available to do external work.

Similarly, Ani. Standing /r shows an overview of a standing-wave refrigerator. The display is basically the same as for Anis. Standing /c and Standing /e discussed above, but now the temperature gradient is less steep, with $dT_m/dx < \nabla T_{crit}$, so that Eq. (5.16) indicates that acoustic power is absorbed. Because of the shallow temperature gradient, the gas parcel's adiabatic temperature oscillation and motion do not match the local solid temperature gradient, so oscillating heat transfer between the gas and the solid occurs, as indicated by the time-dependent red arrows in the magnified view of the stack. The phasing is such that thermal expansion occurs at low pressure and thermal contraction at high pressure, so the pV ellipse circulates counterclockwise, indicating that the parcel consumes acoustic power. The acoustic power thus absorbed by all the parcels of gas in the stack must be supplied by an external agent, such as a loudspeaker.

Example: Standing-Wave Engine
The standing-wave engine of Figs. 1.8 and 1.9 operated just as Ani. Standing /e discussed above, but at 350 Hz instead of the 0.1-Hz frequency of the animation. For a typical operating point, almost 700 W of acoustic power flow was produced in each stack. Examination of Fig. 4.18c, d shows some interesting details of this power production.

The positive slope of $d\mathrm{Re}[U_1]/dx$ and the positive value of $\mathrm{Re}[p_1]$, multiplied together, are responsible for the production of acoustic power in the stack represented by the second term of Eq. (5.8). This term is due to the continuity equation; the $g\, dT_m/dx$ effect is responsible for this production of acoustic power. Both $d\mathrm{Re}[U_1]/dx$ and $\mathrm{Re}[p_1]$ are highest at the hot end of the stack, so that is where the power production is most intense.

At the same time, the positive slope of $d\mathrm{Im}[p_1]/dx$ and the negative value of $\mathrm{Im}[U_1]$ yield the dissipation of acoustic power via the first term of Eq. (5.8). This term is due to viscous dissipation in the momentum equation. Fortunately, the acoustic-power production of the second term is much larger than the acoustic-power dissipation of the second term.

Under these conditions, the temperature gradient in the stack was x dependent, but it was approximately equal to $(T_H - T_0)/\Delta x = 60$ K/cm. This is indeed larger than the critical temperature gradient, which is estimated to be 35 K/cm at the stack midpoint using Eq. (4.44).

5.2   Total Power

Acoustic power is not the only power that is important in thermoacoustics. Total power is perhaps of greater importance. To proceed, we have to consider exactly what we mean by energy in thermoacoustics.[5]

Engineers have long employed the concept of "control volume" for careful thought about energy issues in thermodynamic systems. As reviewed in Sect. 2.2, a control volume is a space of interest, surrounded by a well-defined imaginary boundary, to which the first law of thermodynamics and similar conservation principles can be applied [45].

Figure 5.3 introduces the topic of total power, showing two typical control volumes of interest in thermoacoustics. Consider the thermoacoustic refrigerator shown in Fig. 5.3a, b, driven by a piston or loudspeaker at the left and having thermal insulation around everything except the heat exchangers, so that heat can be exchanged with the outside world only at the two heat exchangers, and work can

Fig. 5.3 A standing-wave refrigerator, insulated everywhere except at the heat exchangers. (**a**) One useful control volume for thermoacoustics, enclosing the *left end*. (**b**) Another useful control volume, enclosing part of the stack. (**c**) Graph of $\dot{E}_2$ and $\dot{H}_2$ in the refrigerator. The discontinuities in $\dot{H}_2$ are the heats transferred at the heat exchangers

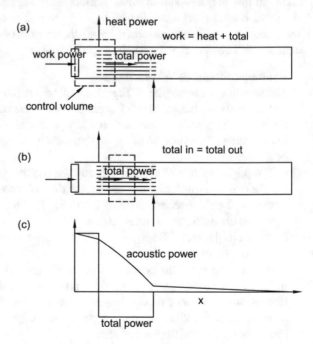

[5]Apologies for the English language: We do not have short, distinct terms for "rate at which work is done" or "rate at which heat is transferred." Many thermodynamics texts avoid this issue by sticking as much as possible with the variables having units of Joules, not Watts, but I prefer to focus on power. So, in the interest of brevity and readability, some of my vocabulary may be awkward or ambiguous. When in doubt, look at the variables: Those with overdots—$\dot{W}$, $\dot{Q}$, $\dot{E}_2$, $\dot{H}_2$, and Chap. 6's $\dot{X}_2$—are powers. For brevity I will refer to $\dot{W}$ as mechanical power or simply as work, and to $\dot{Q}$ as thermal power or simply as heat.

be exchanged with the outside world only at the piston. The principle of energy conservation can be applied to the control volume shown by the dotted line in Fig. 5.3a. In steady state, time-averaged over an integral number of acoustic cycles, the energy inside the control volume cannot change, so the rate at which energy flows into that control volume must equal the rate at which energy flows out. What flows in is clearly the time-averaged mechanical power (which is exactly equal to the acoustic power flowing from the face of the piston into the gas). That must equal the sum of the two outflowing powers, labeled heat power and total power. The heat is easy to think about. It's always least ambiguous to discuss such heat flowing via conduction through a solid, so imagine the control-volume boundary being within the metal surfaces that separate the thermoacoustic gas from the outside world. However, the power flowing down the stack is much more subtle: It's not acoustic power, and it's not heat, and it's not mechanical power—but it's important, because it's what counts in conservation of energy.

Another typical important control volume is shown in Fig. 5.3b, intersecting a stack or regenerator in two places, with thermal insulation around the immovable side walls. Here the only powers flowing are the total powers in and out of the two end surfaces of the control volume. Applying the principle of energy conservation (again, steady state and time averaged) to this control volume shows that total power in equals total power out. So total power cannot depend on x within a stack or regenerator—it has to be constant, independent of x, even while acoustic power depends strongly on x, as discussed in the previous section.

The 1975 experiment of Merkli and Thomann [67] also helps motivate this discussion of total power. As shown in Fig. 5.4, an acoustic resonator was driven at

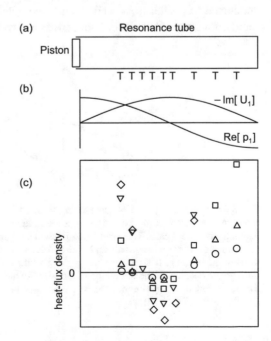

Fig. 5.4 The experiment of Merkli and Thomann, which detected cooling at the velocity antinode of a standing wave. (**a**) Schematic of apparatus. "T" indicates a thermometer on the side wall. (**b**) Half-wavelength standing-wave pattern. (**c**) Local heat-flux density (mks units W/m^2) to or from the wall, inferred from the thermometers. The negative values of heat-flux density near the center of the resonance tube represent cooling

the half-wavelength resonance, using an oscillating piston and employing sensitive thermometers all along the resonator side wall. Merkli and Thomann found that the central thermometers cooled, at the location of greatest dissipation of acoustic power per unit length. That seems surprising, but it is easily explained by the important difference between total power and acoustic power.

So, what is meant by total power? As reviewed in Chap. 2, the enthalpy is the correct energy to consider in moving gases (and liquids) [46]. This is because the energy-flux density in fluid mechanics is the quantity in square brackets on the right side of Eq. (2.33),

$$\left(\rho h + \rho \left|\mathbf{v}\right|^2 /2\right)\mathbf{v} - k\nabla T - \mathbf{v}\cdot\boldsymbol{\sigma}', \tag{5.17}$$

where h is the enthalpy per unit mass, $\rho \left|\mathbf{v}\right|^2 /2$ is the kinetic-energy density, and $\boldsymbol{\sigma}'$ is the viscous stress tensor. The enthalpy itself is the sum of other energies:

$$h = \epsilon + p/\rho, \tag{5.18}$$

where ϵ is the internal energy per unit mass, so the energy-flux density can be thought of as the superposition of five flux densities: kinetic energy, internal energy, "p/ρ," heat, and work done by viscous shear. Looking at Fig. 5.5, it is possible to imagine the first two of these as being attached to the gas and hence being convected along with the gas at velocity $\mathbf{v}$, and to imagine the pressure term representing the rate at which the pressure of the gas "does work" on the gas it is pushing ahead of it. The thermal-conduction term can be imagined as giving the rate at which heat is conducted across the cross section, and the final term might be vaguely imagined as representing work that viscous shear does across the cross section.

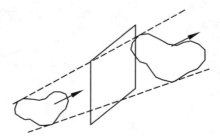

Fig. 5.5 An irregular parcel of gas passing through an imaginary rectangular opening fixed in space. As the parcel of gas passes through the imaginary opening, the energy flux through that opening has several parts. The kinetic and internal energies are convected along with the gas. The pressure does work on the gas ahead as the parcel pushes other gas out of its way in order to pass through the opening. If a temperature gradient in the correct direction exists, heat is conducted through the opening. If the directions of the velocity and its gradients are suitable, viscous shear forces do work on the gas on the other side of the opening. The sum of these forms the energy-flux density given in Eq. (5.17)

To obtain a usable form for the total power $\dot{H}$ carried by thermoacoustic processes, realize first that thermoacoustic total power should be at most second order in the oscillating variables, like acoustic power $\dot{E}_2$, so that the $\rho\,|\mathbf{v}|^2\,\mathbf{v}$ term is negligible. Rott's acoustic approximation to the total power flux in the x direction, time averaged and integrated over the cross-sectional area A of the channel, can then be written

$$\dot{H}_2 = \sum_{j=0}^{2} \int \left[\overline{\rho h u} - \overline{k\,dT/dx} - \overline{(\mathbf{v}\cdot\boldsymbol{\sigma}')_x} \right]_j dA. \tag{5.19}$$

Energy flowing in the x direction in the solid wall of the channel, by means of heat conduction, should also be included.

The first term in Eq. (5.19) has no zeroth-order parts, because $u_m = 0$ (but see Sect. 7.4), and it has no first-order parts because the time average of any first-order variable is zero. The second-order parts are

$$\int \overline{(\rho h u)}_2\,dA = \int \overline{(\rho u)}_2\,h_m\,dA + \int \overline{(\rho u)_1\,h_1}\,dA. \tag{5.20}$$

The first of these two integrals is $\dot{M}_2 h_m$, where $\dot{M}_2$ is the second-order, time-averaged mass flux along the channel. This is zero for most thermoacoustic apparatus—otherwise mass would steadily accumulate in some portions of the apparatus at the expense of other portions (but see Sect. 7.4 for toroidal apparatus). Using again the fact that $u_m = 0$, the second integral simplifies, yielding finally

$$\int \overline{(\rho h u)}_2\,dA = \frac{1}{2}\rho_m \int \mathrm{Re}\,[h_1 \tilde{u}_1]\,dA. \tag{5.21}$$

The second term in Eq. (5.19) has terms of both zeroth and second orders, but only the zeroth-order terms are typically significant. Hence,

$$\int \overline{\left(k\,dT/dx\right)}_0\,dA = (Ak + A_{\mathrm{solid}}k_{\mathrm{solid}})\,dT_m/dx \tag{5.22}$$

suffices, and includes heat flux in the channel wall as well as in the gas within the channel interior.

The third term in Eq. (5.19) includes many terms due to the complicated nature of $\mathbf{v}\cdot\boldsymbol{\sigma}'$. To simplify it, bulk viscosity ζ is neglected, and the assumption that v/u, w/u, $(\partial/\partial x)/(\partial/\partial y)$, and $(\partial/\partial x)/(\partial/\partial z)$ are all of order δ_v/λ eliminates some terms. The five terms remaining are roughly of the order of $\mu\,|u_1|^2\,A/\lambda$. However, Eq. (5.21) is typically of the order of $|p_1|\,|u_1|\,A \sim \rho_m\omega\lambda\,|u_1|^2\,A$. Hence,

$$\frac{\int \overline{(\mathbf{v}\cdot\boldsymbol{\sigma}')}_{x,2}\,dA}{\int \overline{(\rho h u)}_2\,dA} \sim \frac{\delta_v^2}{\lambda^2} \ll 1, \tag{5.23}$$

so the viscous term is usually negligible.

Hence, Eq. (5.19) simplifies to [11, 67, 68]

$$\dot{H}_2(x) = \frac{1}{2}\rho_m \int \text{Re}\,[h_1\widetilde{u_1}]\,dA - (Ak + A_{\text{solid}}k_{\text{solid}})\frac{dT_m}{dx}. \tag{5.24}$$

This acoustic approximation to the total power flowing in the x direction will suffice until streaming is encountered in Chap. 7. The second term is simple conduction of heat, including conduction both in the gas and in whatever solid (e.g., stack or regenerator) is present. The first term is the time-averaged enthalpy flux.[6]

Thomann [69] said that the energy equation is like "a kaleidoscope; each time it is shaken (by rearranging the terms) a different picture is seen." There are many good ways to look at $\dot{H}_2$ given by Eq. (5.24):

First, if $\dot{H}_2$ and work are known, the heat transferred at heat exchangers in thermoacoustic engines and refrigerators can be deduced, as illustrated in Fig. 5.3a. Alternatively, if both work and heat are known, $\dot{H}_2$ can be deduced.

Second, taking enthalpy as a function of temperature and pressure, as in Eq. (2.93),

$$dh = c_p\,dT + \frac{(1-T\beta)\,dp}{\rho},$$

and simplifying to Eq. (2.59) or (3.34),

$$h_1 = c_p T_1,$$

which is true for an ideal gas, Eq. (5.24) becomes

$$\dot{H}_2(x) = \frac{1}{2}\rho_m c_p \int \text{Re}\,[T_1\widetilde{u_1}]\,dA - (Ak + A_{\text{solid}}k_{\text{solid}})\frac{dT_m}{dx}. \tag{5.25}$$

This point of view is useful for calculations of $\dot{H}_2$, and is especially useful for thinking about nearly ideal regenerators, in which $T_1 \simeq 0$. For example, consider the control volume shown by the yellow box in the Stirling refrigerator shown in Ani. Ptr /c. For an ideal regenerator, the heat capacity of the regenerator solid and the good thermal contact in the small pores there maintain temporally isothermal conditions at each location, so $T_1 = 0$ everywhere. If the conduction of heat through the regenerator can be neglected, Eq. (5.25) shows that $\dot{H}_2 = 0$. Hence, the time-averaged cooling power of the refrigerator must equal the time-averaged mechanical power extracted by the cold piston. That is, in suitable units, the area of the pV ellipse for the cold piston must equal the time-averaged width of the red heat arrow at the cold heat exchanger.

---

[6]Notation: Much published literature on pulse-tube refrigerators uses the symbol $\dot{H}$ to represent only the enthalpy flux, adding the conduction term separately and sometimes using another variable to represent the sum.

Using Eqs. (4.67) and (4.53) for T_1 and u_1 and performing the integration[7] in Eq. (5.25) yields

$$\dot{H}_2 = \frac{1}{2}\mathrm{Re}\left[p_1\widetilde{U_1}\left(1 - \frac{f_\kappa - \widetilde{f_\nu}}{(1+\sigma)\left(1-\widetilde{f_\nu}\right)}\right)\right]$$

$$+ \frac{\rho_m c_p |U_1|^2}{2A\omega(1-\sigma^2)|1-f_\nu|^2}\,\mathrm{Im}\left(f_\kappa + \sigma\widetilde{f_\nu}\right)\frac{dT_m}{dx}$$

$$- (Ak + A_{\mathrm{solid}}k_{\mathrm{solid}})\frac{dT_m}{dx}, \tag{5.26}$$

obtained by Rott [13].

If $dT_m/dx = 0$, such as in a water-jacketed resonator component, then Eq. (5.26) gives $\dot{H}_2(x)$. In that case, $d\dot{H}_2/dx$ tells how much energy per unit length is deposited in (or removed from) the water jacket.

Third, taking enthalpy as a function of entropy and pressure, as in Eq. (2.85),

$$dh = T\,ds + dp/\rho,$$

the total power can be written

$$\dot{H}_2 = \frac{1}{2}\mathrm{Re}\left[p_1\widetilde{U_1}\right] + \frac{1}{2}\rho_m T_m \int \mathrm{Re}\left[s_1\widetilde{u_1}\right]\,dA - (Ak + A_{\mathrm{solid}}k_{\mathrm{solid}})\frac{dT_m}{dx}$$

$$\tag{5.27}$$

$$= \dot{E}_2 + \frac{1}{2}\rho_m T_m \int \mathrm{Re}\left[s_1\widetilde{u_1}\right]\,dA - (Ak + A_{\mathrm{solid}}k_{\mathrm{solid}})\frac{dT_m}{dx}. \tag{5.28}$$

The first term is acoustic power. The second term is T_m times the second-order hydrodynamic entropy flux, and the final term is simply conduction of heat.

It can be helpful to intuition to identify the sum of the last two terms of Eq. (5.27) or (5.28),

$$\dot{H}_2 - \dot{E}_2 = \frac{1}{2}\rho_m T_m \int \mathrm{Re}\left[s_1\widetilde{u_1}\right]\,dA - (Ak + A_{\mathrm{solid}}k_{\mathrm{solid}})\frac{dT_m}{dx}, \tag{5.29}$$

as a sort of thermoacoustic heat-pumping power, with the $s_1\widetilde{u_1}$ term generally by far the largest contributor. Many of us who have worked in standing-wave thermoacoustics for a decade loosely refer to this as "thermoacoustic heat flux," and we have been criticized for this practice by careful people who know thermodynamics. Such careful people correctly point out that multiplying a hydrodynamic entropy flux by

---

[7]Tedious.

T does not make a heat flux. The usual description of "bucket-brigade" standing-wave thermoacoustic "heat transport" as shown in Ani. Standing indeed shows heat flowing from plate to gas at one part of the cycle and heat flowing from gas to a different part of the plate 180° later in time. However, careful people correctly point out that it is not accurate to say that the gas "carries" heat from the first location to the second, because there is more to gas thermodynamics than simply heat capacity. I no longer use such careless vocabulary when talking with experts, both to avoid controversy and to avoid a false sense of simple understanding. On the other hand, I continue to use this careless vocabulary often [11], mostly because it allows newcomers to understand the phenomena without getting bogged down in details.

In wide-open isothermal channels, like most resonator components, most of the gas should experience adiabatic oscillations and s_1 should be near zero over most of the area A. Hence, Eq. (5.27) suggests that here the total power and the acoustic power are nearly equal. This contributes to one's ability to deduce a heat exchanger's heat, from knowledge of the total power in the adjacent stack or regenerator and the acoustic power in the adjacent duct.

5.2.1 Traveling Waves

Like a kaleidoscope: Shake the energy equation a little, and you see a new picture. Neglecting ordinary conduction along x, Eqs. (5.25) and (5.28) show that if the total power through a good regenerator is approximately zero, then $\frac{1}{2}\rho_m T_m \int \mathrm{Re}\,[s_1\widetilde{u_1}]\,dA$ must be approximately equal and opposite to the acoustic power $\dot{E}_2 = \frac{1}{2}\mathrm{Re}[\widetilde{p_1}U_1]$ at each location x. This is another way of imagining the "cause" of the Stirling refrigerator's cooling power—as due to the large acoustic power flowing through the regenerator plus the fact that the total power flowing through the regenerator is nearly zero. In Ani. Ptr /r, which shows a close-up view inside the regenerator of an orifice pulse-tube refrigerator, the red arrows indicate heat flow into and out of the parcel of gas. These heat flows cause the oscillating entropy s_1, and its phasing relative to u_1, and hence cause $\mathrm{Re}\,[s_1\widetilde{u_1}]$ to be nonzero.

In a pulse tube or thermal buffer tube, s_1 is due almost entirely to $(u_1/i\omega)\,ds_m/dx$, so $\mathrm{Re}\,[s_1\widetilde{u_1}] \simeq 0$. Hence $\dot{H}_2 \simeq \dot{E}_2$ in a pulse tube or thermal buffer tube.

In the regenerators of traveling-wave systems, $T_1 \simeq 0$, and hence $\dot{H}_2 \simeq 0$. Hence, if a heat exchanger is sandwiched between a regenerator and a pulse tube or thermal buffer tube, or between a regenerator and a resonator component, the heat-exchanger heat and the acoustic power flowing through it must be nearly equal. (In contrast, standing-wave systems have no such constraint on $\dot{H}_2$ and so the heat-exchanger heat can be significantly larger than the acoustic power.)

5.2.2 Standing Waves

For standing-wave phasing with U_1 leading p_1, the inviscid limit of Eq. (5.26),

$$\dot{H}_2 \simeq \frac{1}{2} |p_1| |U_1| \operatorname{Im}[-f_\kappa] \left(\frac{dT_m/dx}{\nabla T_{\text{crit}}} - 1\right) - (Ak + A_{\text{solid}} k_{\text{solid}}) \frac{dT_m}{dx}, \qquad (5.30)$$

is simple and interesting, especially when considered in concert with the standing-wave version of Eq. (5.27):

$$\dot{H}_2 \simeq \frac{1}{2} \rho_m T_m \int \operatorname{Re}[s_1 \widetilde{u_1}] \, dA - (Ak + A_{\text{solid}} k_{\text{solid}}) \frac{dT_m}{dx}. \qquad (5.31)$$

The first term of Eq. (5.30) resembles Eq. (5.16) for $d\dot{E}_2/dx$ derived above: It has the same dependences on f_κ and on dT_m/dx. Hence, aside from viscous effects and the less interesting thermal-conduction term, the total power and the *rate of change of* acoustic power have the same dependences on channel geometry in standing waves. Furthermore, both are zero when $dT_m/dx = \nabla T_{\text{crit}}$, positive when $dT_m/dx > \nabla T_{\text{crit}}$, and negative when $dT_m/dx < \nabla T_{\text{crit}}$.

The first term in Eq. (5.30) must be identified with the hydrodynamic entropy transport represented by the $\operatorname{Re}[s_1 \widetilde{u_1}]$ term in Eq. (5.31). Animation Standing /r illustrates this $\operatorname{Re}[s_1 \widetilde{u_1}]$ contribution to the total power in a standing-wave refrigerator. The parcel of gas absorbs a little heat from the solid walls at the right extreme of its motion, as shown by the red arrows, carries entropy to the left, and gives heat to the solid walls at the left extreme of its motion, returning to the right with less entropy. With x positive to the right, $\dot{H}_2$ is negative. In moving left, the gas has moved up the temperature gradient, so it deposits its heat to the walls at a location of higher temperature than that at which it absorbed heat from the walls. Loosely speaking, the gas moves heat from right to left, up the temperature gradient. So this is a refrigerator. Overall, every parcel of gas in the entire stack, each cycle of the wave, picks up a little heat from the solid, moves a little to the left, deposits heat at a slightly higher temperature, and moves back to the right. The parcels in mid stack are like the middle members of a bucket brigade, passing heat along. At either end are parcels that oscillate between the stack and one of the heat exchangers. At the right end, such parcels absorb heat from the cold heat exchanger and deposit heat in the stack; at the left end, such parcels absorb heat from the stack and deposit heat in the ambient heat exchanger. Overall, the net effect is to absorb heat from the cold heat exchanger and reject waste heat at the ambient heat exchanger.

The process illustrated in Ani. Standing /r occurs in standing-wave refrigerators, for which $dT_m/dx < \nabla T_{\text{crit}}$. If the temperature gradient in the stack is steeper, the signs of all the heat transfers change, as shown for the standing-wave engine of Ani. Standing /e. In this case, even though the gas is heated adiabatically as it moves left, it is not hot enough at the left extreme of its motion to match the local solid temperature when it gets there. So at the left extreme of its motion, at high pressure, it is still cooler than the solid, so heat flows into it, which increases its entropy. Similarly, at the right extreme of its motion, heat flows out of the gas into the solid, decreasing the entropy of the gas. The gas effectively shuttles heat from hot to cold, left to right, down the temperature gradient. With x positive to the right, $\dot{H}_2$ is positive.

5.3 Some Calculation Methods

A broad range of calculational tools is useful for thermoacoustics, representing different trade-offs among ease of use, speed, and accuracy. It is hard to imagine progress in thermoacoustics today without the use of some of these. This section summarizes a few such methods, beginning with the easiest and least accurate.

Among the numerical integration methods, the emphasis in this book is on DELTAEC [61], which we use at Los Alamos. However, many other computer codes, such as SAGE by Gedeon [70], DSTAR by Hofler et al. (Design Simulation for ThermoAcoustic Research, US Naval Postgraduate School, Monterey CA), THERMOACOUSTICA by Tominaga (Institute of Physics, University of Tsukuba, Tsukuba 305, Japan), and REGEN3 used in Radebaugh's group at NIST-Boulder [71], are also accurate for modeling thermoacoustic devices. Although each code has its own strengths and weaknesses, the good ones are in agreement on the fundamentals, and all are traceable to the laws of fluids outlined in Sect. 2.2.

5.3.1 Order-of-Magnitude Estimates

A few simple order-of-magnitude, optimistic estimates can be worthwhile for initial exploration of a totally new thermoacoustic apparatus—perhaps only to rule it out quickly as a hopeless undertaking.

For traveling-wave engines and refrigerators, a good starting point is

$$\dot{E}_2 \sim \frac{1}{2}\,|p_1|\,|U_1| \sim \frac{|p_1|}{p_m}\frac{|\langle u_1 \rangle|}{a}\frac{p_m aA}{2}. \tag{5.32}$$

This provides estimates of the order of magnitude of the acoustic power flowing through the regenerator and the order of magnitude of the heat exchanged at each heat exchanger. The pressure oscillation might plausibly be assumed to be in the range $0.05 \lesssim |p_1|/p_m \lesssim 0.4$, with $|\langle u_1 \rangle|/a$ up to ten times smaller than $|p_1|/p_m$. Experience shows that the efficiency or COP is likely to fall between 20 and 50% of Carnot's efficiency or COP when the entire apparatus is taken into account.

For standing-wave engines and refrigerators, a good starting point is

$$\dot{H}_2 \sim \frac{1}{8}\,|p_1|\,|U_1| \sim \frac{|p_1|}{p_m}\frac{|\langle u_1 \rangle|}{a}\frac{p_m aA}{8}, \tag{5.33}$$

obtained from the inviscid Eq. (5.30) by assuming $\text{Im}[-f_\kappa] \sim 1/2$ and $|(dT_m/dx)/\nabla T_{\text{crit}} - 1| \sim 1/2$. This is the order of magnitude of the heat exchanged at each heat exchanger. The pressure oscillation might plausibly be assumed to be in the range $0.02 \lesssim |p_1|/p_m \lesssim 0.2$, with $|\langle u_1 \rangle|/a$ at least a few times smaller than $|p_1|/p_m$. Experience to date shows that the efficiency or COP is likely to fall between 10 and 25% of Carnot's efficiency or COP when the entire apparatus is taken into account.

For resonator-based systems, size or frequency can be estimated starting from $a = \lambda f$, with the length of the apparatus in the range of $\lambda/4$ to $\lambda/2$. In systems based on free pistons, the moving piston mass and gas-spring effects (see Sect. 3.3.1) set the operating frequency. Channel size in the stack or regenerator can be estimated relative to $\delta_\kappa = \sqrt{k/\pi f \rho_m c_p}$.

5.3.2 Spreadsheet Calculations

Some approximate methods are well suited to spreadsheets. Used skillfully, these methods usually match the performance of real systems to within a factor of 2.

An assumption common to many approximate methods (e.g., [11, 72]) is that the stack or regenerator is short enough and the temperature spanned is small enough that all x- and T_m-dependent variables can be regarded as constant, including dT_m/dx itself. For example, with this approximation the pressure drop across the stack or regenerator is estimated by

$$\Delta p_1 \simeq -\frac{i\omega \rho_m \Delta x\, U_1}{A\,(1 - f_v)} \tag{5.34}$$

[see Eq. (4.54)], where Δx is the length of the stack or regenerator, with ρ_m, U_1, and f_v evaluated at the stack midpoint or mid-temperature.

For further simplification with little additional loss of accuracy for stacks in standing-wave systems, the boundary-layer limit of Eq. (4.56) can be used for f_v and f_κ, and pure standing-wave phasing can be assumed. With these approximations in Eqs. (5.26) and (5.9), simple expressions are obtained for the total power flow through the stack and the acoustic power produced in (or, if negative, consumed by) the stack:

$$\dot{H}_2 \simeq \frac{A\,\delta_\kappa}{4\,r_h}\frac{|p_1||U_1|}{A\,(1+\sigma)\,\Lambda}\left(\Gamma\frac{1+\sqrt{\sigma}+\sigma}{1+\sqrt{\sigma}} - 1 - \sqrt{\sigma} + \frac{\delta_v}{r_h}\right)$$

$$- (Ak + A_{\text{solid}}k_{\text{solid}})\frac{dT_m}{dx}, \tag{5.35}$$

$$\Delta \dot{E}_2 \simeq \frac{A \, \Delta x}{4 r_h} \left[\frac{(\gamma - 1) \, |p_1|^2 \, \delta_\kappa \omega}{\gamma p_m} \left(\frac{\Gamma}{(1 + \sqrt{\sigma}) \, \Lambda} - 1 \right) - \frac{\rho_m \, |U_1|^2 \, \delta_\nu \omega}{A^2 \Lambda} \right],$$

(5.36)

where

$$\Gamma = (dT_m/dx) \, / \, \nabla T_{\mathrm{crit}},$$

(5.37)

$$\nabla T_{\mathrm{crit}} = \omega A \, |p_1| \, / \rho_m c_p \, |U_1|, \quad \text{defined in Eq. (4.44), and}$$

$$\Lambda = 1 - \delta_\nu / r_h + \delta_\nu^2 / 2 r_h^2.$$

(5.38)

Examination of Fig. 4.14 shows that most channel geometries have similar small-r_h behavior, with $\mathrm{Im}[-f]$ approaching 0 quadratically and $\mathrm{Re}[f] \simeq 1$. Picking parallel-plate channels and taking the small-r_h limit, Eqs. (5.26) and (5.9) reduce to the following useful approximate expressions for the regenerators of traveling-wave systems:

$$\dot{H}_2 \simeq -\frac{17}{35} \frac{\rho_m c_p}{A \omega} \frac{r_h^2}{\delta_\kappa^2} |U_1|^2 \frac{dT_m}{dx} - (Ak + A_{\mathrm{solid}} k_{\mathrm{solid}}) \frac{dT_m}{dx},$$

(5.39)

$$\Delta \dot{E}_2 \simeq \dot{E}_{2,0} \left(\frac{T_{(H \, \mathrm{or} \, C)}}{T_0} - 1 \right) - \frac{3 \mu \, \Delta x}{2 A r_h^2} |U_1|^2 - \frac{\omega^2 A r_h^2 \, \Delta x}{6 k T_m} |p_1|^2,$$

(5.40)

where $\Delta \dot{E}_2$ is the acoustic power produced in (or, if negative, consumed by) the regenerator, $\dot{E}_{2,0}$ is the acoustic power entering the ambient end of the regenerator at ambient temperature T_0, and $T_{(H \, \mathrm{or} \, C)}$ is the temperature of the hot end of an engine's regenerator or that of the cold end of a refrigerator's regenerator. The numerical coefficients in Eqs. (5.39) and (5.40) differ for different channel geometries.

Equation (5.34) and the similar approximation to the continuity equation, Eq. (4.70) with $dT_m = 0$, can be used for approximate calculations in resonator components.

5.3.3 DELTAEC's Numerical Integration of Rott's Acoustic Approximation

Chapter 4 developed the momentum equation to relate U_1 and dp_1/dx, as in Eq. (4.54) or (4.72). The momentum equation can be interpreted in two ways, either as an expression of how flow causes a pressure gradient or as an expression of how a pressure gradient causes flow, and both points of view are useful. In steady-flow hydrodynamics, it is usually obvious which point of view is preferable in a given circumstance—the flow of water down a dam's spillway is caused by the gravitational pressure gradient, while the pressure gradients in the oil pumped through the passages of an automobile engine are caused by the volume flow rate coming from the positive-displacement oil pump. In acoustics, the "correct" point of view is more subtle, as the momentum equation relates U_1 and dp_1/dx while simultaneously the continuity equation relates p_1 and dU_1/dx.

Similarly, two interpretations of the thermoacoustic total-power equation are possible, whether Eq. (5.24), (5.25), (5.26), or (5.27). In one interpretation, the temperature gradient dT_m/dx (with all the other relevant variables, such as channel geometry, p_1, and U_1) causes total power $\dot{H}_2$ to flow through a stack or regenerator. In another interpretation, $\dot{H}_2$ (with all the other variables) causes the temperature gradient in the stack or regenerator. Both points of view are useful. For example, if a fixed heater power is delivered into the hot heat exchanger of the standing-wave engine of Fig. 1.9, that power must go through the stack, so it determines the temperature gradient in the stack, in concert with the many additional variables such as those of the acoustic wave. On the other hand, if that heater uses a thermostat set at a fixed T_H, the temperature difference across the stack determines the power flowing through the stack and hence determines the heater power that must be supplied.

The second interpretation of the total-power equation is particularly important because it shows that the total-power equation can be regarded as a first-order differential equation for $T_m(x)$ in a stack or regenerator, if p_1 and U_1 are known. This is evident in the control-volume picture of Fig. 5.3b, which shows that the total power $\dot{H}_2$ must be a constant, independent of x, throughout the stack or regenerator whenever the side walls of the stack or regenerator are insulated. The other variables in Eq. (5.26) are presumably known from the considerations of Chap. 4. Although the temperatures at the ends of a stack or regenerator are often known, in Rott's acoustic approximation there is no better way than Eq. (5.26) to calculate the distribution of temperature at intermediate x, as illustrated in Fig. 5.6.

The numerical integrations in DELTAEC and in similar 1-D acoustic-approximation codes (Hofler, Design Simulation for ThermoAcoustic Research)

Fig. 5.6 The fact that $\dot{H}_2$ is independent of x within an insulated stack or regenerator determines the temperature distribution $T_m(x)$ within the stack or regenerator

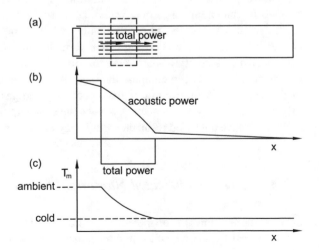

[73] rely on this interpretation of the total-power, momentum, and continuity equations. DELTAEC performs simultaneous integrations along x of the momentum, continuity, and total-power equations, in the forms

$$dp_1 = \mathcal{F}_{\text{mom}}(T_m, U_1; \omega, \text{gas properties, geometry}) \, dx \qquad (5.41)$$

$$dU_1 = \mathcal{F}_{\text{cont}}(T_m, p_1, U_1; \omega, \text{gas properties, geometry}) \, dx \qquad (5.42)$$

$$dT_m = \mathcal{F}_{\text{pow}}(T_m, p_1, U_1; \dot{H}_2, \omega, \text{gas properties, geometry}) \, dx. \qquad (5.43)$$

to find $p_1(x)$, $U_1(x)$, and $T_m(x)$. In Eq. (5.43), $\dot{H}_2$ is taken to be independent of x within each stack or regenerator. Its value may be assigned based on nearby heat-exchanger heats, or its value may be found self-consistently. After $p_1(x)$, $U_1(x)$, and $T_m(x)$ are found, other results such as $\dot{E}_2(x)$ are straightforward.

DELTAEC integrates these one-dimensional equations, in a geometry given by the user as a sequence of segments, such as ducts, compliances, transducers, and thermoacoustic stacks or regenerators. In each segment, DELTAEC uses locally appropriate momentum, continuity, and energy equations, controlled by local parameters (e.g., area or perimeter) and by global parameters (e.g., frequency and mean pressure). Solutions $p_1(x)$, $U_1(x)$, and $T_m(x)$ are found for each segment and are matched together at the junctions between segments to piece together $p_1(x)$, $U_1(x)$, and $T_m(x)$ for the entire apparatus.

It is clear that such a solution $T_m(x)$, $p_1(x)$, $U_1(x)$ is only determined uniquely if essentially five boundary conditions are imposed, because the governing equations comprise one real plus two complex coupled first-order equations in one real plus two complex variables. This is true whether considering a single segment or a one-dimensional string of segments with each joined to its neighbors by continuity of T_m, p_1, and U_1. If all boundary conditions are known at the initial segment, then the integration is utterly straightforward. But usually some of the boundary conditions are known elsewhere. Under such conditions, DELTAEC uses a shooting method, by guessing and adjusting any unknown boundary conditions at the initial segment to achieve desired boundary conditions elsewhere. The user of DELTAEC enjoys considerable freedom in choosing which variables are used as boundary conditions and which are computed as part of the solution. For example, in a simple plane-wave resonator, DELTAEC can compute the input impedance as a function of frequency, or the resonance frequency for a given geometry and gas, or the length required to give a desired resonance frequency, or even the concentration in a binary gas mixture required to give a desired resonance frequency in a given geometry.

5.3.4   More Sophisticated Numerical Integration

Other numerical-integration methods sacrifice speed to improve accuracy by including phenomena "beyond Rott's thermoacoustics" discussed in Chap. 7. For example,

SAGE [70] shares DELTAEC's one-dimensional approach and treats an apparatus as a series of standardized segments, but SAGE does not restrict itself to sinusoidal time dependences—it finds the true steady-state periodic time dependence. In other words, while DELTAEC performs numerical integration only along x, SAGE integrates numerically with respect to both x and t. This provides improved accuracy for high-amplitude devices in which the harmonic content of the waves is high.

REGEN3 [71] also integrates with respect to both x and t, but makes no restrictive assumption that the time dependence must be periodic. This allows REGEN3 to calculate the entire time dependence of transient behavior, such as the initial cool-down time of a refrigerator.

Laborious numerical integration of the fully general laws of fluids reviewed in Sect. 2.2, in all three spatial dimensions and in time, might be performed in order to model a part of a single thermoacoustic component, to improve fundamental understanding. However, such an integration for an entire thermoacoustic apparatus is beyond the capability of today's computers.

5.4 Examples

Example: Standing-Wave Engine

While producing the curves of p_1 and U_1 shown in Fig. 4.18 for the standing-wave engine, DELTAE also computed $T_m(x)$ and the powers. The results are shown in Fig. 5.7. Within the stack, total power $\dot{H}_2$ was constant at 2380 W, because of the sidewall insulation and the first law of thermodynamics. Its magnitude is the same as the heater power supplied at the hot heat exchanger. DELTAE used this value of $\dot{H}_2$ and the total-power equation to determine $T_m(x)$ in the stack. This temperature profile deviates noticeably from a straight line.

The calculated values of acoustic power $\dot{E}_2(x)$ follow directly from the values of $p_1(x)$ and $U_1(x)$ plotted in Fig. 4.18. Notice also that $\mathrm{Re}[p_1]\mathrm{Im}[U_1]$ contributes nothing to $\dot{E}_2$; nonzero acoustic power arises from $\mathrm{Re}[p_1]$ times the small $\mathrm{Re}[U_1]$ and from $\mathrm{Im}[U_1]$ times the small $\mathrm{Im}[p_1]$. A total of 670 W of acoustic power was generated throughout the stack, but especially near the hot end, and 500 W of acoustic power was assumed to flow from right to left at the right end of the figure, coming from the other acoustic engine shown in Fig. 1.9. This total of 1170 W of acoustic power was consumed elsewhere, most dramatically at the side branch to the refrigerators, where 1000 W was delivered. Acoustic power dissipation elsewhere is more subtle, as evidenced by the small negative slope of $\dot{E}_2(x)$ throughout the resonator. This negative slope is somewhat larger in the two heat exchangers. The small negative slope of $\dot{E}_2(x)$ in the hot duct, to the left of the stack and hot heat exchanger, which is almost invisible in the figure except as a slight change in line width between $x = 0$ and $x = 0.05$ m, shows the dissipation of 20 W estimated at the end of Sect. 5.1.1 above.

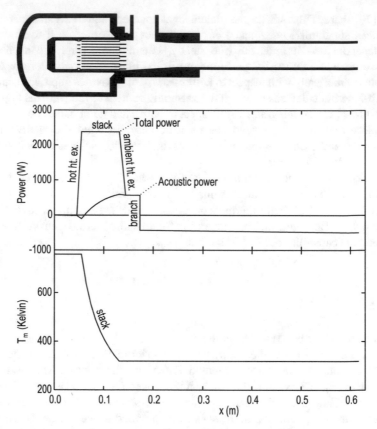

Fig. 5.7 Acoustic power $\dot{E}_2$, total power $\dot{H}_2$, and mean temperature T_m as functions of position x for the standing-wave engine, under the same conditions as Fig. 4.18. ("heat exchanger" is abbreviated "ht. ex.")

Example: Orifice Pulse-Tube Refrigerator

A similar plot of results of DELTAE's numerical integration is shown in Fig. 5.8 for the orifice pulse-tube refrigerator, under conditions similar to those shown in the phasor diagram of Fig. 4.21.

In the regenerator and the pulse tube, $T_m(x)$ was obtained by integration of the total-power equation, for a total power $\dot{H}_2 = 0.48$ kW in the regenerator and 2.92 kW in the pulse tube. In the regenerator, this resulted in only slight deviation of $T_m(x)$ from a straight line. The value of 0.48 kW was obtained self-consistently by the calculation, in order to allow the regenerator to reach from its ambient temperature of 292 K to its cold temperature of 128 K. Both terms in Eq. (5.25) contributed to $\dot{H}_2$ in the regenerator, but their sum is small compared with other powers in the problem. The pulse-tube radius was much greater than δ_κ, so $\dot{H}_2 \simeq \dot{E}_2$ as shown in the figure. The small difference between these two is due primarily to $A_{\text{solid}}k_{\text{solid}} \, dT_m/dx$ in Eq. (5.27). The large changes in $\dot{H}_2$ at

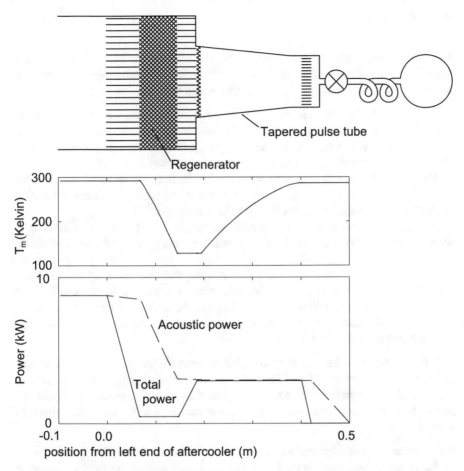

Fig. 5.8 Temperature T_m, total power $\dot{H}_2$, and acoustic power $\dot{E}_2$ throughout the orifice pulse-tube refrigerator of Fig. 1.18. Horizontal dimensions in the sketch above are aligned with corresponding locations in the graphs

each of the three heat exchangers represent the heats exchanged at these heat exchangers. The calculated cooling power at 128 K was $2.92 - 0.48 = 2.44$ kW.

About 2/3 of the 8.77 kW of acoustic power was consumed in the regenerator, and about 1/3 was consumed in the flow resistance of the *RLC* network at the end of the refrigerator. These powers were calculated from the calculated values of $p_1(x)$ and $U_1(x)$ throughout, and are consistent with the phasors in Fig. 4.21 showing p_1 and U_1 at a few locations. Small amounts of acoustic power were also consumed in the two largest heat exchangers, and a negligible amount of acoustic power was consumed in the pulse tube and in the large duct to the left of all the heat-exchange components.

5.5 Exercises

5.1 In the spirit of Fig. 3.6, draw phasor diagrams for several locations of the wave shown in Ani. Wave /k. Discuss how your diagrams are consistent with positive acoustic power flow in the positive x direction, and with $d\dot{E}_2/dx = 0$.

5.2 In a double Helmholtz resonator, how much kinetic energy is in the gas in the neck when it is moving with velocity $|u_1|$? Express your answer in terms of L and U_1. A quarter cycle later, all that kinetic energy is converted to potential energy stored in the compliances. Express that energy in terms of p_1 and C.

Using Eq. (5.11), write down expressions for the time-averaged viscous absorption of acoustic power in the neck, and the time-averaged thermal-relaxation absorption of acoustic power the bulbs. Which is typically bigger?

The quality factor of a resonance is given by $Q = 2\pi$(Stored energy)/(Energy dissipated per cycle). Write down an expression for the Q of this double Helmholtz resonator.

Acoustics experts only: In a simple acoustics textbook, look up the radiation impedance from a circular orifice whose diameter is much smaller than λ. There is an imaginary part, which adds to the inertance of the neck. Estimate how much the inertance "end effect" shifts the resonance frequency of the Helmholtz resonator. How much does it change the Q?

5.3 If your favorite piece of thermoacoustics hardware has a stack or regenerator, estimate the heat deposited and removed from the solid during each half cycle of the oscillation. If you don't have a favorite, estimate the heat deposited and removed during each half cycle of an audio-amplitude sound wave in air, per square meter of a solid boundary.

5.4 Estimate the critical temperature gradient in the stack of your favorite piece of hardware. If you don't have a favorite stack, estimate the critical temperature gradient halfway between the pressure node and adjacent velocity node in a 240 Hz standing wave in air at atmospheric pressure and room temperature. How much of the information given in the preceding sentence was unnecessary to find the answer?

5.5 If your favorite acoustics hardware has a power transducer such as a loudspeaker, estimate the acoustic power it transduces, using its $|\xi_1|$, area, frequency, etc.

5.6 Draw a few control volumes for your favorite thermoacoustics hardware. Discuss thermal insulation, and where energy can cross the boundaries of the control volumes.

5.7 Estimate $\dot{H}_2$ for the stack or regenerator in your favorite piece of thermoacoustics hardware. Black-body radiation is another contributor to total power, which was not included in this chapter. Estimate it for the same component.

5.8 Sketch $\dot{H}_2$ and $\dot{E}_2$ vs x for your favorite thermoacoustics hardware. Don't worry about the actual magnitudes of things, but try to get the overall qualitative picture right: the signs of the powers, the signs and locations of discontinuities, the signs of slopes. Also sketch $|p_1|$ and $|U_1|$.

5.9 Look closely at Ani. Thermal /e. It looks like the power curve shows negative slope at about $y/\delta_\kappa = 4$. This suggests that thermal relaxation in gas at that distance from a solid wall does not dissipate acoustic power—it *generates* acoustic power!—because the imaginary piston at a distance $5\delta_\kappa$ from the wall has to supply *less* power to the gas between it and the wall than does the imaginary piston at a distance $4\delta_\kappa$ from the wall. Was the author sloppy in programming the animation, or does the math confirm this surprising observation? Does something similar occur in boundary-layer viscous dissipation as illustrated in Ani. Viscous?

5.10 Can nonzero acoustic power pass a true pressure node in a standing wave? Can it pass a velocity node? Can nonzero total power pass either a pressure node or a velocity node?

5.11 Consider a large-diameter channel with a nonzero temperature gradient, such as the thermal buffer tube or pulse tube of Figs. 1.23 or 1.18. Show that $\dot{H}_2 \simeq \dot{E}_2$ if you can neglect boundary-layer effects at the walls and ordinary thermal conduction along x (and if streaming, to be discussed in Chap. 7, is negligible). Discuss this result from several points of view: Eqs. (5.25), (5.27), and a Lagrangian picture of moving parcels of gas. With respect to Eq. (5.27), be sure to notice that $s_1 \neq 0$.

5.12 Consider a channel large enough that the boundary-layer approximation is valid. The channel has a spatially uniform T_m, maintained by water flowing in a jacket around the channel. Suppose p_1 and U_1 are known. Use Eq. (5.26) for $\dot{H}_2$ and the principle of energy conservation to derive an expression for the heat $d\dot{Q}$ delivered to (or removed from?) the water in a length dx of the channel. In your answer, use results from Chap. 4 to express dp_1/dx in terms of U_1 and to express dU_1/dx in terms of p_1. Interpret the Merkli-Thomann experiment, Fig. 5.4, from this perspective—are their experimental results consistent with your answer?

5.13 Show that $\dot{H}_2 = \dot{E}_2$ when $dT_m/dx = 0$, $\mu = 0$, and $k = 0$, independent of the size of the channel.

5.14 Simplify Eq. (5.26) using boundary-layer expressions for f_κ and f_ν and assuming standing-wave phasing [11].

5.15 Evaluate $\int \text{Re}\,[s_1\widetilde{u_1}]\,dA$ in boundary-layer approximation. Make a graph, with the vertical axis $A\,|p_1|\,/\rho_m a\,|U_1|$ running from 0 to 10, and the horizontal axis $(dT_m/dx)/\nabla T_{\text{crit}}$ running from 0 to 4. On the graph, for p_1 and U_1 in phase, draw a line indicating where $\int \text{Re}\,[s_1\widetilde{u_1}]\,dA = 0$. Show which side of this line has $\int \text{Re}\,[s_1\widetilde{u_1}]\,dA > 0$ and which side has $\int \text{Re}\,[s_1\widetilde{u_1}]\,dA < 0$. Repeat for phase differences between p_1 and U_1 of $45°, 90°, 135°, \ldots$ Interpret some of these graphs in terms of Ani. Standing.

Chapter 6
Efficiency

Both the raw efficiency and the "fraction of Carnot" serve as useful measures of efficiency for the simplest thermodynamic systems, such as the simple heat engine shown in Fig. 2.5 at the beginning of the book and reproduced here in Fig. 6.1a. The raw efficiency is "what you want divided by what you must spend to obtain it," and Chap. 2 reviewed the well-known bound on raw efficiency given by Carnot's ratio of temperatures, which was quickly derived from the first and second laws of thermodynamics. So it is natural to also cite the fraction of Carnot, which is the ratio of the actual raw efficiency to the maximum raw efficiency allowed by the laws of thermodynamics. The ideal is 100%, but reality always reduces this significantly. Typically, thermoacoustic engines and refrigerators that were designed for high efficiency have operated at or above 20% of Carnot; as of early 2001, the highest reported [37, 38] value is 40% of Carnot.

"Fraction of Carnot" is fine for such simple systems. However, more complex systems need a more sophisticated measure of merit. For instance, consider the machine shown in Fig. 6.1b, which takes air at atmospheric temperature and pressure, and separates and cools it to produce liquid oxygen, liquid nitrogen, liquid argon, solid carbon dioxide, and liquid water. The machine is driven with some work, and it dumps waste heat to ambient temperature. How should the efficiency of a complicated system like this be measured? What one wants to do and what one must spend in order to do it do not even have the same units (e.g., miles per gallon, tons of liquid cryogen per electric kW·h, etc.). How can the limits on the performance of such a system imposed by the two laws of thermodynamics be understood? This chapter introduces some advanced concepts in thermodynamics that can be used to answer such questions.

© Acoustical Society of America 2017
G.W. Swift, *Thermoacoustics*, DOI 10.1007/978-3-319-66933-5_6

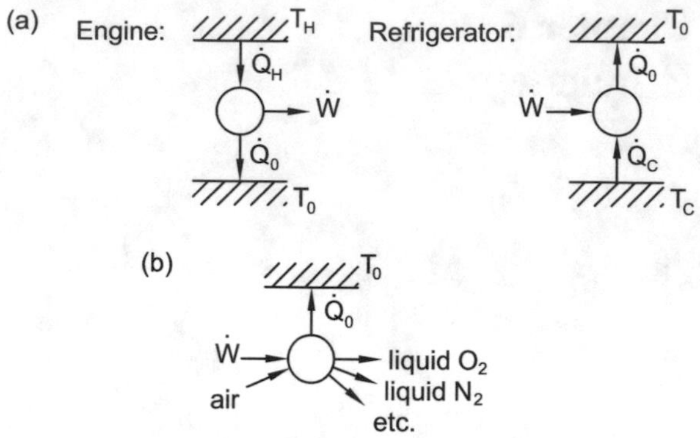

Fig. 6.1 (a) The simplest engine and refrigerator, for which simple measures of efficiency suffice. (b) An air separator–liquefier, requiring a more sophisticated approach to efficiency

6.1 Lost Work and Entropy Generation

In one common and powerful method of accounting for efficiency in complex thermodynamic systems [49, 74], all losses are measured in terms of equivalent lost work, and one temperature T_0 is identified as a special temperature, the environment temperature at which arbitrary amounts of heat of no economic value can be freely exchanged. (If *two* different environmental temperatures existed, and heat of no value could be freely exchanged with both of them, then it would be possible to run a heat engine between them, producing unlimited work—of great value—from heat of no value. Hence, identification of a *single* "ambient" temperature T_0 is not at all artificial.) This section begins with several examples illustrating the use of this method of accounting.

First consider a simple heat engine as shown in Fig. 6.2a. The engine's cold temperature is labeled with the subscript 0, to specify it as the environment temperature where waste heat of no value is rejected. The engine's work can be written as the hot heat times the Carnot efficiency minus something called lost work[1]:

$$\dot{W} = \dot{Q}_H \left(1 - \frac{T_0}{T_H}\right) - \dot{W}_{\text{lost}}. \qquad (6.1)$$

The first term on the right side is the maximum work that an ideal engine could produce, according to the bounds imposed by the first and second laws of thermodynamics, as reviewed in Sect. 2.4. If the engine is not so ideal, the work it actually produces must be lower; the difference is the lost work $\dot{W}_{\text{lost}}$. Similarly, for the simple refrigerator shown in Fig. 6.2b, the work required is the sum of the minimum work that an ideal refrigerator would require plus extra work $\dot{W}_{\text{lost}}$:

---

[1]We will call this "lost work" for brevity, even though it has units of power, not energy.

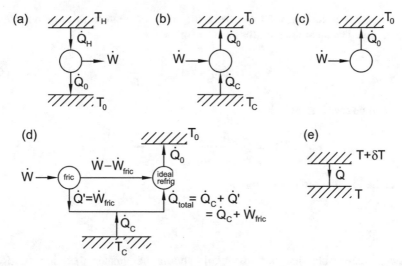

Fig. 6.2 Several examples to illustrate the concept of lost work. (**a**) A simple engine. (**b**) A simple refrigerator. (**c**) Friction at temperature T_0. (**d**) Friction at a temperature $T_C < T_0$ in a refrigerator. (**e**) Heat transfer across a small thermal resistance

$$\dot{W} = \dot{Q}_C \frac{T_0 - T_C}{T_C} + \dot{W}_{\text{lost}}. \tag{6.2}$$

A third example, a very simple example of lost work, is illustrated in Fig. 6.2c: Friction in machinery at ambient temperature dissipates work, all of which is clearly lost, so

$$\dot{W}_{\text{lost}} = \dot{W}_{\text{fric}} . \tag{6.3}$$

The fourth example, illustrated in Fig. 6.2d, is more subtle, and begins to show how challenging and how powerful this point of view can be. Figure 6.2d shows a refrigerator, within which some frictional dissipation occurs at a temperature T_C less than the environment temperature T_0. (Such dissipation could be due to drag on a piston in the cold part of the refrigerator.) How much lost work must be assigned to that frictional dissipation? Suppose that this friction is the only non-ideal feature in the system, so the system can be treated like an ideal refrigerator with Carnot's *COP*, powered by a drive mechanism with friction, with the drive mechanism thermally anchored to the cold temperature T_C, as shown in Fig. 6.2d. Some of the drive power $\dot{W}$ is dissipated into heat at T_C by friction. The remaining power drives an ideal refrigerator, which must pump the heat generated by friction plus the external cooling load $\dot{Q}_C$ from the low temperature to the environment temperature. The ideal refrigerator has Carnot's COP, so

$$\dot{W} - \dot{W}_{\text{fric}} = \left(\dot{Q}_C + \dot{W}_{\text{fric}}\right)\frac{T_0 - T_C}{T_C}, \tag{6.4}$$

which can be rewritten as

$$\dot{W} = \dot{Q}_C \frac{T_0 - T_C}{T_C} + \dot{W}_{\text{fric}}\frac{T_0}{T_C}. \tag{6.5}$$

Hence, the lost work in this system is

$$\dot{W}_{\text{lost}} = \dot{W}_{\text{fric}}\frac{T_0}{T_C}. \tag{6.6}$$

It seems reasonable that the lost work should be greater than the frictional dissipation, by a factor depending on the temperatures, because intuition suggests that dissipating work at low temperature should be worse than dissipating it at room temperature: If work is "lost" at low temperature, pumping the resulting heat up to room temperature costs even more work.

A fifth example—irreversible heat transfer—is shown in Fig. 6.2e. Two thermal reservoirs, at nearly equal temperatures $T + \delta T$ and T, are connected by a thermal resistance through which heat $\dot{Q}$ flows. The work lost in this process can be calculated by considering artificial but *reversible* equipment that accomplishes exactly the same heat flows $\dot{Q}$ for the two reservoirs at T and $T + \delta T$. Suppose that $T > T_0$. Let an ideal heat engine operate between $T + \delta T$ and T_0, and let an ideal heat pump operate between T and T_0. Let the heat removed from the reservoir at $T + \delta T$ by the engine be $\dot{Q}$ and the heat added to the reservoir at T by the heat pump also be $\dot{Q}$. (Ignore the heat exchanged with the reservoir at T_0, because it has no value.) Then the engine produces more work than the heat pump consumes. The work lost in the irreversible process of Fig. 6.2e is the excess work—the work produced by the engine minus that consumed by the heat pump:

$$\dot{W}_{\text{lost}} = \dot{W}_{\text{engine}} - \dot{W}_{\text{pump}} \tag{6.7}$$

$$= \dot{Q}\left(1 - \frac{T_0}{T + \delta T}\right) - \dot{Q}\left(1 - \frac{T_0}{T}\right) \tag{6.8}$$

$$\simeq \dot{Q}\frac{T_0\,\delta T}{T^2}, \tag{6.9}$$

which is small if the temperature difference δT is small. (For illustration here, $T > T_0$ is assumed, but the same conclusion follows from assuming $T < T_0$. See Exercise 6.6.) Imperfect thermal contact in heat exchangers is responsible for this type of lost work.

   These ideas are important because the lost work can be expressed as the product
of the environment temperature T_0 and the sum of all the entropy that is generated
in the system,

$$\dot{W}_{\text{lost}} = T_0 \sum \dot{S}_{\text{gen}} , \tag{6.10}$$

a result known as the Gouy–Stodola theorem [45]. This theorem is presented clearly
in the physics literature [74], but without reference to the decades-earlier work of
Gouy and Stodola.

   The Gouy–Stodola theorem will not be proven here, but some of the examples
of Fig. 6.2 will be used to illustrate it. First: The entropy generated by imperfect
thermal contact in Fig. 6.2e must be the difference between the entropy $\dot{Q}/T$ gained
by the lower reservoir and the entropy $\dot{Q}/(T + \delta T)$ lost by the upper reservoir. The
difference is $\dot{Q}\,\delta T/T^2$ to lowest order in δT, and multiplying by T_0 indeed yields the
lost work given in Eq. (6.9). Second: For friction at a temperature T as illustrated
in Fig. 6.2c but with $T \neq T_0$, the entropy generation must be the frictional heat
divided by T. The frictional heat is exactly $\dot{W}_{\text{fric}}$. Multiplying by T_0, as prescribed
by Eq. (6.10), yields

$$\dot{W}_{\text{lost}} = \dot{W}_{\text{fric}} \frac{T_0}{T}, \tag{6.11}$$

in agreement with both Eqs. (6.3) and (6.6).

   This method and the Gouy–Stodola theorem are important because entropy is
an extensive quantity, so that the sum Σ in Eq. (6.10) can be performed any of a
number of ways. The sum can be performed component by component throughout a
piece of experimental hardware, tabulating how much entropy is generated in each
heat exchanger, stack, resonator, etc. It can even be performed by location within a
component, e.g., to determine whether the hot end of a stack, the middle of a stack,
or the cold end of a stack is the lossiest. This sum can be performed process by
process, tabulating how much entropy is generated by viscous processes, thermal
processes, frictional processes, etc., either within each component or system-wide.
There seems to be no limit to how finely this sum can be subdivided; e.g., one could
even ask how much of the thermal entropy generation at a given location is due to
heat diffusing in the x direction, and how much is due to heat diffusing perpendicular
to x. The sum can be performed with respect to time instead of location or process,
e.g., to learn whether the entropy generated by thermal relaxation in a regenerator
occurs mostly during the compression and expansion phases of the cycle or during
the gas-motion phases of the cycle.

   Taken to an extreme in thermoacoustics, this point of view lets the time-averaged
entropy generation be written as

$$\sum \dot{S}_{\text{gen}} = \frac{\omega}{2\pi} \int_0^{2\pi/\omega} \int_V \rho \dot{s}_{\text{gen}} \, dV \, dt, \tag{6.12}$$

where the instantaneous rate of entropy generation per unit volume is

$$\rho \dot{s}_{\text{gen}} = \frac{\mu}{T}\Phi + \frac{k\,|\boldsymbol{\nabla}T|^2}{T^2} + \frac{j^2}{\alpha T} + \text{friction} + \text{chemistry} + \cdots . \tag{6.13}$$

(Here α is electric conductivity and j is electric current density.) This displays the time-averaged rate of entropy generation as an integral, over time and over the whole volume of an apparatus, of the instantaneous rate of entropy generation, which is itself the sum of effects coming from different processes: viscous, thermal, electrical, frictional, chemical, etc.

Having this in mind lets one think about where and how the inefficiencies arise in a system. Quoting from a clear review [74] of this method: "In accordance with the circumstance that entropy is a quantity having extensive magnitude, it is evident that $[T_0 \dot{S}_{\text{gen}}]$ can be calculated by adding together separate terms of the form $[T_0 \dot{S}_{\text{gen}}]$ for each of the irreversible processes that accompany a total operation. This then allows us to take each of these separate terms ... as a measure of the loss in potential work that results from the corresponding particular irreversible processes in the total operation. This is an important principle, since it allows us to determine the reduction in efficiency resulting from any particular cause without making an analysis of the whole operation."

Source	$\dot{W}_{\text{lost}}$ (macroscopic)	$\rho \dot{s}_{\text{gen}}$ (microscopic)
Electrical	$I^2 R T_0/T$	$j^2/\alpha T$
Friction	$\dot{W}_{\text{fric}} T_0/T$	$v\sigma_{\text{fric}}/T$ (per unit area)
Viscosity	$U\,\Delta p\,T_0/T$	$\mu\Phi/T + \zeta\,(\boldsymbol{\nabla}\cdot\mathbf{v})^2/T$, with Φ given by Eq. (2.43)
Imperfect heat transfer ($\delta T \ll T$)	$\dot{Q} T_0\,\delta T/T^2$	$k\,|\boldsymbol{\nabla}T|^2/T^2$
Heat leak	$\dot{Q} T_0\,(T_H - T_C)/T_H T_C$	$k\,|\boldsymbol{\nabla}T|^2/T^2$
Free expansion	$\dot{M} c_p T_0 \log\,[p_{\text{final}}/p_{\text{initial}}]$	
Thermal mixing	$\dot{M} c_p T_0 \log\left[\frac{x+\tau(1-x)}{\tau^{1-x}}\right]$, where $\dot{M} = \dot{M}_1 + \dot{M}_2$, $x = \dot{M}_1/\dot{M}$, and $\tau = T_2/T_1$	
Mass mixing		
Mass diffusion		
Chemical reaction		
...etc.!		

The table above gives expressions for lost work and entropy generation for some dissipative processes encountered in thermoacoustics.[2] For example, if dc electric

---

[2]Watch out for factors of two if you use the macroscopic column for time-averaged acoustic situations.

power is dissipated by resistance, the dissipated electric power is I^2R, and the lost work is T_0/T times the dissipated power. Microscopically, the entropy generation is given by the heat generation rate j^2/α per unit volume, divided by the temperature at which the heat is generated. Integrating that entropy generation over the whole volume of a resistor, and multiplying by T_0 to convert from entropy generation to lost work, yields the macroscopic expression for lost work. For a second example, the lost work accompanying viscous flow down a pipe can be computed with the macroscopic expression, if microscopic details such as whether the flow is turbulent or laminar are of no concern. If instead one is concerned with microscopic details, perhaps to understand the spatial distribution of dissipation within a channel of a stack, then the microscopic expression can be used. This list of entropy-generating processes could be extended—many mechanisms contribute to the increase of the entropy of the universe!

Imperfect heat transfer is one of the most important loss mechanisms in engines and refrigerators, so consider the $k|\nabla T|^2/T^2$ term in more detail. If $k = 0$ — perfect insulation—there is no entropy generation. In the other extreme, $k = \infty$, there is also no entropy generation, because in that case ∇T is zero. Only the in-between case is lossy. Unfortunately, all real gases and materials have nonzero, finite thermal conductivity, so all are lossy! In standing-wave systems, the situation is severe, because standing-wave engines and refrigerators *depend* on such imperfect thermal contact to get useful thermodynamic behavior, with heat being transferred irreversibly across a thermal penetration depth, throughout the stack, every cycle of the wave. For this reason, standing-wave systems have sometimes been called "intrinsically irreversible" [75].

In our experience at Los Alamos, the important types of processes that are responsible for irreversibility in thermoacoustic engines and refrigerators are:

(1) Thermal relaxation occurs within the stack in any standing-wave system, because imperfect thermal contact characterized by $\text{Im}[f_\kappa] \neq 0$ in the stack is necessary for nonzero power in standing-wave systems. In traveling-wave systems, those losses are in principle and in practice smaller, giving traveling-wave systems an inherent efficiency advantage over standing-wave systems.
(2) Viscous drag in stacks and regenerators is important in both standing- and traveling-wave systems. Some viscous loss is unavoidable, because the Prandtl number is not much less than unity for anything except liquid metals [60], so the viscous penetration depth is of the same order as the thermal penetration depth.
(3) Ordinary conduction of heat in the x direction from hot to cold along stacks and regenerators is conceptually a very simple loss. However, it is unavoidable, especially in the wall of the pressure vessel.
(4) Thermal-relaxation loss and viscous loss occur not only in stacks and regenerators but also on resonator walls and in heat exchangers.
(5) Serious thermal bottlenecks can occur elsewhere in heat exchangers, in the solid metal parts and in the fluids such as water flowing through them carrying heat to or from the system.

(6) If used, electroacoustic transducers contribute significantly to system loss, through electric dissipation and sometimes through piston friction.
(7) Finally, there are things that are not yet understood. Our thermoacoustic systems never perform as well as we think they should. And effects beyond Rott's acoustic approximation add to the severity of many of the items in this list. For example, turbulence in resonators adds to the viscous losses there.

6.2 Exergy

Consider the device shown in Fig. 6.3, processing a mass flow $\dot{M}$ of fluid from one temperature and pressure to another temperature and pressure. Liquefaction of nitrogen or methane is typical of such a process. The system consumes work $\dot{W}$, and waste heat $\dot{Q}_0$ is rejected at temperature T_0. The first law of thermodynamics indicates that the sum of the powers going in must equal the sum of the powers going out:

$$\dot{W} + \dot{M}h_{\text{in}} = \dot{Q}_0 + \dot{M}h_{\text{out}}. \tag{6.14}$$

Note the use of enthalpy here. Enthalpy is the correct energy to use in the first law for open systems, as discussed in Chaps. 2 and 5. The second law of thermodynamics indicates that the entropy leaving the system must equal the sum of the entropy entering the system plus the entropy generation within the system:

$$\frac{\dot{Q}_0}{T_0} + \dot{M}s_{\text{out}} = \dot{M}s_{\text{in}} + \sum \dot{S}_{\text{gen}}. \tag{6.15}$$

The sources $\dot{S}_{\text{gen}}$ inside the system are viscous, frictional, and other dissipative processes as discussed in the previous section.

Combining these two equations by eliminating the uninteresting quantity $\dot{Q}_0$ (i.e., the heat of no economic value) yields

$$\dot{W} = \dot{M}\left[(h - T_0 s)_{\text{out}} - (h - T_0 s)_{\text{in}}\right] + T_0 \sum \dot{S}_{\text{gen}}. \tag{6.16}$$

This shows that $h - T_0 s$ is another important, special energy in thermodynamics problems. If the system is ideal—if there are no entropy-generating processes in the

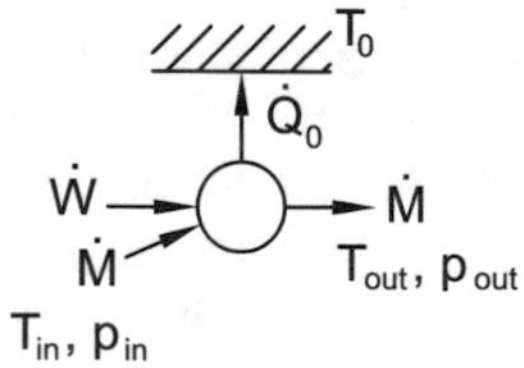

Fig. 6.3 A thermodynamic device using work to process a flowing stream of material $\dot{M}$ from an input condition to an output condition. The device rejects waste heat at T_0

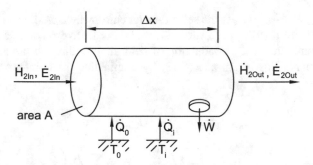

Fig. 6.4 A generalized portion of thermoacoustic apparatus. The x direction is horizontal, and this portion of the apparatus, with length Δx, has acoustic power $\dot{E}_{2\mathrm{In}}$ and total power $\dot{H}_{2\mathrm{In}}$ flowing into the left open end and acoustic power $\dot{E}_{2\mathrm{Out}}$ and total power $\dot{H}_{2\mathrm{Out}}$ flowing out of the right open end. In between, the portion produces (or receives) time-averaged mechanical power $\dot{W}$, absorbs (or rejects) thermal power $\dot{Q}_0$ from ambient at T_0, and absorbs (or rejects) one or more other thermal powers $\dot{Q}_i$ from other temperatures T_i. The sign conventions are indicated by the *arrows*

system, so $\Sigma \dot{S}_{\mathrm{gen}} = 0$ —then the work needed to process unit mass of the fluid is given by the difference between the outgoing and incoming $h - T_0 s$. Any real system must use at least that much work to perform this process.[3] This energy per unit mass $b \equiv h - T_0 s$ is called the flow availability [45], well known to mechanical engineers and especially to chemical engineers. It is invaluable for calculating the minimum work required to perform a complex process such as that shown in Fig. 6.1b.

A similar approach leads to a derivation of an acoustic approximation to the Gouy–Stodola theorem, and to the introduction of the important concept of exergy X, a form of "energy" that accounts for the ability to do useful work when a thermal reservoir at temperature T_0 is freely accessible. Consider the schematic of a generalized portion of a thermoacoustic system shown in Fig. 6.4. If the portion shown is in steady-state operation, the first law of thermodynamics shows that the difference between energy fluxes going out and going in must be zero:

$$0 = \dot{H}_{2\mathrm{Out}} + \dot{W} - \dot{H}_{2\mathrm{In}} - \dot{Q}_0 - \sum_i \dot{Q}_i. \qquad (6.17)$$

The second law of thermodynamics says that the difference between the entropy fluxes going out and going in must equal the entropy generated within the portion:

$$\sum \dot{S}_{\mathrm{gen}} = \frac{\dot{H}_{2\mathrm{Out}} - \dot{E}_{2\mathrm{Out}}}{T_{m\mathrm{Out}}} - \frac{\dot{H}_{2\mathrm{In}} - \dot{E}_{2\mathrm{In}}}{T_{m\mathrm{In}}} - \frac{\dot{Q}_0}{T_0} - \sum_i \frac{\dot{Q}_i}{T_i}. \qquad (6.18)$$

---

[3]Physicists note: This is *not* the Gibbs energy $h - Ts$.

In Eq. (6.18), $(\dot{H}_2 - \dot{E}_2)/T_m$ is the second-order entropy flux, as shown in Eq. (5.29). The first term of Eq. (5.29) is T_m times the second-order hydrodynamic entropy flux and the second term is the conduction of heat along x through the gas and solid parts. Combining Eqs. (6.17) and (6.18) by eliminating the uninteresting heat $\dot{Q}_0$ easily yields

$$T_0 \sum \dot{S}_{\text{gen}} = \frac{T_0}{T_{m\text{In}}} \dot{E}_{2\text{In}} + \left(1 - \frac{T_0}{T_{m\text{In}}}\right) \dot{H}_{2\text{In}}$$

$$- \frac{T_0}{T_{m\text{Out}}} \dot{E}_{2\text{Out}} - \left(1 - \frac{T_0}{T_{m\text{Out}}}\right) \dot{H}_{2\text{Out}} - \dot{W} + \sum_i \dot{Q}_i \left(1 - \frac{T_0}{T_i}\right). \qquad (6.19)$$

This is written in terms of exergy as

$$T_0 \sum \dot{S}_{\text{gen}} = \dot{X}_{2\text{In}} - \dot{X}_{2\text{Out}} - \dot{X}_W + \sum_i \dot{X}_{Q_i}, \qquad (6.20)$$

where the thermoacoustic approximation to the time-averaged exergy flux $\dot{X}_2(x)$ in the x direction in a channel is

$$\dot{X}_2 = \frac{T_0}{T_m} \dot{E}_2 + \left(1 - \frac{T_0}{T_m}\right) \dot{H}_2. \qquad (6.21)$$

Similarly, the exergy flux $\dot{X}_W$ associated with work is the work $\dot{W}$ itself,

$$\dot{X}_W = \dot{W}, \qquad (6.22)$$

and the exergy flux $\dot{X}_Q$ associated with heat $\dot{Q}$ flowing from temperature T is

$$\dot{X}_Q = \dot{Q}(1 - T_0/T). \qquad (6.23)$$

Each of these last three equations for exergy flux gives the ability of the associated power to do work, in the presence of a thermal reservoir at T_0. If $\sum \dot{S}_{\text{gen}} = 0$, Eqs. (6.20)–(6.23) prescribe how mechanical power, thermal power, acoustic power, and thermoacoustic total power could be freely interchanged within the bounds of the first and second laws of thermodynamics.

Reminiscent of the second-order equation for total power discussed in Chap. 5, the second-order equation for exergy flux, Eq. (6.21), is also like a kaleidoscope, giving many different views depending on how it is shaken. A few will be examined:

In a stack, regenerator, or thermal buffer tube, laterally insulated and bounded so that no exergy can flow out perpendicular to x, a thermoacoustic approximation to the Gouy–Stodola theorem results from letting $\Delta x \to dx$ in Eq. (6.20):

$$-\frac{d\dot{X}_2}{dx} = T_0 \frac{d \sum \dot{S}_{\text{gen}}}{dx}. \qquad (6.24)$$

This equation expresses the combined first and second laws of thermodynamics in thermoacoustics, showing that

$$\frac{d\dot{X}_2}{dx} \leq 0 \qquad (6.25)$$

always, just like $d\dot{H}_2/dx = 0$ expresses the first law in a stack, regenerator, or thermal buffer tube. Exergy is conserved if and only if no entropy is generated.

If $T_m = T_0$, then Eq. (6.21) shows that $\dot{X}_2 = \dot{E}_2$. In words, acoustic power at T_0 is exactly the ability to do work, in the strictest thermodynamic sense. This is consistent with the interpretation near Eq. (5.6) of $\dot{E}_2$ in terms of work done by a piston. In this simplest situation, lost work and entropy generation are related in the most straightforward manner,

$$-\frac{d\dot{X}_2}{dx} = T_0 \frac{d\sum \dot{S}_{\text{gen}}}{dx} = -\frac{d\dot{E}_2}{dx}. \qquad (6.26)$$

Now suppose that $T_m \neq T_0$, and consider an ideal thermal buffer tube or pulse tube, for which $\dot{E}_2 = \dot{H}_2$. Then $\dot{X}_2 = \dot{E}_2$, independent of T_m. In this case, Eq. (6.26) holds, with $d\dot{E}_2/dx = 0$ only if $\sum \dot{S}_{\text{gen}} = 0$. Again, $\dot{E}_2$ represents the ability to do work.

Continuing to suppose that $T_m \neq T_0$, consider an ideal regenerator with $\dot{H}_2 = 0$. Then $\dot{X}_2 = \dot{E}_2 T_0/T_m$. In other words, acoustic power in a temporally isothermal space, such as a regenerator, at temperature $T_m < T_0$ has an ability to do work that is greater than $\dot{E}_2$ itself. Similarly, acoustic power in a temporally isothermal space at temperature $T_m > T_0$ represents a reduced ability to do work. The full $\dot{E}_2$ cannot be converted directly to $\dot{W}$. This surprising result can be understood by consideration of Fig. 6.5b. Imagining the process shown in the figure—the extraction of work from the oscillation in an isothermal space (i.e., a regenerator) at T_m—is problematic, if not paradoxical. For the control volume shown, with short extent along the x direction, the gas's U_1 must equal the piston's u_1 times its area, and the force on the piston must equal the gas pressure p_1 times the piston area, so $\dot{W} = \dot{E}_2$. However, the first law seems to prohibit such a simple process, because it would suggest $\dot{W} = \dot{H}_2$ while $\dot{E}_2 \neq \dot{H}_2 = 0$ in a perfect regenerator. The resolution of this paradox is that work can only be extracted from an isothermal space if there is a net, time-averaged heat flux $\dot{Q} = \dot{W} - \dot{H}_2 = \dot{W}$ into the control volume. (In the bucket-brigade picture of "heat transport" through a regenerator, $\dot{Q}$ equals the heat removed by a member of the bucket brigade adjacent to the piston, at the right end of its motion.) This heat at $T_m \neq T_0$ carries exergy—exactly the amount needed to add to that carried by $\dot{E}_2$ in order to total the exergy removed by $\dot{W}$.

These points are illustrated in Fig. 6.6 for an ideal orifice pulse-tube refrigerator, assumed to be thermally insulated except at the three heat exchangers. The total power must be piecewise constant because of the insulation, with discontinuities at the heat exchangers equal to the heats transferred there. The acoustic power must be continuous, because p_1 and U_1 are. It decreases through the regenerator not because

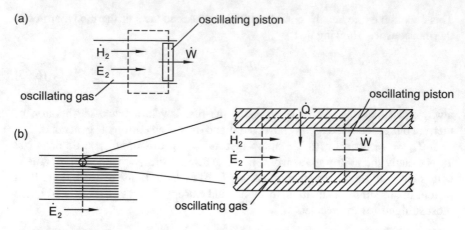

Fig. 6.5 Conversion of acoustic power to mechanical power. In both parts, the *dashed boxes* outline a control volume. (**a**) Direct conversion of acoustic power to mechanical power in an ideal adiabatic channel, where $\dot{H}_2 = \dot{E}_2$, presents no conceptual problem. (**b**) Direct conversion of acoustic power to mechanical power in an ideal isothermal channel, where $0 = \dot{H}_2 \neq \dot{E}_2$, is surprisingly complicated. The *right illustration* is a magnified view of a small portion of the *left illustration*

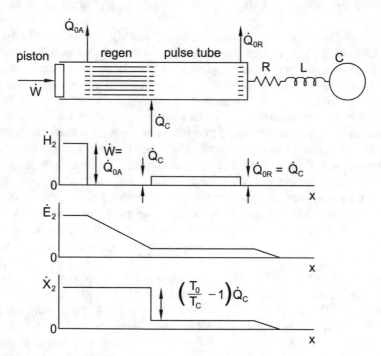

Fig. 6.6 Schematic of an ideal orifice pulse-tube refrigerator, showing the total power, acoustic power, and exergy flux as functions of position. The apparatus is insulated except at the heat exchangers, so the piston work $\dot{W}$, aftercooler heat $\dot{Q}_{0A}$, refrigeration load $\dot{Q}_C$, and final waste heat $\dot{Q}_{0R}$ are the only energies entering or leaving the apparatus

of viscosity but rather because U_1 decreases as T_m decreases, as shown by Eq. (5.14). It decreases again in the flow resistance R because of viscosity. The exergy flux is constant throughout the regenerator. It decreases at the cold heat exchanger by $\dot{Q}_C(1 - T_0/T_C)$, which is the amount of exergy carried out of the system by the heat load $\dot{Q}_C$. It is constant again along the pulse tube, finally dropping to zero smoothly in the flow resistance R, which is the only location with nonzero $\sum \dot{S}_{\mathrm{gen}}$.

Using Eq. (6.21) to write an explicit expression for $d\dot{X}_2/dx$ in a stack, regenerator, pulse tube, or thermal buffer tube, in which $d\dot{H}_2/dx = 0$, yields

$$\frac{d\dot{X}_2}{dx} = \frac{T_0}{T_m}\left[\frac{d\dot{E}_2}{dx} + \left(\dot{H}_2 - \dot{E}_2\right)\frac{1}{T_m}\frac{dT_m}{dx}\right]. \tag{6.27}$$

In the inviscid limit, $\mu = 0$, and using appropriate expressions from Chaps. 4 and 5, Eq. (6.27) becomes

$$\frac{T_m}{T_0}\frac{d\dot{X}_2}{dx} = -\frac{1}{2r_\kappa}\,|p_1|^2\left[1 + 2\frac{dT_m/dx}{\nabla T_{\mathrm{crit}}}\sin\phi_{pU} + \left(\frac{dT_m/dx}{\nabla T_{\mathrm{crit}}}\right)^2\right]$$

$$- (Ak + A_{\mathrm{solid}}k_{\mathrm{solid}})\frac{(dT_m/dx)^2}{T_m}, \tag{6.28}$$

which is always ≤ 0, for any values of dT_m/dx and ϕ_{pU}, as expected from Eq. (6.25). The angle ϕ_{pU} is the phase difference by which p_1 leads U_1.

The interpretation of the second term (i.e., second line) of Eq. (6.28) is trivial based on the discussion of heat-transfer irreversibility in the previous section. The first term (i.e., right side of first line) is more interesting. It can be reduced to zero by making $r_\kappa = \infty$, as is approximately the case for traveling-wave systems. However, the intrinsic irreversibility of standing-wave systems is apparent here, because $1/r_\kappa \propto \mathrm{Im}[-f_\kappa]$ must be nonnegligible for standing-wave systems; otherwise, they develop no power, as shown by the discussion near Eq. (5.30). In this case, the only way to make the first term in Eq. (6.28) zero is by setting $\sin\phi_{pU} = -1$ [i.e., pure standing waves, with the sign convention discussed near Eq. (5.16)] and $dT_m/dx = \nabla T_{\mathrm{crit}}$, exactly the situation for which standing-wave systems have zero power.[4]

---

[4]Thus far I have been unable to grind through the algebra to find a similarly pretty version of $d\dot{X}_2/dx$ for $\mu \neq 0$. I had hoped it would depend cleanly on r_κ, r_ν, and $(dT_m/dx)/\nabla T_{\mathrm{crit}}$, because the only sources of irreversibility are viscous shear, imperfect thermal contact, and the trivial term arising from ordinary conduction of heat along x.

6.3   Examples

Example: Orifice Pulse-Tube Refrigerator

The orifice pulse-tube refrigerator of Figs. 1.17 and 1.18 provides a realistic example similar to the ideal orifice pulse-tube refrigerator considered above. The flows of exergy, acoustic power, and total power are shown in Fig. 6.7, for the same operating point as that used to generate the power plots in Fig. 5.8 and the phasors in Fig. 4.21. The exergy curve was calculated using Eq. (6.21), using DELTAE's values for $\dot{H}_2$, $\dot{E}_2$, and T_m. Incoming from the left, the acoustic power, total power, and exergy flux are all nearly equal at 8.8 kW in the large-diameter entrance duct. The ambient-temperature aftercooler removes 8.3 kW of heat at T_0, which reduces the total power by that amount but has no effect on the exergy because heat at T_0 carries no ability to do work. Both r_v and r_κ cause a small reduction in acoustic power in the aftercooler, and the exergy decreases in step with this acoustic power, as required in a channel at temperature T_0. Next, in the regenerator the acoustic power drops considerably, but only part of this drop represents a loss in exergy (mostly due to viscosity). Much of this

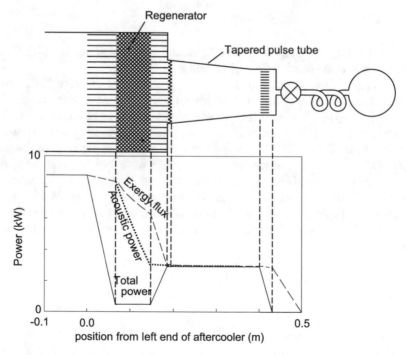

Fig. 6.7 Schematic of the orifice pulse-tube refrigerator of Figs. 1.17 and 1.18, with lengths along x to scale, and diagram showing acoustic power, total power, and exergy flux at a representative operating point. We assume that all components are thermally insulated, so the three heat exchangers and the open left end are the only locations where energy can enter or leave the system

decrease in acoustic power would occur even for $\mu = 0$, as a result of the mean temperature gradient. The exergy flux drops through the cold heat exchanger, but this represents no loss—only the flow of exergy out of the thermoacoustic part of the system and into the refrigeration load. Through the pulse tube, exergy decreases only a little, due largely to the heat leak down the walls of the pulse tube. At the end of the apparatus, exergy is destroyed in viscous flow in the orifice, while thermal insulation around the orifice, inertance, and compliance ensure that the energy associated with that exergy destruction shows up at the adjacent heat exchanger. Overall, the exergy losses in the orifice and in the regenerator are comparable, and are larger than that in the aftercooler. The fact that almost half the exergy loss in this system occurs in the orifice motivated our research [76] into methods of feeding acoustic power from the hot end of the pulse tube back to the left of the heat exchange components instead of dissipating it in the orifice.

Example: Standing-Wave Engine
The engine of Figs. 1.8 and 1.9 serves as an additional illustration of the use of exergy. In Fig. 6.8, an exergy curve calculated using Eq. (6.21) has been added to the energy plot of Fig. 5.7. At the hot heat exchanger, 2380 W of heat was added, causing 2380 W of total power to flow down the stack. This heat, at a temperature of 760 K, carried only 1440 W of exergy into the system. Some 720 W of this exergy was lost in the stack, due to viscosity, imperfect thermal contact, and conduction along x, in the process of generating 670 W of acoustic power. A full 1000 W of exergy flowed into the refrigerator at the side branch, in the form of 1000 W of acoustic power at T_0.

At another operating point for the standing-wave engine, one which received both experimental and calculational attention, a total of 3000 W of electric power was supplied to the heaters of the two engines, and 490 W of acoustic power was delivered to the pulse-tube refrigerator. The lost work is the difference between

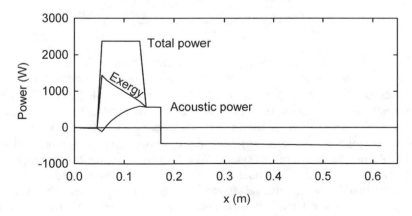

Fig. 6.8 Powers for the standing-wave engine, at the same operating point as in Fig. 5.7, with an exergy curve added

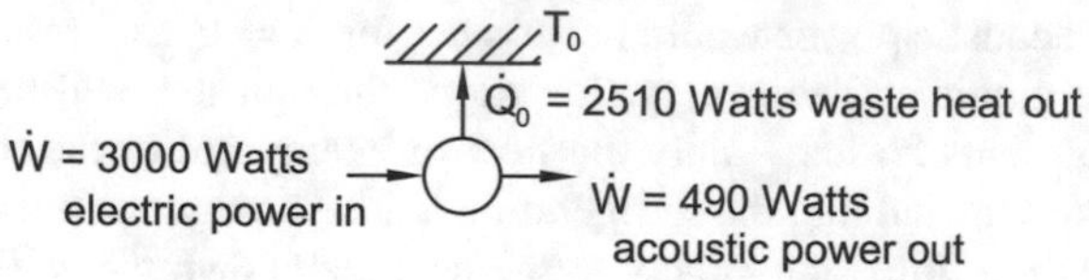

Fig. 6.9 At one experimental operating point, the standing-wave engine accepted 3000 W of input electric power and rejected 2510 W of waste heat at T_0. Hence, 2510 W of work was lost. Details are displayed in the table

these: 2510 W. Although we did not measure it, we would infer that 2510 W of heat passed into the room-temperature cooling-water streams at T_0, as illustrated in Fig. 6.9.

To gain understanding of the sources of irreversibility in the system at this operating point, I made an arbitrary choice of how to break up this lost work: mostly by component, but with some subdividing according to process. Making intelligent estimates based partly on measurement and partly on a corresponding DELTAE model of this system generated this table.

Source	$\dot{W}_{lost}$
I^2R	730 W
Room heat leak	300 W
Hot heat exchangers, acoustics	20 W
Hot heat exchangers, δT	10 W
Stacks	950 W
Stacks pressure vessel heat leak	150 W
Cold heat exchangers, acoustics	10 W
Cold heat exchangers, δT	70 W
Resonator	35 W
Unknown	235 W
Total:	2510 W

Such a table can serve as one guide when contemplating improvements to a thermoacoustic system, to help decide which components are responsible for losses, and maybe to help decide how to spend time and money most effectively to improve the overall system efficiency. In this case, consider the sources of lost work starting with the largest: The stacks contributed 950 W. We believed that we optimized the gaps, lengths, and positions of these parallel-plate stacks well during the design of this system, so I doubt that rebuilding the stacks with different gaps or lengths would improve performance much. However, most of that 950 W was thermal-relaxation loss, not viscous losses. The thermal-relaxation loss was large in this system because of the size constraint imposed by the project's sponsor. If we were free to make the system bigger, we could operate closer to the critical temperature gradient and thus have a more efficient thermoacoustic engine. Hence, this table strongly suggests that we should re-negotiate the size constraint with the sponsor. We might consider trying to build a pin-array stack, because computer calculations show that a pin-array stack should perform about 20% better than a parallel-plate stack. However, building such an intricate pin array, with supports for the pins that do not block the sound wave badly, seems daunting (i.e., expensive).

The second-largest loss in the table is the I^2R loss in the NiCr heaters in the two thermoacoustic engines. What does this mean, that 730 W was lost to I^2R, when 3000 W was converted from electricity to heat in the heaters? We were turning 3000 W of electric power into 3000 W of heat at temperature T_H. The formula in the previous table shows that the lost work in that process was $I^2R T_0/T_H$. "Work was lost" in this process because the first and second laws allow for a more efficient way to deliver 3000 W of heat to T_H, in principle: This 3000 W of heat could have been delivered to the engine at $T_H = 1230$ K by spending only 2270 W of electric power, driving a heat pump with Carnot's *COP* operating between T_0 and T_H. The difference, 3000–2270 W, is the lost work. Reducing lost work from this source also sounds expensive, given the sponsor's insistence on an electrically heated engine: The only way to lose less work in this electric heater would be to operate it at higher temperature, which would require more expensive materials for the stack, the pressure vessel, and the heater itself.

The next largest sources seem more promising. Heat leak from T_H to the room contributed 300 W, which might be reduced simply by more insulation, if the sponsor's size constraint is not compromised. Unknown sources contributed 235 W, which might serve as justification for asking the sponsor for more funds for fundamental research in thermoacoustics. The heat leaks along the two pressure vessels enclosing the two stacks contributed 150 W, which suggests that we might look for a reasonable-cost alloy with greater strength or lower thermal conductivity than the Incolloy 800H used here.

6.4 Exercises

6.1 A simple heat engine operates between temperatures T_H and T_0. Its work output is used to drive a simple refrigerator, which pumps heat from a load at temperature T_C; it rejects waste heat at temperature T_0. The efficiency of this composite system is defined to be the ratio of heat pumped from T_C to heat consumed from T_H. What is the limit on efficiency imposed by the first and second laws of thermodynamics?

6.2 An ideal gas, with temperature-independent specific heat c_p, flowing with mass flow rate $\dot{M}$, must be cooled from T_0 to T_C. In case 1, the cooling is provided by a Carnot refrigerator removing heat from T_C and rejecting waste heat to T_0. In case 2, the cooling power is provided by two Carnot refrigerators rejecting waste heat to T_0. The gas stream first encounters the first refrigerator, whose cold end is at $(T_0 + T_C)/2$, and then encounters the second refrigerator, whose cold end is at T_C.

Case 3 is similar to case 2, except that the colder refrigerator rejects its waste heat at $(T_0 + T_C)/2$ and the first refrigerator must absorb that heat in addition to the heat from the gas stream. Calculate the work required in all three cases, and compare it to the minimum work given by Eq. (6.16). How might you design cooling hardware that would approach the minimum work more closely?

6.3 Demonstrate the Gouy–Stodola theorem for the system of Fig. 6.2d, by showing that the difference between the entropy increase of the reservoir at T_0 and the entropy decrease of the reservoir at T_C equals the lost work given by Eq. (6.6) divided by T_0.

6.4 Investigate $\rho \dot{s}_{\mathrm{gen}}$ for the simplest thermal-penetration-depth problem, with $u_1 = 0$ and boundary-layer approximation. Make a plot of the time average of the entropy generation per unit volume, as a function of distance from the wall. How does this plot compare with the plot of power dissipated per unit volume shown in Ani. Thermal? Does this make sense? Compare $\int \rho \dot{s}_{\mathrm{gen}} \, dy$ with the results of Exercise 6.8, to verify the Gouy–Stodola theorem in this context.

6.5 Pick one process, one component, or one process within one component of your favorite thermoacoustic system, and estimate (order of magnitude) the lost work. What fraction of the total lost work for the whole system does this represent?

6.6 Show that Eq. (6.9) is true even if $T < T_0$.

6.7 On a trans-Pacific airplane flight, you are surprised to see your rich uncle, a banker, whom you haven't met in 10 years. He tells you that he is about to invest heavily in a natural-gas recovery project that will make him unimaginably wealthy.

The project involves a huge, high-pressure natural-gas reservoir under the Arctic Ocean. The gas will arrive at the floating platform on the ocean surface in a pipe, at a pressure of 100 bar and a temperature of $0\,^{\circ}\mathrm{C}$, which is the temperature of the ocean. He says that the brilliant scientist in whom he will be investing has invented two special machines. One machine will use the gas pressure to drive a special turbine, to generate power from the 100 bar gas. As it does work on the turbine, the gas will expand to atmospheric pressure, and cool. He doesn't remember how much it will cool. The second machine will use some of the power from the turbine to further cool the gas and liquefy it. There will be power left over, to run the lights, water pumps, radios, computers, heaters, etc. for the workers on the platform. The liquefied gas will be carried in cryogenic tankers to Tokyo.

He is certain that the planned platform will require no external source of power, nor will it need to burn any of the gas. It will be completely powered by the expansion of the gas from 100 bar. He remembers that one or both of the two machines will have a large heat exchanger connected to the ocean water.

You remember that methane liquefies at 112 K at atmospheric pressure, and that the latent heat is approximately equal to the total heat that must be removed to cool gaseous methane from $0\,^{\circ}\mathrm{C}$ to 112 K. You assume that the gaseous methane, CH_4, can be treated as an ideal gas with molar mass 0.016 kg/mole. You remember that changes in enthalpy of an ideal gas are given by $dh = c_p \, dT$, and that changes in entropy of an ideal gas are given by $ds = (c_p/T) \, dT - (1/\rho T) \, dp$. You know that a methane molecule has more degrees of freedom than a nitrogen molecule has, so you are willing to assume that $\gamma \simeq 1.3$, a little less than the value for air that you remember.

Evaluate the feasibility of this scheme using the first and second laws of thermodynamics. Convince your uncle that the first and second laws of thermodynamics are applicable to this scheme, and that you can decide that his investment plan is unwise without understanding any of the details of the two special machines.

6.8 Derive the boundary-layer expression for $d\dot{X}_2/dx$ in a stack. Express your answer using the same variables as in Eq. (6.28) plus r_ν and σ. Then assume standing-wave phasing to simplify your result, showing that it is ≤ 0.

6.9 Find the error in this chain of reasoning: For an ideal gas, $\oint pu\,dt = \oint (p/\rho)\rho u\,dt = R\oint T\rho u\,dt$. In a perfect regenerator, T is independent of time, so it can be pulled outside the integral, leaving $RT\oint \rho u\,dt$. But $\oint \rho u\,dt = 0$ because there is no net time-averaged mass flow through a perfect regenerator. Hence, $\oint pu\,dt = 0$. So acoustic power cannot flow through a perfect regenerator. (If you get stuck, wait until Chap. 7.)

6.10 NASA probes to the outer planets require electric power generators that operate for a decade with extremely high reliability. Currently, this electricity is generated with the thermoelectric effect in a solid. The high-temperature heat is supplied to the thermoelectric component by radioactive decay of a plutonium isotope, and waste heat is carried away from the thermoelectric component by a heat pipe and finally rejected by means of black-body radiation to space. The efficiency, i.e., the ratio of electric power to plutonium heat, is 7%.

Consider a thermoacoustic engine as an alternative to the thermoelectric component, perhaps with piezoelectric or electrodynamic transduction of acoustic to electric power. Your assignment: Decide whether this idea might be worth pursuing. Make some plausible assumptions, and make a few rough estimates using various approximate relations throughout the book, to see if a thermoacoustic system might have an efficiency greater than 7%, while maintaining extremely high reliability. If you decide this idea might be worth pursuing, write a letter to your friend the program manager at NASA, explaining why you think it's worth pursuing—try to convince her that her office should provide you with financial support to attempt a meaningful design. If you decide the idea is *not* worth pursuing, write a letter to her explaining why she should reject any half-baked thermoacoustics proposals she receives.

6.11 A simple thermoacoustic refrigerator, with *COP* equal to 0.3 times Carnot's *COP*, lifts heat from 276 K and rejects waste heat at $T_0 = 300$ K. The refrigerator is used to cool a stream of 10 g/s of water from 300 to 276 K. This cooling load is the only load on the refrigerator.

In addition to the obvious irreversibility within the thermoacoustic refrigerator, the heat transfer in the cold heat exchanger must be irreversible, because this heat exchanger is at 276 K but the water enters it at 300 K. (Assume water's heat capacity is 4.2 J/g·K, independent of temperature.)

How much heat must be removed from the water stream? How much work power is used to run this refrigerator? Calculate the lost work in the refrigerator and the lost work in the irreversible heat transfer. Compare the appropriate sum/difference of these three numbers with the minimum work required to cool this stream, using Eq. (6.16). How much exergy is added to the water stream? (Comment on the sign of this exergy term, regarding the ability of the cold water stream to do work.)

To reduce the irreversibility in the heat exchanger, it is proposed to use two refrigerators instead of one. The first will operate at 288 K and remove half of the heat load from the water stream and the second will operate at 276 K and remove the other half of the heat load from the water stream. (Hence, each refrigerator will have half of the cooling power of the original refrigerator.) Assume that both refrigerators reject heat to 300 K, and both have $COP/COP_{Carnot} = 0.3$. What is the total work power required by this system? Tabulate the lost work in four locations: the two refrigerators, and the two cold heat exchangers.

Now suppose the colder of the two refrigerators rejects its waste heat at 288 K instead of 300 K, so that the warmer of the two refrigerators must pump this heat load in addition to the heat load due to the water stream. Again calculate the total work required, and tabulate the lost work in the four locations. Discuss some of this in terms of exergy; for example, where is exergy destroyed, where does it flow into and out of the apparatus, etc.

6.12 Show that heat exergy $Q(1 - T_0/T)$ indeed represents the ability to do work, by considering a Carnot engine operating between T and T_0 when $T > T_0$.

6.13 A machine can absorb heat at a rate $\dot{Q}_C$ at temperature T_C. Show that a Carnot engine attached to this machine could produce work at a rate $\dot{W} = (T_0/T_C - 1)\dot{Q}_C$. Interpret this result in terms of the heat exergy associated with $\dot{Q}_C$.

6.14 Superficially, we sometimes say that acoustic power is dissipated, turned into heat, in the orifice of an orifice pulse-tube refrigerator. Think about this issue carefully, considering Fig. 6.6 or 6.7. Write a paragraph or short essay, describing carefully what actually happens. Use the word "heat" only to mean $T\Delta S$, where ΔS is an entropy flow across a gas–solid interface, through a stationary solid, or across a solid–liquid interface [21, 77, 78].

6.15 In the study of thermodynamics, confusion arises because the overdot on a variable can have two or more fundamentally different meanings, all of which share the "units" of inverse time. For example, $\dot{S}$ can mean the rate at which a mass stream carries entropy across the boundary of a control volume, or the rate at which the total entropy inside a control volume increases, or the rate at which irreversible entropy generation occurs within a control volume. Where such ambiguity exists, de Waele's group at TU Eindhoven sometimes uses different symbols for these different meanings; e.g., an asterisk, instead of a dot, above a variable indicates the first meaning [21]. With pens or highlighters of different colors, mark all variables having overdots in Chap. 6, according to which of these meanings is relevant. Discuss any difficult cases or additional cases if you find them.

6.16 In the same style as the derivation of Eq. (5.24), provide an alternative derivation of Eq. (6.21) by considering the x component of the availability-flux density plus the exergy-flux density due to ordinary conduction of heat in the x direction.

Chapter 7
Beyond Rott's Thermoacoustics

Chapter 6 completed a review of the foundations of thermoacoustics, based mostly on the work of Nikolaus Rott. Chapter 4, summarized in Fig. 4.15, showed that any thermoacoustic apparatus can be regarded as a series of short channels (possibly with internal structure, such as in a stack), each of which can be analyzed using the momentum and continuity equations. This allows a short length of arbitrary thermoacoustic component to be treated as a combination of only five impedance elements: series inertance, series viscous resistance, parallel compliance, parallel thermal-relaxation resistance, and the parallel current-controlled current source that appears when a nonzero temperature gradient exists along the channel. In this representation, each dynamical variable, such as pressure, is assumed to have a sinusoidally oscillating part, with the amplitude and phase of the oscillation the only variables of interest. Once oscillating pressure and oscillating velocity are understood in this context, other oscillating variables such as temperature and density follow easily. Then, as described in Chaps. 5 and 6, the acoustic power $\dot{E}_2$, the second-order total power $\dot{H}_2$, the time-averaged heats $\dot{Q}$ transferred at heat exchangers, and the second-order exergy flux $\dot{X}_2$ can be calculated without controversy from these oscillating variables when the first and second laws of thermodynamics are brought into consideration.

If an engine or refrigerator is built according to these principles, and no mistakes are made (see "Common pitfalls," Appendix A), it will work as designed—but only in the limit of low amplitudes. Figure 7.1 illustrates this typical behavior, showing data obtained during early debugging of the standing-wave engine of Figs. 1.8 and 1.9, using only one engine and no load on the resonator. Heater power, displayed on the horizontal axis, was applied, resulting in measured pressure oscillations and hot temperatures as shown by the circles. The lines are calculations using DELTAE, based on Rott's acoustic approximation and hence consistent with the discussion in Chaps. 4 and 5. At low power, measurements and calculations agree, but at the highest powers the measured and calculated temperatures differ by almost 200 °C and the measured and calculated pressure amplitudes differ by almost 25%. Thermoacoustic engines and refrigerators with practical levels of power per unit

© Acoustical Society of America 2017
G.W. Swift, *Thermoacoustics*, DOI 10.1007/978-3-319-66933-5_7

Fig. 7.1 Pressure amplitude and hot temperature as functions of heater power, during early debugging of the standing-wave engine of Figs. 1.8 and 1.9, with the side branch to the refrigerators blocked, and using one spiral stack with its heat exchangers in one end of the resonator but no stack or hot heat exchanger in the other end of the resonator. Helium, 3 MPa, 370 Hz. The *points* represent measurements, with a 300-W heat leak to the room subtracted from the heater power. The *lines* are calculations using DELTAE, assuming laminar flow in the resonator

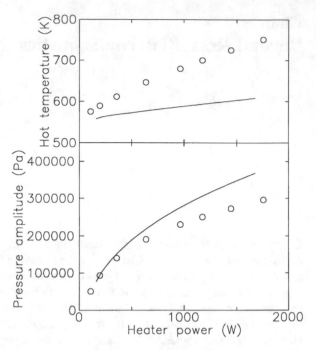

volume and per unit mass must operate at high amplitudes such as these, where actual behavior deviates significantly from Rott's acoustic approximation.

In the early days of thermoacoustics research, we were impressed that Rott's acoustic approximation came so close to the truth at high amplitudes, but our standards are much more demanding now: We hope to understand such deviations quantitatively. Those of us who come from an acoustics background must go well beyond our acoustics-based knowledge and intuition. For example, we must learn about the high-Reynolds-number phenomena encountered in other branches of hydrodynamics, such as aerodynamics and pipeline hydraulics.

This chapter examines some of the gas-dynamics phenomena leading to high-amplitude deviations from Rott's acoustic approximation. The relevant fundamental physics is well known: The gas dynamics and thermodynamics are believed to be included in the momentum, continuity, and energy equations reviewed in Chap. 2,

$$\frac{\partial \rho}{\partial t} + \nabla \cdot (\rho \mathbf{v}) = 0, \tag{7.1}$$

$$\rho \left[\frac{\partial \mathbf{v}}{\partial t} + (\mathbf{v} \cdot \nabla) \mathbf{v} \right] = -\nabla p + \nabla \cdot \boldsymbol{\sigma}', \tag{7.2}$$

$$\frac{\partial}{\partial t} \left(\rho \epsilon + \frac{1}{2} \rho |\mathbf{v}|^2 \right) = -\nabla \cdot \left[-k \nabla T - \mathbf{v} \cdot \boldsymbol{\sigma}' + \left(\rho h + \frac{1}{2} \rho |\mathbf{v}|^2 \right) \mathbf{v} \right], \tag{7.3}$$

with appropriate boundary conditions, and in the properties of the gas

$$p = \rho \Re T, \tag{7.4}$$

$$\mu = \mu(p, T), \tag{7.5}$$

$$k = k(p, T), \tag{7.6}$$

$$\gamma = \text{constant}. \tag{7.7}$$

The time-averaged total power [46] flowing in the x direction,

$$\dot{H} = \int \left[\overline{\rho u \left(v^2/2 + h \right)} - k \frac{\overline{dT}}{dx} - \overline{(\mathbf{v} \cdot \boldsymbol{\sigma}')_x} \right] dA, \tag{7.8}$$

and the time-averaged exergy flux [45] in the x direction,

$$\dot{X} = \int \left[\overline{\rho u \left[v^2/2 + (h - h_0) \right]} - T_0 \overline{\rho u \left(s - s_0 \right)} \right.$$

$$\left. - k \left(1 - \frac{T_0}{T} \right) \frac{\overline{dT}}{dx} - \overline{(\mathbf{v} \cdot \boldsymbol{\sigma}')_x} \right] dA, \tag{7.9}$$

are also unambiguous, where h_0 and s_0 are the enthalpy and entropy per unit mass in the so-called dead state [45]. However, knowing that all the fundamental physics involved is captured in these deceptively simple equations is of little use in the day-to-day practice of high-amplitude thermoacoustics, because personal computers and human brains are incapable of processing such complicated equations and truths quickly. The challenge is to find the relevant aspects of Eqs. (7.1)–(7.9) and to distill them into comprehensible, compact forms and into usable design procedures and analysis procedures. "Rott's acoustic approximation" of Chaps. 4–6 is one such distillation, but data like those displayed in Fig. 7.1 demonstrate that this approximation omits some important phenomena.

While Chaps. 4–6 of this book are built on a well-established foundation, many of the topics in this chapter are at the frontiers of current understanding. If this chapter seems inelegant or confused, it may be due to the fact that some of these issues are indeed complicated, and are not well understood. Also likely is that some of these issues are well understood by people outside the present, small thermoacoustics community, and that our rushed, "close-enough" hardware-development approach at Los Alamos has led us to overlook relevant information published somewhere else in the vast hydrodynamics literature. I will not be surprised if some of this chapter turns out to be wrong.

The approach here will be to build upon and extend Rott's acoustic approximation, adding (one might say kludging) various phenomena onto it. A strength of this approach: It builds on a firm foundation. A weakness of this approach: Most extensions seem to help only with the momentum equation—not with heat transfer. A danger of this approach: It will not find a dramatically improved point of view, no matter how interesting or important that might potentially be.

7.1 Tortuous Porous Media

In Rott's acoustic approximation, the dynamics and thermodynamics of oscillating flow are theoretically well understood in simple geometries such as arrays of circular tubes, but the regenerators of many Stirling engines and refrigerators comprise screen beds, randomly stacked spheres, and other tortuous porous geometries, for which computational methods are still evolving and fundamental data are scarce. Organ [16] points out many of the shortcomings of the current state of the art. However, given today's imperfect foundations, additional approximations are useful whenever they improve computational ease dramatically while only slightly reducing accuracy.

To understand the notation commonly used to describe the geometry of tortuous porous media, consider a regenerator of length Δx and cross-sectional area A. The porosity ϕ is the ratio of gas volume to total regenerator volume, so $\phi A \, \Delta x$ is the total volume of gas in the regenerator. The hydraulic radius r_h of the regenerator is the ratio of total gas volume to gas–solid interface surface area (commonly referred to as "wetted area").

Expressions for the hydraulic radius of stacked-screen regenerators (Fig. 7.2) as a function of porosity and wire diameter are derived by Organ [16]. Although the individual wire segments in the woven screen can be modeled as sinusoids, it usually turns out that expressions for straight, circular rods suffice. The definitions of hydraulic radius and porosity yield

$$r_h = D_{\text{wire}} \frac{\phi}{4\,(1-\phi)} \qquad (7.10)$$

for a collection of straight, circular rods, no matter how they are arranged in a volume, as long as the surface area of the rods' ends is negligible compared with the surface area of their sides. If the rods are arranged in touching layers, with each layer having n wires per unit width, then

$$\phi = 1 - \frac{\pi n D_{\text{wire}}}{4}, \qquad (7.11)$$

which is also commonly used for modeling woven screen.

Chapter 4 mentioned that screen-bed regenerators can be incorporated into the point of view summarized in Fig. 4.15, having effective equivalent f_ν, f_κ, etc. [63]. To force the complicated viscous and thermal properties of screen-bed geometries

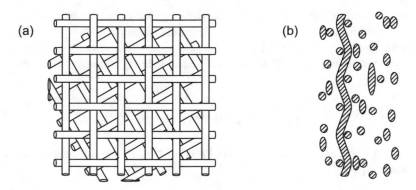

Fig. 7.2 A screen bed is the most common example of a tortuous porous medium for Stirling engines and refrigerators. (**a**) A view of two layers of plain-weave screen, with the x direction normal to the page. (**b**) A cross section through a screen bed, with the x direction *from left to right*

and other porous media into this formalism, all oscillating variables are assumed to be small and to have simple-harmonic time dependence $e^{i\omega t}$.

At the same time, an additional, important assumption is made: The flow at any instant of time has no memory of its recent history. In other words, the gas flows through the pores so vigorously that viscous drag and heat transfer between the gas and the porous solid are determined only by the *instantaneous* velocity, with no dependence on the velocity at earlier times. This can occur if a typical parcel of gas moves through many layers of screen in a time that is much shorter than the period of the oscillation, so its velocity is essentially constant during the time that the parcel moves through those many layers of screen. This condition is equivalent to $r_h \ll |\xi_1| = |U_1|/\phi A\omega$, which can be rewritten as $r_h \ll \sqrt{|N_{R,1}|\delta_v}$, where

$$N_{R,1} = \frac{4\langle u_1 \rangle r_h \rho_m}{\mu} = \frac{4U_1 r_h \rho_m}{\phi A \mu} \qquad (7.12)$$

is the complex Reynolds-number amplitude.[1] With this assumption, *steady-state* measurements of viscous drag and heat transfer provide the necessary fundamental information. For screen-bed regenerators, the friction-factor and heat-transfer data of Kays and London [64] can be used for viscous and thermal-relaxation effects. However, this point of view allows incorporation of any steady-state friction-factor and heat-transfer correlations, for screen beds or for any other tortuous porous geometry [79], into a harmonic analysis in a self-consistent manner that uses the full velocity-dependent information in the correlations.

Conversion of the Kays and London correlations to forms appropriate for oscillating flow is described fully in [63]. Here, a small part of that derivation is

---

[1]The standard convention is followed here: The Reynolds number uses hydraulic diameter $4r_h$ as length scale.

repeated, showing how to impose the sinusoidal approximation on the x component of the momentum equation to obtain r_v. With the assumption that $|\xi_1| \gg r_h$, the instantaneous pressure gradient should be related to the instantaneous velocity in the same way that pressure gradient is related to velocity in steady, unidirectional flow. Thus

$$\frac{\partial p(x,t)}{\partial x} = -\frac{1}{2}\rho(t)\,|\langle u(t)\rangle|\,\langle u(t)\rangle\,\frac{f(t)}{r_h}, \qquad (7.13)$$

where f is the friction factor, which is plotted in Fig. 7.9 of [64] as a function of Reynolds number

$$N_R = 4\,|\langle u\rangle|\,r_h\rho/\mu \qquad (7.14)$$

and porosity ϕ. In the simple harmonic approximation, the goal is to derive an expression for the temporal fundamental Fourier component of Eq. (7.13).

To make further progress, f is written

$$f \simeq \frac{c(\phi)}{N_R} + C(\phi), \qquad (7.15)$$

a reasonable approximation to Fig. 7.9 of [64]. Using Eqs. (7.15) and (7.14) to write $f(t)$ more explicitly, Eq. (7.13) becomes

$$\frac{\partial p(x,t)}{\partial x} = -\frac{\mu(t)}{r_h^2}\,\langle u(t)\rangle\left(\frac{c(\phi)}{8} + \frac{C(\phi)r_h\,|\langle u(t)\rangle|\,\rho(t)}{2\mu(t)}\right). \qquad (7.16)$$

The variables μ and ρ can be expressed by first-order Taylor series expansions:

$$\mu(t) = \mu_m + (\partial\mu/\partial T)[T(t) - T_m] + (\partial\mu/\partial p)[p(t) - p_m] \qquad (7.17)$$

and similarly for ρ. Using these expansions in Eq. (7.16), substituting expressions of the form of Eqs. (4.45)–(4.51) for $\langle u(t)\rangle$, $p(t)$, and $T(t)$, multiplying the entire equation by $e^{-i\omega t}/\pi$, and integrating the resulting expression with respect to ωt from 0 to 2π generates the fundamental Fourier component.

On the left side of the equation, this procedure produces simply dp_1/dx, the desired result. On the right side, the first of the two terms reduces easily to $[\mu_m c(\phi)/8r_h^2]\langle u_1\rangle$. In the second term, T_1 and p_1 are neglected because their contributions are third or fourth order in the small oscillating amplitudes. With this approximation, the final result is

$$\frac{dp_1}{dx} = -\frac{\mu_m}{r_h^2}\left(\frac{c(\phi)}{8} + \frac{C(\phi)\,|N_{R,1}|}{3\pi}\right)\langle u_1\rangle. \qquad (7.18)$$

Hence, in the notation of Chap. 4, the viscous resistance per unit length is

$$r_v = \frac{\mu_m}{r_h^2 \phi A} \left(\frac{c(\phi)}{8} + \frac{C(\phi)\,|N_{R,1}|}{3\pi} \right). \tag{7.19}$$

This result depends, of course, on the specific choice of Kays-and-London data for $f(N_R)$. However, the method—deriving the complex fundamental Fourier component of dp_1/dx from steady-state friction-factor data—is general enough to use for any $f(N_R)$, for screens or any other tortuous regenerator geometry.

Proper accounting for the time-averaged solid thermal conductance along x in a porous medium is another important challenge. In stacked screens, the tiny points of contact between the wires of adjacent layers serve to reduce the thermal conductance well below the most naive estimate, $k_{solid}A(1 - \phi)/\Delta x$, which would hold if the same total mass of metal were arranged in columns of x-independent cross section. Lewis et al. [80] have measured this reduction factor in screen regenerators under circumstances of interest in cryocoolers, finding a reduction factor of typically 0.1.

Example: Thermoacoustic-Stirling Heat Engine

The regenerator in the thermoacoustic-Stirling heat engine shown in Figs. 1.22 and 1.23 was a pile of stainless-steel screens. The diameter of the screen wire was 65 μm, and the screen was plain-weave, "120 mesh" (meaning 120 wires per inch in each of the two perpendicular directions of the weave). Equation (7.11) predicts $\phi = 0.76$, and in fact the pile was compressed somewhat more tightly than this, achieving $\phi = 0.72$ as built. Using this as-built value, Eq. (7.10) predicts $r_h = 42\,\mu m$, considerably smaller than δ_κ, which ranges from about 150 μm at ambient temperature to 450 μm at the temperature of the hot end of the regenerator.

The length of the regenerator (along x) was 7.3 cm. At a high-power operating point, the gas displacement amplitude $|\xi_1| = |U_1|/\omega\phi A = 1.4$ cm, and the magnitude of the oscillating Reynolds number was $|N_{R,1}| \simeq 300$.

In DELTAE design calculations, we usually account for the axial thermal conductance of the pressure vessel around the regenerator by adjusting the screen longitudinal-thermal-conductance reduction factor mentioned above. While the recommended value [80] of this factor is about 0.1, we found that the pressure-vessel contribution in this engine brought this factor up to about 0.3.

7.2 Turbulence

Figure 7.3 is the well-known friction-factor diagram [50] for steady incompressible flow in a circular pipe. [Here "steady" means that U is independent of time, although $u(x, y, z, t)$ may fluctuate rapidly.] The figure summarizes practical knowledge about pressure drops in such flows. Equation (7.2) includes all the physics displayed in Fig. 7.3, but Eq. (7.2) is too far removed from the engineering reality of turbulent steady flow to be useful in ordinary engineering work.

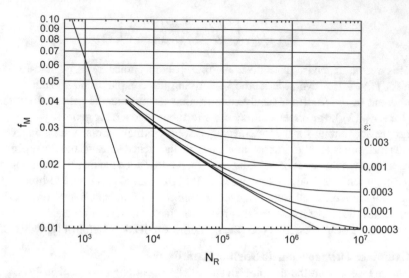

Fig. 7.3 Friction factor vs Reynolds number in steady flow in a circular pipe. This plot appears in most fluid-mechanics texts, such as [50]. The different curves in the turbulent regime are for different surface roughnesses in the pipe

Note the use of dimensionless groups of variables in Fig. 7.3. Engineers have long realized that correlating data using such dimensionless groups provides the only practical, compact approach for sharing data. The Reynolds number

$$N_R = \frac{\langle u \rangle D \rho}{\mu} \tag{7.20}$$

is a dimensionless measure of the average velocity, where D is the diameter of the pipe. The vertical axis is the dimensionless Moody friction factor f_M, used to calculate the pressure drop across a length L of the pipe[2]:

$$\Delta p = f_M \frac{L}{D} \frac{1}{2} \rho \langle u \rangle^2. \tag{7.21}$$

The dimensionless relative roughness ε is the roughness height on the pipe's inner surface divided by the pipe diameter. Such dimensionless groups will often be used in this chapter, and in the final section a formal approach to the selection of dimensionless groups for thermoacoustic phenomena will be outlined.

Figure 7.3 shows that steady flow in a circular pipe has two regimes. Below $N_R \sim 2000$, the flow is laminar. In this regime, the $(\mathbf{v} \cdot \nabla) \mathbf{v}$ term on the left side of

---

[2]Be alert for factor-of-two or -four differences in the definition of friction factor in different circumstances and with different authors; e.g., compare definitions in Sects. 7.1 and 7.2 in this book.

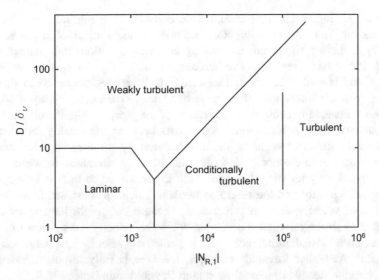

Fig. 7.4 Approximate boundaries between regimes of oscillating flow, in a smooth circular pipe, as a function of peak Reynolds number $|N_{R,1}|$ and the ratio of pipe diameter D to viscous penetration depth δ_v. Adapted from [82]

Eq. (7.2) can be neglected, and the equation is simple enough for analysis leading to a closed-form solution for the friction factor,

$$f_M = 64/N_R, \tag{7.22}$$

which is the straight line in the left part of Fig. 7.3. Above $N_R \sim 2000$ the flow is turbulent. Roughness $\varepsilon > 0$ gives higher friction factors in the turbulent regime.

For such steady incompressible flow in a circular pipe, two dimensionless parameters—N_R and ε—are sufficient to span the space of all possible flows, so the friction factor of *any* steady flow can be obtained from Fig. 7.3 by specifying these two parameters. In oscillating flow, however, one additional parameter is needed, which must be related to the frequency of oscillations. The viscous penetration depth is a familiar frequency-dependent variable, so D/δ_v provides one convenient dimensionless form for this third parameter. (The Womersley [81] number $D/\sqrt{2}\delta_v$ and the Reynolds number based on viscous penetration depth, $\langle u \rangle \, \delta_v \rho/\mu$, are also commonly used for the third dimensionless parameter.) One slice through the three-dimensional parameter space of oscillating flow is shown in Fig. 7.4 for a circular pipe. This slice is for $\varepsilon = 0$, and shows the regimes of behavior as a function of *peak* Reynolds number

$$|N_{R,1}| = \frac{|U_1| D \rho_m}{A \mu} \tag{7.23}$$

and D/δ_v.

Figure 7.4 represents more complicated behavior than Fig. 7.3 represents, and mapping out and understanding this behavior is the subject of much research [83–87], including significant uncertainty about the locations and natures of the boundaries between regimes. The laminar regime of Fig. 7.4 was the subject of Chap. 4, and is well understood. The weakly turbulent regime seems to share the laminar regime's mathematically simple behavior in the boundary layer, with the turbulence essentially confined to the center of the pipe, not significantly affecting the boundary layer. A transitional zone exists between the weakly turbulent and conditionally turbulent regimes. In the conditionally turbulent regime, hot-wire anemometer measurements [84] show that the flow alternates between weakly turbulent and fully turbulent behavior, with the transition to turbulence occurring at the peak velocity and the return to weak turbulent flow occurring at the zero crossings of velocity. (This is remarkable, because one might have expected the transition to strong turbulence occurring when the velocity rises past a threshold, and the return to weak turbulence occurring when the velocity falls below the same threshold.) At higher Reynolds numbers, the flow is fully turbulent, essentially resembling the steady turbulent flow at high Reynolds number in Fig. 7.3.

The mathematics developed in Chaps. 4 and 5 is probably inapplicable in the conditionally turbulent and fully turbulent regimes, because in these regimes the turbulence disturbs the boundary layer deeply and violently.

At the high amplitudes necessary to achieve high power density, various components in thermoacoustic engines and refrigerators operate in all regimes shown in Fig. 7.4, and it is important to calculate peak Reynolds numbers at typical locations in each component to get a rough idea whether deviations from Rott's acoustic approximation might be expected. Pulse tubes and thermal buffer tubes typically operate in the weakly turbulent regime, but may enter the conditionally turbulent zone. Resonator components are often fully turbulent or conditionally turbulent. Inertances in pulse-tube refrigerators are often conditionally or fully turbulent. For other components having non-circular cross sections but with parallel walls, one can only hope that Fig. 7.4 gives reasonable guidance, using $D \sim 4r_h$. With this criterion, standing-wave stacks are often in the high-Reynolds-number part of the laminar regime, sometimes with their cold ends extending into the conditionally turbulent regime, and heat exchangers may fall in any regime except fully turbulent. High-amplitude operation of standing-wave stacks typically falls near the intersection between the laminar, weakly turbulent, and conditionally turbulent regimes in Fig. 7.4, where new research [83] suggests that Fig. 7.4 significantly underestimates the Reynolds number at which the laminar-to-turbulent transition occurs. As discussed in the previous section, the screen beds of regenerators typically have $r_h/\delta_\nu \ll 1$ and $|N_{R,1}| \sim 500$, but Fig. 7.4 offers *no* guidance for the microscopically tortuous geometry of screen beds.

Chapters 4–6 showed the usefulness of expressions for dp_1/dx, dU_1/dx, $d\dot{E}_2/dx$, and $\dot{H}_2$ everywhere in thermoacoustic engines and refrigerators. The present state of knowledge is very incomplete in this regard. These expressions are only certain in the laminar regime in Fig. 7.4 (and even there entrance effects, to be discussed later this chapter, confuse the issue). The mathematical expressions developed in Chap. 4

are probably applicable in the weakly turbulent regime, because the weak turbulence probably does not penetrate significantly into the boundary layer, and because some experimental evidence [88, 89] suggests that laminar boundary-layer mathematics predicts phenomena accurately. Everywhere else in Fig. 7.4, present understanding is too incomplete to give quantitatively accurate predictions for thermoacoustics.

However, in the fully turbulent regime, "Iguchi's hypothesis" gives partial, perhaps quantitatively dependable guidance. In this regime, the displacement amplitude $|\xi_1|$ of the gas is much larger than both δ_ν and D, so Iguchi [82] suggested that the flow at any instant of time should have little memory of its past history: The flow at each instant of time should be identical to that of fully developed steady flow with that same velocity, represented by Fig. 7.3. This hypothesis must be excellent in the low-frequency limit, in which D/δ_ν is not too large, a limit that is approached in the inertances of many pulse-tube refrigerators. (This assumption was used with confidence in Sect. 7.1, where $r_h/\delta_\nu \ll \sqrt{|N_{R,1}|}$ in *tortuous* channels.) The validity of the assumption for large D/δ_ν, which is of interest in many resonators, is unknown.

To incorporate Iguchi's hypothesis into the impedance framework of Chap. 4, an expression for dp_1/dx can be derived, modified to account for the turbulence. If the volume flow rate and hence Reynolds number N_R vary sinusoidally in time, then the instantaneous friction factor $f_M(t)$ obtained from Fig. 7.3 has a complicated time dependence. This time dependence is approximated by using a Taylor-series expansion around the peak Reynolds number:

$$f_M(t) \simeq f_M + \frac{df_M}{dN_R} |N_{R,1}| \left(\frac{\left| \text{Re} \left[U_1 e^{i\omega t} \right] \right|}{|U_1|} - 1 \right), \tag{7.24}$$

where f_M and the derivative on the right side are evaluated at the peak Reynolds number. It is then straightforward to integrate the instantaneous power dissipation over a full cycle, obtaining for the time-averaged dissipation of acoustic power per unit length

$$\frac{d\dot{E}_2}{dx} = -\frac{32\rho_m |U_1|^3}{3\pi^3 D^5} \left[f_M - \left(1 - \frac{9\pi}{32} \right) |N_{R,1}| \frac{df_M}{dN_R} \right], \tag{7.25}$$

where, again, f_M and df_M/dN_R are evaluated at the peak Reynolds number. (Note that df_M/dN_R is negative.)

Equation (5.9) shows that viscous power dissipation in a channel is $d\dot{E}_2/dx = -r_\nu |U_1|^2/2$, suggesting that Eq. (7.25) should be expressed as

$$r_{\nu,\text{turb}} = \frac{64\rho_m |U_1|}{3\pi^3 D^5} \left[f_M - \left(1 - \frac{9\pi}{32} \right) |N_{R,1}| \frac{df_M}{dN_R} \right]. \tag{7.26}$$

When this is compared to the equivalent result for laminar flow, Eq. (4.74), it is apparent that this model of turbulence multiplies the power dissipation $d\dot{E}_2/dx$ by a factor

$$m = \frac{\delta_\nu^2 \, |N_{R,1}|}{6A} \frac{[f_M - (1 - 9\pi/32) \, |N_{R,1}| \, df_M/dN_R]}{\mathrm{Im}\,[-f_\nu] \,/\, |1 - f_\nu|^2}. \tag{7.27}$$

Both f_M and df_M/dN_R as functions of N_R and ε can be evaluated directly from Fig. 7.3, or by using the iterative expression

$$\frac{1}{\sqrt{f_M}} = -2 \, \log_{10} \left(\frac{\varepsilon}{3.7} + \frac{2.51}{N_R \sqrt{f_M}} \right), \tag{7.28}$$

which is a remarkably good approximation to the friction factor [50].

At low enough velocities, $m \rightarrow 1$ and the flow is laminar. According to Eq. (7.27), the $m = 1$ boundary between laminar and turbulent zones occurs roughly at

$$|N_{R,1}| \simeq 2000 \text{ for } D/\delta_\nu < 4, \tag{7.29}$$

$$\frac{|N_{R,1}|}{D/\delta_\nu} \simeq 500 \text{ for } D/\delta_\nu > 4, \tag{7.30}$$

which is in rough agreement with the transitions from laminar or weakly turbulent to conditionally or fully turbulent in Fig. 7.4.

In our limited experience with resonator components and inertances at high Reynolds numbers, using $\varepsilon \simeq 5 \times 10^{-4}$ in this algorithm frequently gives reasonable agreement with experimental pressure drops, even when the actual pipe roughness would suggest a much lower value of ε. (See Fig. 9.12.)

Turbulence must also affect the other elements of the five-parameter impedance model of Chap. 4: the inertance per unit length l, the compliance per unit length c, the thermal-relaxation resistance r_κ, and the source/sink factor g. As turbulence increases r_ν, it probably changes the average velocity gradient at the wall, so the boundary-layer thickness probably changes, changing the effective area in the center of the duct that is available to carry the plug flow responsible for most of the inertance. Turbulence must also allow the influence of the isothermal boundary condition at the wall to reach farther into the gas, making more of the channel appear isothermal and hence increasing c. With more volume affected by thermal relaxation, r_κ probably decreases, and g may increase.

This analysis inspires little confidence. Experimental confirmation exists only for r_ν [82], and only in a limited regime. Even in the fully turbulent regime, where Iguchi's quasi-steady hypothesis is most credible, no expression exists for $\dot{H}_2$ and there are only vague ideas for the corrections to inertance, compliance, and thermal-relaxation resistance. Experiments and careful analysis are needed to improve the situation. Accurate estimates or experiments are needed for $T_1(y)$ in and near the boundary layer.

Example: Standing-Wave Engine

Figure 7.1 showed some data obtained during debugging of the standing-wave engine shown in Figs. 1.8 and 1.9, with lines representing numerical integrations

Fig. 7.5 Similar to Fig. 7.1, but with the addition of the *dashed lines*, showing calculations that account for turbulence according to the method described in this section

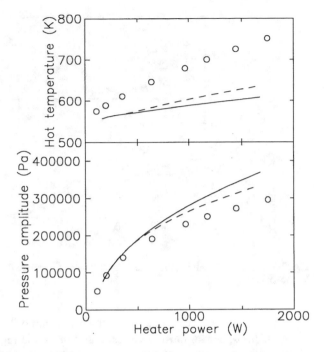

based on the considerations of Chaps. 4 and 5. Incorporating the algorithm just described to account for turbulence in the resonator of the engine changes the numerical integrations significantly, as shown by the dashed lines in Fig. 7.5. This brings the calculations significantly closer to the measurements. Turbulence has the greatest effect on the calculations at the center of the resonator, where $|N_{R,1}| = 10^6$ and Eq. (7.27) gives $m = 2.0$ for the highest-amplitude end of the plotted curve in Fig. 7.5. With $D/\delta_\nu = 800$, this is in the turbulent regime in Fig. 7.4.

We do not know whether the remaining disagreement between calculation and measurement is due to inadequacy of the turbulence calculation algorithm or to other phenomena beyond Rott's acoustic approximation.

7.2.1 Minor Losses

In flow at high Reynolds number, additional pressure drops are associated with the transitions between channels, and with changes in the direction of a channel. These effects are known as "minor losses" because in long piping systems such transitions are indeed minor contributors to overall pressure drop. However, in thermoacoustic engines and refrigerators operating at high amplitudes, these so-called minor losses can be major, as in the thermoacoustic-Stirling heat engine of Figs. 1.22 and 1.23, where minor losses are severe in the tee and in the 180° and 90° bends.

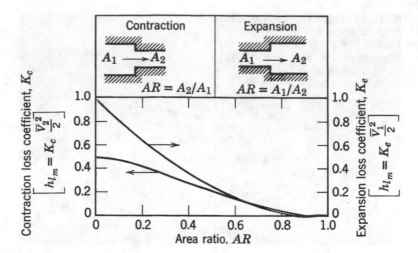

Fig. 7.6 Typical chart giving minor-loss coefficient K for steady flow. Reproduced from Fox and McDonald [50]

We have done a very small amount of work on minor loss in oscillating flow [90] at Los Alamos, but for most of our estimates of oscillating minor losses we rely on steady-flow data. For steady flow, the pressure drop that arises from minor losses is characterized by the dimensionless minor-loss coefficient K,

$$\Delta p = K \frac{1}{2} \rho u^2 = K \frac{\rho U^2}{2A^2}, \tag{7.31}$$

with tables and charts of K, as illustrated in Fig. 7.6, available for a wide variety of geometries [91, 92]. This "head loss" pressure difference is in addition to the dissipationless "Bernoulli" pressure difference $\Delta p = \Delta(\rho u^2/2)$ accompanying any change in velocity u. Minor losses arise from turbulence and other types of "secondary" flow, as illustrated in Figs. 7.7 and 7.8.

For flow of such high amplitude that the displacement amplitude $|\xi_1|$ is far larger than all other dimensions in the vicinity, the quasi-steady hypothesis might be applicable (especially during those phases of the oscillations in which velocity and acceleration have the same sign [93]). Then Eq. (7.31) is assumed to hold at each instant of time-dependent flow, so that[3]

---

[3]Note that the sign of the pressure drop is usually taken to be "obvious" in steady flow, so that fluid-mechanics textbooks usually use no minus signs in equations such as Eqs. (7.31), (7.21), and (7.37). Oscillating flow demands more careful notation, as the sign of Δp oscillates while the sign of $|U_1|^2$ remains nonnegative. In this book, beginning in Chap. 4, U_1 is taken to be positive in the positive x direction and Δp_1 is taken to be positive when p_1 increases with increasing x, consistent with the sign conventions of differential calculus. These choices lead to the unusual forms of Eqs. (7.32) and (7.13).

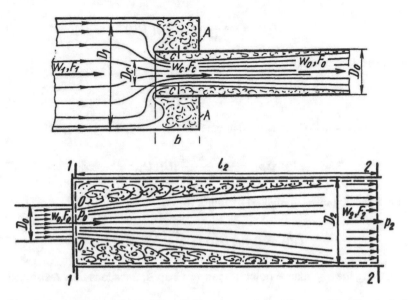

Fig. 7.7 Illustrations of turbulent eddies at abrupt contractions and expansions, for steady flow. Reproduced from Idelchik [92]

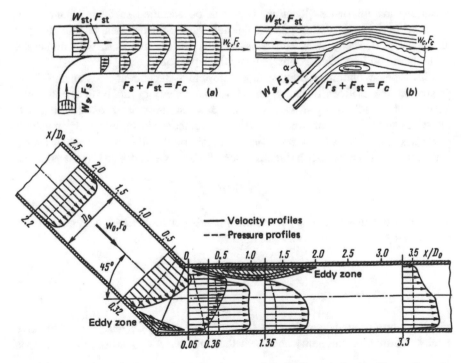

Fig. 7.8 Illustrations of sources of turbulent losses and dissipation at unions and in an elbow, for steady flow. Reproduced from Idelchik [92]

$$\Delta p(t) = -\frac{K}{2A^2} \rho(t) \, |U(t)| \, U(t). \tag{7.32}$$

Using sinusoidal velocity

$$U(t) = |U_1| \sin \omega t \tag{7.33}$$

and constant K, the first-order fundamental Fourier component of $\Delta p(t)$ is

$$\Delta p_1 = -\frac{4}{3\pi} \frac{K \rho_m}{A^2} |U_1| \, U_1, \tag{7.34}$$

and the acoustic power dissipated by minor loss is

$$\Delta \dot{E}_2 = \overline{\Delta p(t) \, U(t)} = \frac{2}{3\pi} \frac{K \rho_m}{A^2} |U_1|^3 . \tag{7.35}$$

Hence, using the impedance point of view of Chap. 4, a viscous flow resistance

$$R_v = \frac{4}{3\pi} \frac{K \rho_m}{A^2} |U_1| \tag{7.36}$$

exists at each minor-loss location, if K can be assumed to be independent of velocity.[4] In the DELTAEC file for the thermoacoustic-Stirling heat engine included in Appendix B, many such minor-loss resistances are included, with numerical values based on Idelchik's extensive tabulations [92] of K.

This type of nonlinear flow, with an amplitude-dependent resistance, is also present in the orifice (typically an adjustable valve) of an orifice pulse-tube refrigerator. Handbooks give values for the dimensionless head-loss coefficient K for valves and fittings, but American valve manufacturers usually specify a valve's properties in terms of a "flow coefficient" C_V. Although catalogs give values for C_V without units, it is not really a dimensionless number: C_V is defined by the equation

$$\Delta p = \rho U^2 / C_V^2 \tag{7.37}$$

for steady flow, using gallons per minute, pounds per square inch, and gm/cm^3 for U, Δp, and ρ respectively, so C_V has units of

$$\frac{\text{gal}}{\text{min}} \sqrt{\frac{\text{gm/cm}^3}{\text{psi}}}, \tag{7.38}$$

which is a unit of area equal to 24×10^{-6} m^2. The use of C_V instead of K in valve catalogs eliminates ambiguity about whether the area A appearing in Eq. (7.31) is

---

[4]See "Gedeon streaming" in this chapter for a discussion of velocity-dependent K.

the inside diameter of the pipes entering the valve, the smallest cross section inside the valve itself, or the "nominal" pipe size of the valve. To convert among K, C_V, and R_v for *steady* flow, use any of

$$K = \frac{2A^2}{[(24 \times 10^{-6} \, \text{m}^2)C_V]^2},$$ (7.39)

$$R_v = \frac{\rho U}{[(24 \times 10^{-6} \, \text{m}^2)C_V]^2} = \frac{\sqrt{\rho \, \Delta p}}{[(24 \times 10^{-6} \, \text{m}^2)C_V]},$$ (7.40)

$$R_v = \frac{K\rho U}{2A^2} = \sqrt{\frac{K\rho \, \Delta p}{2A^2}}.$$ (7.41)

For sinusoidal flow, use any of

$$K = \frac{2A^2}{[(24 \times 10^{-6} \, \text{m}^2)C_V]^2},$$ (7.42)

$$R_v = \frac{8\rho_m |U_1|}{3\pi \, [(24 \times 10^{-6} \, \text{m}^2)C_V]^2} = \frac{\sqrt{8\rho_m \, |\Delta p_1| / 3\pi}}{[(24 \times 10^{-6} \, \text{m}^2)C_V]},$$ (7.43)

$$R_v = \frac{4K\rho_m |U_1|}{3\pi A^2} = \sqrt{\frac{4K\rho_m \, |\Delta p_1|}{3\pi A^2}}.$$ (7.44)

Equations (7.42)–(7.44) are not particularly accurate. For most valves, C_V depends on the direction of flow, but manufacturers specify it only for one direction, their "intended" direction of flow. Furthermore, these equations incorporate only the incompressible approximation for C_V.

I hope that Eq. (7.36) and its variants incorporate the most important features of minor losses for thermoacoustics, but present understanding is very incomplete. When minor-loss locations are close together, such as the tee and elbow of Fig. 1.22, it is not known how their minor-loss phenomena interact, so it is not clear what effective R_v to assign. Sometimes minor loss is large even when $|\xi_1|$ is small enough that the quasi-steady hypothesis may not be applicable and hence Eq. (7.36) is questionable. It is not known whether to assign to each minor-loss location some correction for inertance, compliance, or thermal-relaxation resistance in addition to the correction to viscous resistance given by Eq. (7.36). And when both $U(t)$ and $\Delta p(t)$ contain harmonics, Eq. (7.36) is much too simple.

Example: Standing-Wave Engine

Short gradual cones, with an area reduction of a factor of 0.64 and a full included angle of 7° (i.e., 3.5° between center line and side wall), joined the cold heat exchangers to the central resonator tube in the standing-wave engine shown in Figs. 1.8 and 1.9. At a typical operating point, $|N_{R,1}| \sim 5 \times 10^5$ at the small end in each cone, so turbulence and minor losses might not have been negligible.

Handbooks show $K \simeq 0.15$ for outflow through such a cone, and smaller values for inflow, so we guessed that $K \simeq 0.1$ is a good estimate for minor loss in this component. Then Eq. (7.35) predicts that approximately 4 W of acoustic power is dissipated in each cone at a typical operating point. This is small, but not negligible, compared to the ordinary boundary-layer thermoviscous dissipation (Chap. 5) in each cone, 12 W. It is larger than the turbulence correction [Eq. (7.27)] predicted by Iguchi's quasi-steady hypothesis, $\sim$2 W. These powers are small compared with the net power of 500–1000 W that was supposed to be delivered to the refrigerators by this engine system at typical operating points.

Example: Thermoacoustic-Stirling Heat Engine

The large 90° tee in the thermoacoustic-Stirling heat engine, shown at the bottom of Fig. 1.23, was probably a source of significant minor losses, due to eddies and shears qualitatively similar to those shown for the tees at the top of Fig. 7.8.

Idelchik's handbook [92] tabulates losses for steady flows through tees, for a variety of trunk-to-side-branch velocity ratios and tee angles. Using the numerical values from Idelchik, we estimated that the minor losses in the thermoacoustic-Stirling heat engine's tee amounted to 70 W at a typical operating point—5% of the acoustic power delivered to the resonator. This must be regarded as only a rough estimate, because (a) the displacement amplitudes were not very much larger than the tee dimensions, so the use of steady-flow data from Idelchik is not well justified, and (b) even if the displacement amplitudes had been much larger, the trunk and side-branch velocities had different time phases, and we did not undertake a detailed time integration of this phenomenon.

Nevertheless, the rough estimate serves as a warning that this component needs improvement, and examination of Idelchik's extensive tables suggests that redesigning the tee with an angle closer to 30° instead of 90° and with areas chosen so that velocities are nearly equal would lead to improved performance.

7.3 Entrance Effects and Joining Conditions

One of the unspoken assumptions of Rott's acoustic approximation is that the gas displacement amplitude $|\xi_1|$ is much smaller than all other relevant dimensions in the x direction. This assumption is violated in thermoacoustic engines and refrigerators operating at high amplitudes, where the displacement amplitude is typically comparable to the entire length of heat exchangers and may be one-tenth of the length of a stack or regenerator.

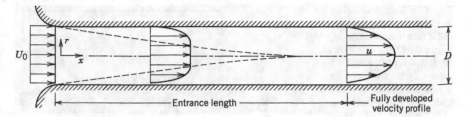

Fig. 7.9 Illustration of entrance effects for steady laminar flow, reproduced from Fox and McDonald [50]. The flow has its "fully developed" parabolic profile only at distances greater than Δx_{entr} downstream of the entrance to the channel. At shorter distances, the flow retains memory of its profile upstream of the entrance

7.3.1   Entrance Effects

Figure 7.9 illustrates one aspect of this situation for steady flow. Flow with a flat velocity profile enters a channel, from wide-open space to the left. It takes time for viscosity to diffuse the influence of the zero-velocity boundary condition at the wall into the flow, and time is equivalent to distance along the flow. For laminar flow, the entrance length [50] required to approach a fully developed velocity profile is given by

$$\Delta x_{\mathrm{entr}} \simeq 0.06\,N_R\,D. \tag{7.45}$$

Some calculations have been made of entrance effects for oscillating flow [94–96], including both viscous and thermal effects, but experimental work is lacking and the phenomena are not well understood. However, a rough estimate is very illuminating. If the entrance length for oscillating laminar flow is comparable to that for steady laminar flow, and if

$$D = 4r_h, \tag{7.46}$$

$$|N_{R,1}| = \frac{\rho\,|u_1|\,D}{\mu} = \frac{8r_h\,|\xi_1|}{\delta_\nu^2}, \tag{7.47}$$

then the result is

$$\Delta x_{\mathrm{entr}} \simeq 2\left(\frac{r_h}{\delta_\nu}\right)^2 |\xi_1|. \tag{7.48}$$

Hence, in a stack where typically $r_h \sim \delta_\nu$, the entrance length may be of the order of the displacement amplitude. At the 10% pressure amplitudes typical of the standing-wave engine of Figs. 1.8 and 1.9, roughly 10% of the length of the stack may be affected. In this region, the assumptions underlying the derivations of Chaps. 4 and 5 may be violated.

Fig. 7.10 Temperature
profile in the pulse tube of a
pulse tube refrigerator,
reproduced from Storch et al.
[97]. Naively one might
expect a profile without the
temperature overshoots at the
ends, whose magnitudes are
of the order of the adiabatic
temperature oscillation
amplitude and whose spatial
extents are of the order of the
gas displacement amplitude

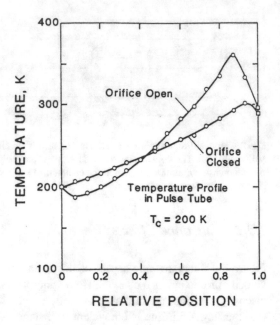

7.3.2 Joining Conditions

Some thermoacoustic computation algorithms have used continuity of complex
pressure amplitude p_1, complex volume-flow-rate amplitude U_1, and mean gas
temperature T_m as "joining conditions," i.e., boundary conditions to match solutions
across the interface between two adjacent thermoacoustic components (e.g., the
interface between a stack and a heat exchanger, or between a heat exchanger and
a pulse tube). For example, unless instructed otherwise, DELTAEC [61] uses these
joining conditions to pass from the end of one "segment" to the beginning of the
next.

There are many situations for which these traditional acoustic joining conditions
are obviously inadequate to describe real thermoacoustic phenomena. As discussed
above, continuity of p_1 is inappropriate if minor losses are important. Storch et
al. [97] observed a distorted temperature profile near the ends of a pulse tube, as
shown in Fig. 7.10, which can be attributed to an effective mismatch in T_m between
the end of the pulse tube and the adjacent heat exchanger. The spatial extent of
the "discontinuity" is of the order of the gas displacement amplitude. Swift [98]
observed a similar discontinuity in T_m between a hot heat exchanger and the adjacent
duct in a standing-wave engine; the discontinuity in T_m was proportional to $|p_1|$.

As a hypothetical example of the inadequacy of the traditional joining conditions,
consider two adjacent heat exchangers, with separation (along the direction of
oscillating gas motion) much less than the gas displacement amplitude, and each
having hydraulic radius much smaller than the thermal penetration depth δ_κ. If the
two heat exchangers are at equal temperatures, then U_1 is continuous across the

interface between them. However, if the temperatures of the two heat exchangers differ by δT_m, continuity of first-order mass flow $\rho_m U_1$ (with ρ_m the mean gas density) more accurately describes flow across the interface. Hence, U_1 has a discontinuity $\delta |U_1| \sim |U_1|\, \delta T_m / T_m$.

As an alternative to continuity of U_1, Rott [12, 99] proposed continuity of

$$\psi_1 = \frac{\omega}{i} \exp\left[-\int \frac{(f_\kappa - f_\nu)\, dT_m}{(1-\sigma)(1-f_\nu) T_m}\right] U_1 = \frac{\omega}{i} U_1 e^{-\int g\, dx}. \qquad (7.49)$$

The plausibility of ψ_1 as a joining variable is easily examined for some simple cases. For example, Eq. (7.49) becomes

$$\psi_1 = \frac{\omega}{i}\left[1 - \int \frac{f_\kappa\, dT_m}{T_m}\right] U_1 \qquad (7.50)$$

for an inviscid ideal gas, if the integral is much smaller than unity. Then, in the adiabatic limit ($f_\kappa \to 0$) of traditional acoustics, continuity of ψ_1 and of U_1 are identical, so the ψ_1 joining condition reduces to the traditional acoustic U_1 joining condition. In contrast, in the isothermal limit ($f_\kappa \to 1$), Eq. (7.50) becomes

$$\psi_1 = \frac{\omega}{i}\left[1 + \frac{\delta \rho_m}{\rho_m}\right] U_1, \qquad (7.51)$$

so that in this case continuity of ψ_1 and of first-order mass flow $\rho_m U_1$ are identical. As argued in the preceding paragraph, this is the correct joining condition between two heat exchangers at different temperatures, each with $r_h \ll \delta_\kappa$.

Although continuity of ψ_1 appears plausible in these two limiting situations, it is probably not always correct. To demonstrate this, joining conditions will be derived for the interface between an isothermal space ($f_\kappa = 1$, such as a good heat exchanger) and an adiabatic space ($f_\kappa = 0$, such as a pulse tube, thermal buffer tube, or resonator duct), as shown in Fig. 7.11. To lowest order, a first-order expression for δT_m and second-order expressions for δU_1 and δp_1 are needed. The flow in the adiabatic space will be assumed to be stratified, as it would be for a pulse tube or buffer tube with adequate flow straightening (see "Jet-driven streaming" below).

7.3.2.1   Mean Temperature

Closely following Smith and Romm [100] (see also Kittel [101], de Waele et al. [78] or Steijaert [21], Bauwens [102], and Romm and Smith [103]), consider the interface, illustrated in Fig. 7.11, between an isothermal heat exchanger at temperature T_x and an open space in which the gas is stratified (such as in an ideal pulse tube). Neglect viscosity, and let the frequency be low so that the acoustic wavelength is much larger than any other distance of interest. Suppose that

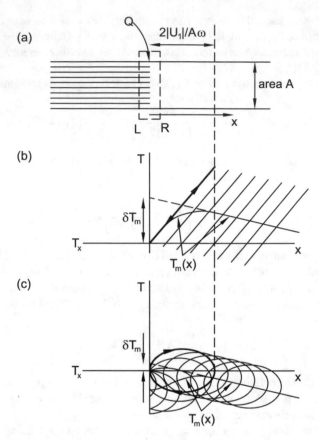

Fig. 7.11 A guide to the discussion of joining conditions. (**a**) The geometry under consideration. The interface between an isothermal heat exchanger (*left*) at temperature T_x and an adiabatic duct (*right*) of area A is at $x = 0$, with x increasing to the right. In the duct, the peak-to-peak gas displacement is $2|U_1|/\omega A$. Locations "L" and "R" are just to the left and right, respectively, of the interface. Somewhere far to the right, another heat exchanger imposes another temperature boundary condition, thereby contributing to the temperature gradient here. (**b**) For $\theta = -90°$, the gas motion is essentially that of a standing wave, with pressure and displacement in phase. The "particular" slice of gas has both $\xi(t)$ and $T(t) - T_x$ proportional to $1 - \cos \omega t$. The *heavy line* shows $T(t)$ vs $x = \xi(t)$ for this slice. Slices to the right of the "particular" slice have similar T vs x traces, shown as *parallel lines*. Slices to the left follow a portion of the "particular" slice's line while $\xi(t) > 0$, and have $T \equiv T_x$ while $\xi(t) < 0$. (**c**) For $\theta = 0$, the gas motion is essentially that of a traveling wave, with pressure and velocity in phase. The "particular" slice of gas has $\xi(t) \propto 1 - \cos \omega t$ and $T(t) - T_x \propto \sin \omega t$. The *heavy ellipse* shows T vs x for this slice. Slices to the right of the "particular" slice follow similar T vs x ellipses; slices to the left follow *truncated ellipses* while $x(t) > 0$ and $T \equiv T_x$ while $x(t) < 0$. In both (**b**) and (**c**), the resulting Eulerian $T_m(x)$ is shown as the *solid curve*; extrapolating it from large x to $x = 0$ gives the effective discontinuity δT_m in mean temperature

pistons or other external means cause the mass flow (positive to the right) across the interface at $x = 0$ to be exactly $\omega M_a \sin \omega t$ and the pressure to be exactly $p_m + |p_1| \sin (\omega t + \theta)$. Assume that the gas displacement amplitude is much greater than δ_κ.

Because the gas in the open cylinder is stratified, each differential slice of gas dM can be followed as it moves, considering its temperature, position, etc. as functions of time t. Begin by considering $T(t)$ for the slice of gas that is at $x = 0$ when $t = 0$. This particular slice touches the heat exchanger only at $t = 0, 2\pi/\omega, 4\pi/\omega,$ $\ldots$; slices to the right of it never touch the heat exchanger; slices to the left of it spend a nonzero fraction of their time inside the heat exchanger. When $t = 0$, this particular slice has temperature T_x, because at that moment it is in thermal contact with the heat exchanger. Thereafter, its temperature evolves isentropically in response to the changing pressure. Exercise 2.18 shows that $p/T^{\gamma/(\gamma-1)}$ is a constant during an isentropic process. Applying this relation here yields

$$T(t) = T_x \left[\frac{p_m + |p_1| \sin (\omega t + \theta)}{p_m + |p_1| \sin \theta} \right]^{(\gamma-1)/\gamma}. \tag{7.52}$$

To first order in p_1, Eq. (7.52) becomes

$$T(t) = T_x \left(1 + \frac{\gamma - 1}{\gamma} \frac{|p_1|}{p_m} [\sin (\omega t + \theta) - \sin \theta] \right). \tag{7.53}$$

Meanwhile, the position of that slice of gas is

$$\xi(t) = \frac{M_a}{\rho_m A} (1 - \cos \omega t) = \frac{|U_1|}{\omega A} (1 - \cos \omega t) \tag{7.54}$$

to first order, where A is the cross-sectional area of the open space. The path $T = T(t)$, $x = \xi(t)$ traced out by this particular slice of gas is shown as the heavy solid line with attached arrowheads in Fig. 7.11b, c for two choices of θ. Its average temperature is $T_x - [(\gamma - 1) |p_1| /\gamma p_m] T_x \sin \theta$, and its average location is $|U_1|/\omega A$.

Slices of gas to the right of the particular slice essentially share its oscillating motion and oscillating temperature, but with different average values. For $x > 2|U_1|/\omega A$, the slope dT_m/dx of their average temperatures is determined by weak phenomena outside the scope of this section (typically, heat conduction in the gas). Slices of gas to the left of the particular slice have complicated temperature histories, with constant temperatures for times when $\xi(t) < 0$, and temperatures evolving according to trigonometric functions of t (to first order) when $\xi(t) > 0$. The net effect of all such slices is to produce an average temperature profile $T_m(x)$, shown in Fig. 7.11b, c, which is unremarkable for $x > 2|U_1|/\omega A$ but which is rather complicated for $0 < x < 2|U_1|/\omega A$. Experimental evidence of such a complicated $T_m(x)$ in a pulse tube is shown in Fig. 7.10.

A suitable joining condition for T_m would allow convenient matching of the $x < 0$ and $x > 2|U_1|/\omega A$ solutions, the latter extrapolated to $x = 0$. Examination of Fig. 7.11b, c shows that this occurs if T_m is given a discontinuity δT_m at $x = 0$:

$$T_{m,R} - T_{m,L} = \delta T_m = -T_m \frac{\gamma - 1}{\gamma} \frac{|p_1|}{p_m} \sin\theta - \frac{|U_1|}{\omega A} \frac{dT_m}{dx}. \tag{7.55}$$

The discontinuity is first order in the acoustic amplitudes, and it can have either sign. Equation (7.55) is in reasonable agreement with the measurements of Storch [97] and Swift [98].

The signs in Eq. (7.55) work whether the open cylinder is on the right [as in Fig. 7.11a] or on the left, as long as these conventions are followed: θ is the phase angle by which p_1 leads U_1, U_1 is positive in the $+x$ direction, and dT_m/dx is positive if T_m increases with x.

7.3.2.2 Oscillating Volume Flow Rate

To deduce a suitable joining condition for volume flow rate, the slices to the left of the "particular" slice must be considered in greater detail. Each such slice is labeled by t^*, the time at which it crosses $x = 0$ from right to left. Thus $\pi \leq \omega t^* \leq 2\pi$ includes all slices, and the "particular" slice of the previous section has $\omega t^* = 2\pi$. Each slice crosses $x = 0$ from left to right at time t^{**} given by $\omega t^{**} = 2\pi - \omega t^*$. While each slice enjoys adiabatic conditions at $x > 0$, its pressure changes from $p_m + |p_1| \sin(2\pi - \omega t^* + \theta)$ to $p_m + |p_1| \sin(\omega t^* + \theta)$, a net pressure change of $2|p_1|\cos\theta \sin\omega t^*$. Hence, just before that slice enters the heat exchanger, its temperature must be $T_x\{1 + 2[(\gamma - 1)|p_1|/\gamma p_m]\cos\theta \sin\omega t^*\}$. As it enters the heat exchanger, returning to temperature T_x, its volume dM/ρ_m changes abruptly by an amount given by $-2[(\gamma - 1)|p_1|/\gamma p_m]\cos\theta \sin\omega t^* \, dM/\rho_m$. The discontinuity in volume flow rate is obtained by dividing by dt^*:

$$\delta U(t^*) = 2\frac{\gamma - 1}{\gamma p_m}\frac{\omega M_a}{\rho_m}|p_1|\cos\theta \sin^2\omega t^*, \qquad \pi \leq \omega t^* \leq 2\pi,$$

$$\delta U(t^*) = 0, \qquad 0 \leq \omega t^* \leq \pi. \tag{7.56}$$

The fundamental Fourier component of this function of t^* is

$$\delta U_{fFc}(t^*) \cong -\frac{8}{3\pi}\frac{(\gamma - 1)}{\gamma p_m}\frac{\omega M_a}{\rho_m}|p_1|\cos\theta \sin\omega t^*$$

$$= -\frac{8}{3\pi}\frac{(\gamma - 1)}{\gamma p_m}|p_1||U_1|\cos\theta \sin\omega t^*. \tag{7.57}$$

Hence, the discontinuity of volume flow rate contains a fundamental Fourier component δU_1. This discontinuity is in phase with the volume flow rate itself, so there is no discontinuity in the phase of U_1. The discontinuity in $|U_1|$ is

$$|U_1|_R - |U_1|_L = \delta\,|U_1| = -\frac{8}{3\pi}\frac{\gamma-1}{\gamma}\frac{|p_1|\,|U_1|}{p_m}\cos\theta \qquad (7.58)$$

$$= -\frac{16}{3\pi}\frac{\gamma-1}{\gamma}\frac{\dot{E}_2}{p_m}, \qquad (7.59)$$

where $\dot{E}_2 = \frac{1}{2}|p_1|\,|U_1|\cos\theta$ is the acoustic power. This expression may be used whether the open cylinder is on the right or on the left, as long as $\dot{E}_2$ is taken to be positive when power flows in the $+x$ direction. Note that $\delta\,|U_1|$ is second order in the acoustic amplitudes, and it is zero for standing-wave phasing. Hence, it is larger for pulse-tube and Stirling refrigerators than for low-amplitude standing-wave thermoacoustic engines and refrigerators.

The overall form of Eq. (7.58) is not surprising. In the geometry under consideration, there is a volume of order $|U_1|/\omega$ in the open cylinder, adjacent to the heat exchanger, that might naively be considered adiabatic. However, the gas occupying this volume actually behaves as near to isothermally as to adiabatically. Hence, its response to changing pressure must be corrected in rough proportion to the difference between the isentropic and isothermal compressibilities:

$$\delta V \sim \frac{|U_1|}{\omega}\,|p_1|\left(\frac{1}{\gamma p_m}-\frac{1}{p_m}\right). \qquad (7.60)$$

Multiplication by ω to convert from δV to $\delta\,|U_1|$, and by $\cos\theta$ to eliminate the effect for standing-wave phasing, recovers the form of Eq. (7.58).

Note that $\theta = 0$ and $dT_m/dx = 0$ makes $\delta T_m = 0$ in Eq. (7.55). In that case, examination of Eq. (7.49) or (7.50) shows that continuity of ψ_1 implies continuity of U_1, yet Eq. (7.57) or (7.58) shows a maximum (with respect to θ) discontinuity in $|U_1|$. Hence, the joining condition derived here is fundamentally different from continuity of ψ_1. Perhaps these two joining conditions are applicable in different circumstances.

7.3.2.3 Oscillating Pressure

Consistent with the previous section, a pressure discontinuity

$$\delta p \sim U_1\,\delta U_1 \qquad (7.61)$$

might be expected at the interface, due to the $(\mathbf{v}\cdot\nabla)\mathbf{v}$ term in the momentum equation. Since this pressure discontinuity would apparently be third order in the acoustic amplitude, it is neglected. The second-order minor-loss pressure discontinuity is given by Eq. (7.34).

7.3.2.4   Acoustic Power Dissipation and Entropy Generation

In addition to the minor-loss dissipation of acoustic power given in Eq. (7.35), a
third-order dissipation of acoustic power is caused by the second-order discontinuity
in volume flow rate given in Eq. (7.58):

$$\delta \dot{E}_2 = \dot{E}_{2,L} - \dot{E}_{2,R} = \frac{1}{2\pi} \int_0^{2\pi} p(t) \, \delta \, |U_1| \sin \omega t \, d(\omega t)$$

$$= \frac{4}{3\pi} \frac{\gamma - 1}{\gamma p_m} |p_1|^2 \, |U_1| \cos^2 \theta. \tag{7.62}$$

Naturally, this expression is never negative. It represents an inherent irreversibility
at such an interface due to irreversible heat transfer in slices of gas approaching
$x = 0$ from the right with temperatures different from T_x. This irreversibility was
discussed by Smith and Romm [100] from the point of view of entropy generation.
The "adiabatic model" explained by Urieli and Berchowitz [15] incorporates this
irreversibility into Stirling-engine analysis. Combining this book's notation with
the results of Smith and Romm, the interface dissipation per cycle is $2\pi \, \delta \dot{E}_2/\omega =
T_x S_{gen}$. Here S_{gen} is the entropy generation per cycle, obtained from the second law
of thermodynamics,

$$S_{gen} = \int s_R \, dM - \int s_L \, dM - \frac{1}{T_x} \int dQ, \tag{7.63}$$

where s_R and s_L are the entropies per unit mass just right and left of the interface and
$dQ = c_p(T_R - T_L) \, dM$ is the heat absorbed by mass dM crossing the interface. Smith
and Romm numerically integrated Eq. (7.63) for a monatomic ideal gas ($\gamma = 5/3$),
without making Rott's acoustic approximation, and they display a graph of their
results as a function of θ and $|p_1|/p_m$. Equation (7.62) is indistinguishable from
their results at $|p_1|/p_m \sim 0.1$, and disagrees with their more accurate results by at
most 10% at $|p_1|/p_m \sim 0.3$.

This loss mechanism is usually negligible in standing-wave systems, where
$\cos \theta$ is very small. However, it can represent a significant loss in high-amplitude
traveling-wave systems, as

$$\frac{\delta \dot{E}_2}{\dot{E}_2} \sim \frac{8}{3\pi} \frac{\gamma - 1}{\gamma} \frac{|p_1|}{p_m} \cos \theta. \tag{7.64}$$

In this regard, diatomic and polyatomic gases are more efficient than monatomic
gases.

(An expression similar to Eq. (7.62), but with $\theta = 0$, was derived incorrectly by
Swift [104]. The error in that derivation arises on page 4162, where the equation
in the second line should be $\delta S = \delta Q \, \delta T / 2T^2$. This error propagated through that
analysis, to Eqs. (27) and (28), which should have been divided by two.)

7.3.2.5   Summary

This rather intricate derivation provides higher-order joining conditions between an isothermal channel (some heat exchangers, some flow straighteners) and a stratified adiabatic channel (pulse tubes, thermal buffer tubes, some resonator ducts). It lacks careful experimental confirmation, and it provides no guidance for joining conditions between other types of channels encountered throughout thermoacoustic engines and refrigerators. Some calculations (e.g., [94, 95]) have been made that point the way for future understanding of these phenomena.

Example: Thermoacoustic-Stirling Heat Engine
The interface between the top of the thermal buffer tube and the bottom of the regenerator in the thermoacoustic-Stirling heat engine of Figs. 1.22 and 1.23 approximates the interface between an isothermal and open space, if the electric hot heat exchanger is regarded as so acoustically transparent that the interface phenomena occur at the end of the locally isothermal regenerator. At a typical high-amplitude operating point, Eq. (7.62) predicts 70 W of dissipated acoustic power from this joining irreversibility, roughly 5% of the net power delivered by the engine to the resonator at the tee. The gas displacement amplitude $|\xi_1| \sim 1$ cm in this vicinity, and the adiabatic temperature oscillation has an amplitude of 40 °C.

7.4   Mass Streaming

In acoustics, the term "streaming" usually refers to a steady mass-flux density or velocity, usually of second order, that is superimposed on the larger first-order oscillating acoustic mass-flux density or velocity and is driven by the first-order oscillations [105]. When streaming exists, it is useful to think of the motion of gas during each cycle as something like 102 steps forward and 98 steps backward, equivalent to the superposition of a steady drift forward of 4 steps during each cycle of an oscillating motion with a peak-to-peak amplitude of 100 steps. In thermoacoustic engines and refrigerators, streaming is important because it is a mechanism for convective heat transfer. Carrying heat, streaming can be either an undesirable loss mechanism or an essential heat-transfer method.

Pulse tubes and thermal buffer tubes are usually oriented vertically to avoid gravity-driven convection, which is one form of time-independent mass-flux density that could be superimposed on the oscillations. This section will present four other undesirable steady flows, which can be called streaming because they are driven by the first-order acoustic phenomena. These generally undesirable types of streaming are illustrated in Fig. 7.12. "Gedeon streaming" [106] is a net time-averaged mass

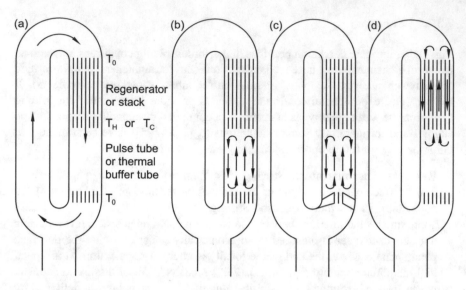

Fig. 7.12 Illustration of some types of mass streaming, generally harmful to thermoacoustic engines and refrigerators. *Arrows* indicate the time-averaged mass-flux density, which is superimposed on the much larger oscillating flow discussed in Chap. 4. (**a**) Gedeon streaming. (**b**) Rayleigh streaming. (**c**) Jet-driven convection or streaming. (**d**) Streaming within a regenerator or stack

flow along x through a regenerator, pulse tube, etc.[5] "Rayleigh streaming" here refers to a time-averaged toroidal circulation within a pulse tube or thermal buffer tube, driven by boundary-layer effects at the side walls [88, 108, 109], similar to the standing-wave streaming described for $\delta_v \ll r_h$ in Rayleigh's book [105, 110]. "Jet-driven streaming" is a third type, which is also a toroidal circulation within a pulse tube or buffer tube, but which is driven by inadequate flow straightening at an end of the tube. A fourth type, "internal streaming" within a regenerator or stack, also occurs under some circumstances.

However, perhaps such superimposed steady flow can be used advantageously in some circumstances, to deliberately transfer heat. This steady flow could be either parallel or perpendicular to x, as illustrated in Fig. 7.13. The purpose of such flow is to carry heat, which is a second-order quantity in thermoacoustics, so these flows will be treated as second order, referred to as streaming, and treated in this section.

To discuss streaming, the perturbation-series expansion of relevant variables must be extended one step beyond the first-order acoustic approximation that was

---

[5]I refer to this as Gedeon streaming because David Gedeon wrote an early paper [106] on this subject in the context of engines and refrigerators. This streaming is commonly called "dc flow" in the pulse-tube-refrigerator literature [107], but it seems to me that "dc flow" sounds as general as "streaming" and hence might be interpreted as referring to any of the second-order steady flows described here.

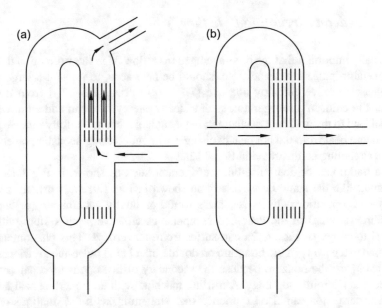

Fig. 7.13 Illustration of mass streaming that can be beneficial. Note the elimination of a heat exchanger in each case. (**a**) Flow parallel to x, through a stack or regenerator. (**b**) Flow perpendicular to x, across one end of a stack or regenerator

introduced in Chap. 3. Here any time-dependent variable $\mathcal{F}$ must be written in the more accurate form

$$\mathcal{F}(t) = \mathcal{F}_m + \mathrm{Re}\left[\mathcal{F}_1 e^{i\omega t}\right] + \mathcal{F}_{2,0} + \mathrm{Re}\left[\mathcal{F}_{2,2} e^{2i\omega t}\right]. \tag{7.65}$$

The subscript "2,0" identifies the streaming term, which is independent of time. (The second-order oscillating term, which oscillates at 2ω, is of no interest in this section, but will be considered in "Harmonics" below.) The second-order time-averaged mass-flux density $\dot{m}_{2,0}(x, y, z)$ is of primary interest, because it convects a time-averaged enthalpy-flux density $\dot{m}_{2,0} c_p T$. As illustrated in Figs. 7.12 and 7.13, the second-order time-averaged mass flow

$$\dot{M}_2 = \int \dot{m}_{2,0}\, dA \tag{7.66}$$

in the x direction may be either zero or nonzero, depending on the circumstances. Associated gradients in the second-order time-averaged pressure $p_{2,0}$ are also important, and can be either large or small, depending on the impedance of the space through which $\dot{M}_2$ flows.

7.4.1 Gedeon Streaming ("dc flow")

It is usually important that the time-averaged mass flow $\dot{M}$ in the x direction through
a regenerator, pulse tube, stack, etc. should be near zero, to prevent a large time-
averaged convective enthalpy flux $\dot{M}c_p(T_H - T_0)$ or $\dot{M}c_p(T_0 - T_C)$ from flowing
from hot to cold. In a refrigerator, such a steady energy flux can add an unwanted
thermal load to the cold heat exchanger; in an engine, it can wastefully remove high-
temperature heat from the hot heat exchanger without creating acoustic power. This
type of streaming is illustrated in Fig. 7.12a.

In a traditional orifice pulse-tube refrigerator such as shown in Fig. 1.18 or in
something like the standing-wave engine shown in Fig. 1.9, $\dot{M}$ is exactly zero in
steady state operation: Otherwise, mass would gradually accumulate (or deplete,
depending on sign) in the dead-end components. However, any system with the
toroidal topology of Fig. 7.12a can suffer from nonzero $\dot{M}$. This phenomenon is
discussed extensively in the literature on double-inlet [111] pulse-tube refrigerators,
in which $\dot{M}$ can be nonzero because the secondary orifice opens up a path having
the necessary toroidal topology. A Stirling machine with an imperfect seal around
the displacer piston can also experience such streaming, with the small piston gap
completing the toroidal path. Asymmetry in the flow impedance of the double-inlet
valve or synchronous displacer-gap wobble can pump mass through such paths.

To enforce $\dot{M}c_p \,\Delta T = 0$, every time-averaged term in the perturbation expansion
of $\dot{M}$ must be zero. In particular, $\dot{M}_2 \equiv \dot{M}_{2,0} = 0$ in Eq. (7.65). In terms of density
and volume flow rate,

$$\dot{M}_2 = \frac{1}{2}\mathrm{Re}\,[\widetilde{\rho_1} U_1] + \rho_m U_{2,0}. \tag{7.67}$$

Hence, the two terms on the right side of Eq. (7.67) *must* turn out to be equal and
opposite in systems lacking the toroidal topology, such as traditional orifice pulse-
tube refrigerators or standing-wave thermoacoustic engines or refrigerators. The two
first-order factors ρ_1 and U_1 are "known" from the considerations of Chap. 4, so
setting $\dot{M}_2 = 0$ in Eq. (7.67) determines what $U_{2,0}$ *must* be.[6]

The second-order time average of the momentum equation can be written in the
form

$$0 \simeq -\frac{dp_{2,0}}{dx} + r_{v,2,0}U_{2,0} \tag{7.68}$$

in a regenerator having a large enough resistance per unit length to second-order
flow, $r_{v,2,0}$, that the $\rho_1 u_1$ and $(\mathbf{v}_1 \cdot \nabla)\mathbf{v}_1$ terms arising from the left side of Eq. (2.25)
are negligible. [*Keeping* those terms while *neglecting* dissipation gives a very

---

[6]When reading or writing about streaming, be careful to distinguish between the time-averaged
velocity at a point, $u_{2,0}$, and the time-averaged velocity of a particle, $u_{2,0} + \mathrm{Re}\,[\widetilde{\rho_1} u_1]/2\rho_m$.
Ambiguously, both of these have been called "the streaming velocity" in the literature.

different, well-known result [46, 112] for $p_{2,0}(x)$.] Hence, a nonzero $dp_{2,0}/dx$ will generally also exist in a regenerator or stack whenever $\dot{M}_2 = 0$ through it. It is helpful to think of the $\Delta p_{2,0}$ that exists across the regenerator as causing the viscous flow of $U_{2,0}$ through the regenerator. In a traditional orifice pulse-tube refrigerator, this small time-averaged pressure difference appears automatically, because the topology of the hardware imposes $\dot{M}_2 = 0$.

However, as shown in Fig. 7.12a, any system with the topology of a torus, such as the thermoacoustic-Stirling heat engine of Fig. 1.23, can have $\dot{M}_2 \neq 0$ if the two terms on the right side of Eq. (7.67) are not in balance. To estimate how severe the time-averaged convective enthalpy flux carried by such streaming might be in a cryogenic refrigerator if $\dot{M}_2 \neq 0$, Gedeon [106] showed that

$$\frac{1}{2}\mathrm{Re}\,[\widetilde{\rho_1} U_1] \simeq \rho_m \dot{E}_2 / p_m \qquad (7.69)$$

in a regenerator, where $\dot{E}_2 = \frac{1}{2}\mathrm{Re}[\widetilde{\rho_1} U_1]$ is the usual acoustic power passing through the regenerator. Hence, $\frac{1}{2}\mathrm{Re}[\widetilde{\rho_1} U_1]$ *must* be nonzero in traveling-wave engines and refrigerators. In the context of open (adiabatic) spaces, Bradley [113, 114] refers to this streaming term as "squeeze flow" mass transport. Setting Eq. (7.67) equal to zero shows that efficient regenerator operation requires

$$U_{2,0} = -\frac{1}{2}\mathrm{Re}\,[\widetilde{\rho_1} U_1]\,/\rho_m = -\dot{E}_2/p_m. \qquad (7.70)$$

The consequences of ignoring this requirement can be severe indeed. If $\dot{M}_2 > 0$, an undesired, streaming-induced enthalpy flux

$$\dot{Q}_{\mathrm{loss}} \sim \dot{M}_2 c_p \,(T_0 - T_C) \qquad (7.71)$$

flows from ambient to cold through the regenerator. (If $\dot{M}_2 < 0$, such enthalpy flows from ambient to cold through the pulse tube, with equally harmful effect.) For $U_{2,0} = 0$, the ratio of $\dot{Q}_{\mathrm{loss}}$ to the ordinary regenerator loss $\dot{H}_{2,\mathrm{reg}}$ is of the order of

$$\frac{\dot{Q}_{\mathrm{loss}}}{\dot{H}_{2,\mathrm{reg}}} \sim \frac{\gamma}{\gamma-1}\frac{(T_0-T_C)}{T_0}\frac{\dot{E}_2}{\dot{H}_{2,\mathrm{reg}}} \sim \frac{\gamma}{\gamma-1}\frac{(T_0-T_C)}{T_C}\frac{\dot{Q}_{C,\mathrm{gross}}}{\dot{H}_{2,\mathrm{reg}}}, \qquad (7.72)$$

where $\dot{Q}_{C,\mathrm{gross}}$ is the gross cooling power, equal to $\dot{E}_2$ in the cold heat exchanger [see Chap. 5]. In the final expression in Eq. (7.72), each of the three fractions is > 1 for cryocoolers. Hence, their product is $\gg 1$, and the unmitigated streaming-induced heat load would be much greater than the ordinary regenerator loss.

A toroidal topology, such as shown in Fig. 7.12a, ensures that the path integral of ∇p around the torus must be zero, so that the pressure p at any location will be uniquely defined. At second order, this means that $\oint (dp_{2,0}/dx)\,dx = 0$ around the torus. Equation (7.68) shows what $\Delta p_{2,0}$ must exist across the regenerator to obtain the desired $U_{2,0}$ in the regenerator. Hence, the other components of the loop must be designed so that their aggregate $\Delta p_{2,0}$ is equal and opposite to the $\Delta p_{2,0}$ desired across the regenerator.

The desired $\Delta p_{2,0}$ can be estimated using the low-Reynolds-number limit of Fig. 7–9 of Kays and London [64]

$$\frac{dp}{dx} \simeq -\frac{6U\mu}{A r_h^2} \tag{7.73}$$

for the pressure gradient in a screen bed of cross-sectional area A and hydraulic radius r_h. (The numerical factor depends weakly on the volumetric porosity of the bed.) With Eq. (7.70) for U, this yields

$$\Delta p_{2,0,\text{reg}} \simeq \frac{6}{A r_h^2 p_m} \int_{\text{reg}} \mu_m(x)\dot{E}_2(x) \; dx \tag{7.74}$$

for the pressure difference across the regenerator when $\dot{M}_2 \equiv 0$. (The x dependence of the viscosity is due to the temperature gradient.) For the devices shown in Figs. 1.22, 1.23 and 1.17, 1.18 and for other traveling-wave devices having p_m of tens of bar and $|p_1|/p_m \sim 0.1$, $\Delta p_{2,0,\text{reg}}$ given by Eq. (7.74) is typically of the order of a few hundred Pa.

In the limit of low viscosity or large tube diameter and in the absence of turbulence, $p_{2,0}(x)$ would be described by some acoustic version of the Bernoulli equation. This suggests that an acoustically ideal path around the torus from one end of the regenerator to the other would impose across the regenerator a pressure difference of the order of $\Delta\left[\,\rho_m u_1 \widetilde{u}_1\,\right]$. (Such an ideal loop might include a pulse tube or thermal buffer tube, inertance or transmission line, and compliance, without heat exchangers or other components with small passages. It would also be necessary to avoid all minor losses in such a loop—a significant challenge.) This pressure difference is typically much smaller than the $\Delta p_{2,0,\text{reg}}$ given in Eq. (7.74) that is required for $\dot{M}_2 = 0$.

Hence, to produce the required $\Delta p_{2,0,\text{reg}}$, an additional physical effect or structure is needed in the torus, relying on turbulence, viscosity, or some other physical phenomenon not included in the Bernoulli equation.

Asymmetry in minor-loss effects can be used to produce this required $\Delta p_{2,0,\text{reg}}$ [37, 38, 76]. In a gently tapered transition between a small-diameter tube, where $|u_1|$ is large, and a large-diameter tube, where $|u_1|$ is small, turbulence would be avoided and Bernoulli's equation would hold. At the opposite extreme, the oscillating pressure drop across an abrupt transition between a small-diameter tube and a large-diameter tube exhibits direction-dependent minor losses, leading to what can be called a gas diode. (There was some discussion of this idea in the late 1990s at some of the cryocooler conferences, in the context of "dc flow" in double-inlet pulse tube refrigerators. See *Cryocoolers 10*, Plenum, 1999, edited by R. G. Ross, Jr.)

In the example shown in Fig. 7.14, a small tube is connected to an essentially infinite open space. When the gas (at velocity u inside the tube) flows out of the

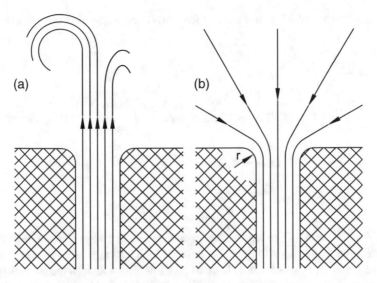

Fig. 7.14 Asymmetry of high-Reynolds-number flow at the transition between a small tube and a wide-open space. (**a**) For outflow, a jet extends far into the open space, and downstream turbulence dissipates kinetic energy. (**b**) For inflow with a well-rounded entrance lip, there is little dissipation

tube, a jet occurs, and kinetic energy is lost to turbulence downstream of the jet; $K_{out} = 1$. In contrast, when gas flows into the tube, the streamlines in the open space are widely and smoothly dispersed; K_{in} varies from 0.5 to 0.04, with smaller values for larger radius r of rounding of the edge of the entrance. For such asymmetric transitions, K must have time dependence in Eq. (7.32), which becomes

$$\Delta p(t) = -\frac{1}{2A^2} K(t)\rho(t) \, |U(t)| \, U(t). \tag{7.75}$$

If $U = |U_1| \sin \omega t$, the time-averaged pressure drop is obtained by integrating Eq. (7.75) in time:

$$\overline{\Delta p_{ml}} = -\frac{\omega}{2\pi A^2} \left(\int_0^{\pi/\omega} K_{out} \frac{1}{2}\rho \, |U_1|^2 \sin^2 \omega t \, dt \right.$$

$$\left. - \int_{\pi/\omega}^{2\pi/\omega} K_{in} \frac{1}{2}\rho \, |U_1|^2 \sin^2 \omega t \, dt \right)$$

$$= -\frac{1}{8A^2}\rho \, |U_1|^2 \, (K_{out} - K_{in}). \tag{7.76}$$

Such easy control of $\dot{M}_2$ is not penalty free, however. Acoustic power is dissipated at a rate

$$
\begin{aligned}
\Delta\dot{E}_2 &= -\frac{\omega}{2\pi}\int_0^{2\pi/\omega}\Delta p\, U\, dt \\
&= \frac{\omega}{2\pi A^2}\left(\int_0^{\pi/\omega}K_{\text{out}}\frac{1}{2}\rho\,|U_1|^3\sin^3\omega t\, dt - \int_{\pi/\omega}^{2\pi/\omega}K_{\text{in}}\frac{1}{2}\rho\,|U_1|^3\sin^3\omega t\, dt\right) \\
&= \frac{1}{3\pi A^2}\rho\,|U_1|^3\,(K_{\text{out}}+K_{\text{in}}) \tag{7.77}\\
&= -\frac{8}{3\pi}\overline{\Delta p_{ml}}\,|U_1|\,\frac{K_{\text{out}}+K_{\text{in}}}{K_{\text{out}}-K_{\text{in}}}, \tag{7.78}
\end{aligned}
$$

where A is the area of the small tube. Equation (7.78) shows that the best way to produce a desired $\overline{\Delta p_{ml}}$ is to insert the device at a location where $|U_1|$ is small, and to shape it so that $K_{\text{out}} - K_{\text{in}}$ is as large as possible. Even though the acoustic power dissipation given in Eqs. (7.77) and (7.78) is formally of only third order, it is large if u_1 is large enough to generate a substantial $\overline{\Delta p_{ml}}$.

Our measurements at Los Alamos provide qualitative evidence for most of the features discussed here, but quantitative agreement between Eqs. (7.76) and (7.74) is poor. Once again, there is something important that is not understood well.

Example: Thermoacoustic-Stirling Heat Engine
The adjustable "jet pump" labeled in Fig. 1.23 employed asymmetry of high-Reynolds-number flow to suppress Gedeon streaming. (The terminology "synthetic jet" [115] and "gas diode" is also used.) Temperature measurements in the center of the regenerator and measurements of $\dot{Q}_H$ required to maintain engine operation at constant amplitude both agreed that Gedeon streaming was stopped, but only when the jet area A was 2–3 times smaller than would be predicted by equating Eqs. (7.76) and (7.74) and setting $K_{\text{out}} - K_{\text{in}} \simeq 0.9$.

7.4.2   Rayleigh Streaming

Ideally, the gas in a pulse tube or thermal buffer tube acts as a long (and slightly compressible) piston, transmitting pressure and velocity oscillations from one end to the other. The gas should also thermally insulate the ends of the tube from each other. Unfortunately, convection *within* the tube [109] can carry enthalpy from one end to the other, adding a heat load to the cold heat exchanger in a refrigerator or consuming heat at the hot heat exchanger of an engine. If Gedeon streaming is eliminated, there can be no *net* mass flow $\dot{M}_2$ through the tube, but the possibility remains that enthalpy can be carried by mass-flux density $\dot{m}_{2,0}(r)$ streaming upward near the side walls of the tube and downward in the central portion of the tube, or vice versa. Such streaming can be due to a jet (or jets) caused by inadequate flow straightening at either end of the tube, as described in the next subsection, but this subsection summarizes Rayleigh streaming, which is forced convection confined within the tube and driven by viscous and thermal boundary-layer phenomena at the side walls of the tube, as illustrated in Fig. 7.12b.

The boundary-layer calculation [88] of streaming in such a geometry, based on the earlier work of Rott [68] and incorporating variable cross-sectional area $A(x)$, yields a prediction for the side-wall taper that suppresses this streaming:

$$
\frac{1}{A} \frac{dA}{dx} = \frac{\omega A |p_1|}{\gamma p_m |U_1|} \left[\left(1 + \frac{2 (\gamma - 1) (1 - b) \sqrt{\sigma}}{3 (1 + \sigma)} \right) \sin \theta \right.
$$
$$
\left. - \left(1 + \frac{2 (\gamma - 1) (1 - b \sigma^2)}{3 \sigma (1 + \sigma)} \right) \cos \theta \right] - \frac{(1 - b) (1 - \sqrt{\sigma})}{3 (1 + \sigma) (1 + \sqrt{\sigma})} \frac{1}{T_m} \frac{dT_m}{dx}
$$
$$
\tag{7.79}
$$

$$
\simeq (0.64 \sin \theta - 0.75 \cos \theta) \frac{\omega A |p_1|}{p_m |U_1|} - 0.0058 \frac{1}{T_m} \frac{dT_m}{dx}. \tag{7.80}
$$

where θ is the phase of the acoustic impedance. (The sign of θ here is opposite to the unfortunate sign convention adopted in [88, 89, 108].) In Eq. (7.80), numerical values for helium gas are used: $\gamma = 5/3$, $\sigma = 0.69$, and $b = (T/\mu) \, d\mu/dT = 0.68$.

When Eq. (7.79) or (7.80) is satisfied, a large number of side-wall boundary-layer effects are in delicate balance, with some trying to cause streaming up along the side walls and down through the center, while others try to cause streaming down along the side walls and up through the center.

One such effect is easily imagined and is illustrated in Fig. 7.15c. Consider a small parcel of gas oscillating up and down along the wall, at a distance from the wall of the order of the relevant boundary layer: the viscous penetration depth δ_ν. On average, the gas between the parcel and the wall will have a different temperature during the parcel's upward motion than during its downward motion, due to imperfect thermal contact with the wall's temperature gradient and due to the temperature oscillations that have time phasing depending on oscillating motion and oscillating pressure. The moving parcel will experience a different amount of viscous drag during its upward motion than during its downward motion because the viscosity depends on temperature, so it will undergo a different displacement during its upward motion than during its downward motion. After a full cycle, the parcel does not return to its starting point: It experiences a small net drift which contributes to $\dot{m}_{2,0}$ at its location. This process is represented by terms proportional to the product of b, T_1, and u_1 in the nonlinear equations of the derivation, and is responsible for the presence of b in Eq. (7.79).

The effect of a taper on streaming can also be imagined, with reference to Fig. 7.15d. In general, a gas parcel close to the wall will be farther from the wall during, say, its upward motion than during its downward motion, due to, for example, the compressibility of the gas in the boundary layer and the phasing

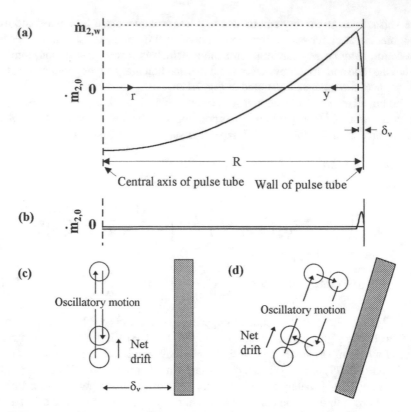

Fig. 7.15 The net drift near the wall, illustrated in (**c**) and (**d**), affects the entire tube, and can cause the offset parabolic mass-flux-density profile $\dot{m}_{2,0}(r)$ shown in (**a**) in the laminar regime [108]. (**a**) For the signs chosen here for illustration, the gas moving downward in the center of the tube is hotter than the gas moving upward around it, so that heat is carried downward. The coordinate r measures the distance from the center of the tube, whose radius is $\mathcal{R}$. The calculations are for $\delta_v \ll \mathcal{R}$. (**b**) Mass-flux density for the case where the streaming is suppressed by tapering the tube. Although there is still some streaming near the wall and a correspondingly small offset mass flux in the rest of the tube, they carry negligible heat. (**c**) Illustration of net drift caused by one process within δ_v of the tube wall. A parcel of gas is shown in three consecutive positions. Here we imagine that the temperature dependence of the viscosity is the only important effect, and assume that the temperature is lower during upward motion than during downward motion. (**d**) Illustration of net drift caused by a different process, as discussed in the text

between oscillating motion and pressure. Hence, again the moving parcel will experience a different amount of viscous drag during its upward motion than during its downward motion, and so the parcel will again fail to return to its starting point after a full cycle. This effect is represented in the derivation's starting equations by terms proportional to the product of u_1 and v_1. However, the boundary-layer continuity equation couples v_1 and du_1/dx, while the y-averaged continuity equation $i\omega A \langle \rho_1 \rangle + \rho_m d \langle A u_1 \rangle /dx = 0$ couples du_1/dx to dA/dx. Hence, the process shown in Fig. 7.15d is controlled in part by the taper dA/dx, and is responsible for the presence of dA/dx in Eq. (7.79).

Including all such second-order streaming effects in the calculations allows determination of the conditions under which they all add to zero, represented by Eq. (7.79). Note that all the variables in the right side of Eq. (7.79) or (7.80) should be known during the design of a pulse tube refrigerator and are experimentally accessible.

The effort necessary to obtain expressions for p or u at second order, describing streaming, is far greater than the effort of Chap. 4 to obtain first-order expressions for these variables. For example, Olson's calculation [88] begins with the x component of the general momentum equation (2.25)

$$\rho \left[\frac{\partial u}{\partial t} + (\mathbf{v} \cdot \nabla) u \right] = -\frac{\partial p}{\partial x} + \frac{\partial}{\partial x} \left[\mu \left(\frac{4}{3} \frac{\partial u}{\partial x} - \frac{2}{3} \frac{\partial v}{\partial y} \right) \right] + \frac{\partial}{\partial y} \left[\mu \left(\frac{\partial u}{\partial y} + \frac{\partial v}{\partial x} \right) \right],$$

(7.81)

which becomes, to second order,

$$\frac{1}{2} \text{Re} \left[-i\omega \rho_1 \widetilde{u_1} + \rho_m \widetilde{u_1} \frac{\partial u_1}{\partial x} + \rho_m \widetilde{v_1} \frac{\partial u_1}{\partial y} \right]$$

$$= -\frac{dp_{2,0}}{dx} + \mu_m \frac{\partial^2 u_{2,0}}{\partial y^2} + \frac{1}{2} \text{Re} \left[\frac{\partial}{\partial y} \left(\mu_1 \frac{\partial \widetilde{u_1}}{\partial y} \right) \right]$$

(7.82)

when time averaged. This is far more complicated than the first-order x component of the momentum equation, Eq. (4.52). In particular, to obtain correct answers at second order, viscosity must be kept inside the derivative on the right side of Eqs. (7.81) and (7.82). This aspect of the present problem differs from almost all other problems in acoustics and fluid dynamics.

The fact that Rayleigh streaming in a pulse tube can be suppressed so simply and conveniently is the result of a remarkable series of fortunate coincidences. First, there is no *a priori* guarantee that tapering the tube would have a large enough effect on streaming to cancel streaming's other causes. Second, it might have turned out that rather large taper angles (say, greater than $10°$ or $20°$) were required; in this case, jet-like flow separation of the high-Reynolds-number first-order velocity from the tube wall would have invalidated the entire laminar, boundary-layer approach. Third, it is fortunate that most pulse tubes operate in the "weakly turbulent" regime of oscillating flow (see Fig. 7.4), and with tube surface roughness much smaller than the boundary-layer thickness, so that laminar analysis is adequate in the boundary layer.

Fourth, the perturbation expansion upon which this calculation is based is only valid for zero or extremely weak mass streaming—the very situation of most interest. This point is subtle. Strong streaming (but nevertheless with the streaming velocity small compared to the oscillating velocity) distorts the axial temperature profile of the pulse tube significantly, contradicting the fundamental assumption that the time-averaged temperature, density, etc. in the boundary layer are well approximated by their zero-oscillation values $T_m(x)$, $\rho_m(x)$, etc. This fundamental assumption requires that the streaming be so weak that the temperature profile in the pulse tube is unperturbed by the streaming—or, equivalently, that the streaming is so weak that it carries negligible heat! Hence, the calculation self-consistently predicts the conditions of zero streaming, but it cannot be relied on to accurately predict the magnitude of strong streaming.

Fifth, there are numerous other fourth-order enthalpy-flux terms in addition to $\dot{m}_{2,0}c_pT_{2,0}$ that would in principle have to be considered to obtain a formally correct fourth-order result. Fortunately, the only other large term, $\rho_{2,0}c_p\mathrm{Re}[T_1\widetilde{u_1}]$, is zero at the same taper angle that makes $\dot{m}_{2,0}c_pT_{2,0}$ zero, while the remaining terms, such as those involving products of first- and third-order quantities, are small for all angles. Hence, the suppression of $\dot{m}_{2,0}$ is sufficient to suppress all fourth-order enthalpy transport.

Finally, as a practical matter, it is very convenient that the streaming-suppression taper is independent of oscillation amplitude and is only weakly dependent on temperature gradient.

In the future, perhaps someone will use a tapered pulse tube in a low-frequency "GM-type" or "valved" pulse-tube refrigerator, in which high harmonic content must contribute additional terms to $\overline{m}$ near the wall. Although Eqs. (7.79)–(7.80) would no longer be valid, it seems likely that some taper (which could be discovered experimentally) would suppress Rayleigh streaming outside the boundary layer.

Example: Orifice Pulse-Tube Refrigerator
When the pulse-tube refrigerator of Figs. 1.17 and 1.18 was being designed, Eq. (7.80) showed that a taper angle of $1.3°$ (with the cold end larger than the ambient end) would suppress Rayleigh streaming in the pulse tube at the design operating point, so the pulse tube was fabricated with that taper and care was taken to keep internal surface roughness much smaller than δ_κ and δ_ν. When the refrigerator was in operation [89], measurements (described in Chap. 9; see Fig. 9.15) showed that $\dot{H}_2/\dot{E}_2 = 0.96$ in the pulse tube at the design operating point, so the "heat leak" down the pulse tube was only 4% of the acoustic power flowing up the pulse tube into the RLC acoustic impedance network. This value was extremely close to the design calculation, which included only the metallic thermal conductance of the tube wall ($\gtrsim 2\%$) and the second-order boundary-layer hydrodynamic entropy flux

$$\frac{1}{2}\rho_mT_m\int \mathrm{Re}\,[s_1\widetilde{u_1}]\,dA \tag{7.83}$$

described in Chap. 5 ($\lesssim 1\%$). Adjusting the acoustic-network orifice valves R_s and R_p to change the phase of the complex acoustic impedance Z in the pulse tube just $5°$ from the design point, in either direction, caused the measured values of $\dot{H}_2/\dot{E}_2$ to drop below 0.9, as Rayleigh streaming convected enthalpy along the pulse tube.

7.4.3 Jet-Driven Streaming

Ideally, the gas in a pulse tube or thermal buffer tube would do nothing more than oscillate in plug flow, transmitting $\dot{E}_2 \equiv \dot{H}_2$ from one end to the other. In reality, several nonidealities can cause $\dot{E}_2 \neq \dot{H}_2$, "carrying heat" down the temperature gradient: ordinary conduction along x (discussed in Chap. 5), ordinary thermoacoustic boundary-layer phenomena carrying entropy along x near the wall, bucket-brigade fashion (discussed in Chap. 5), Gedeon streaming (discussed in this chapter), Rayleigh streaming (discussed in this chapter), gravity-driven convection, and jet-driven streaming (discussed here).[7] Simultaneous suppression of all these heat-transport mechanisms is not trivial.

If either end of the tube has a small-diameter entrance, a jet may blow into the tube when gas enters [115], driving time-averaged convection within the tube, as illustrated in Figs. 7.12c and 7.16. A rough estimate indicates how easily jet-driven mass streaming (or any other time-averaged flow internal to a pulse tube or thermal

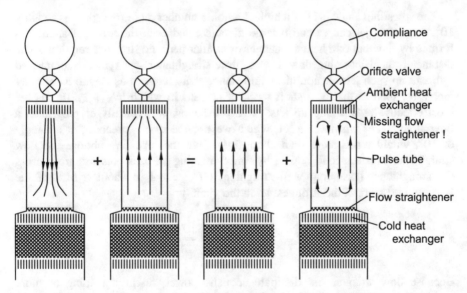

Fig. 7.16 Illustration of jet-driven streaming in a pulse tube, caused by a missing flow straightener at the top of the pulse tube. Downflow shown in the *first frame* is concentrated in the center of the tube, while upflow shown in the *second frame* is broadly distributed. These two flows can be regarded as the superposition of a broadly distributed oscillating flow and a time-averaged toroidal circulation, as shown in the *third and fourth frames*

---

[7]Jet-driven time-averaged mass flux often turns out to be linear (not quadratic) in the acoustic amplitude. The perturbation expansion Eq. (7.65) must be used with care, if at all, in this circumstance. Such first-order steady motion is often called "nonlinear streaming" (although some people reserve the term "streaming" for second-order motions only).

buffer tube) can ruin $\dot{E}_2 \equiv \dot{H}_2$. In a tube with a reasonably large diameter, such as $100\delta_\kappa$, there is little radial thermal contact in the gas, so any time-averaged flow $\dot{m}_{2,0}$ from end to end carries a full ΔT load of heat the length of the tube. For a 98% effective pulse tube, this heat must be no more than 2% of $\dot{E}_2$:

$$\dot{m}_{2,0} c_p \Delta T \sim (0.02) \frac{1}{2} |p_1| |u_1| . \tag{7.84}$$

Using $\rho c_p = \gamma p/(\gamma - 1)T$ and solving for $\dot{m}_{2,0}/\rho_m$ yields

$$\dot{m}_{2,0}/\rho_m \sim 10^{-2} \frac{\gamma - 1}{\gamma} \frac{T}{\Delta T} \frac{|p_1|}{p_m} |u_1| . \tag{7.85}$$

Hence, in a typical case, the maximum allowable steady velocities are roughly $\dot{m}_{2,0}/\rho_m \sim 1$ cm/s in the presence of $|u_1| \sim 10$ m/s. In terms of displacements, this means that gas displacements in the $+x$ and $-x$ directions must match to within 10^{-3}.

The spreading angle [91] of a high-Reynolds-number jet in free space is roughly $10°$, so such a jet can extend a large distance, and even the array of small jets formed by flow out of a heat exchanger or similar periodic structure requires some distance to heal into plug flow. Hence, flow straighteners are typically employed to break up such jets. A nonlinear impedance that increases as Reynolds number increases, such as that of a short stack of screens in which $|N_{R,1}| \gtrsim 10^3$, is the most effective at breaking up jets. At Los Alamos, we typically employ such a flow straightener whenever a jet, which we assume to be spreading at an angle of $10°$, would reach more than about $2|\xi_1|$ into the tube. We choose the flow straightener's resistance $R_{straightener}$ by simply making the peak pressure drop across the straightener (assuming well-straightened U_1) equal to about a fifth of the Bernoulli pressure of the strongest jet in the vicinity:

$$|U_1| R_{straightener} \sim \frac{1}{10} \rho \left[\frac{|U_{1,jet}|}{A_{jet}} \right]^2 . \tag{7.86}$$

Because flow straighteners dissipate acoustic power, research leading to more quantitative design methods for flow straighteners will help improve the efficiency of thermoacoustic engines and refrigerators. This seems particulary important for orifice pulse-tube refrigerators, to prevent jet-driven streaming in the pulse tube from consuming cooling power.

The "imaginary piston" of gas in the tube should execute plug flow, but it is a very fragile object. Only a few subtle effects work to maintain its integrity. First, gravity and the mean density gradient tend to keep this fragile imaginary piston stratified. With a typical mean-density difference of only a few kg/m^3 from top to bottom, only $\Delta p \sim \Delta \rho g \Delta x \sim 1$ Pa is available, in the presence of oscillations typically at least 10^4 times larger. Viscous shear forces caused by gradients in $u_{2,0}$

also tend to reduce $\dot{m}_{2,0}$, but the pressure differences available from this source are only of order $\mu \, \Delta x \, u_{2,0}/\mathcal{R}^2$, which is typically much smaller than 1 Pa.

Good radial thermal contact in the tube can prevent $\dot{m}_{2,0}$ from carrying the full ΔT from end to end, essentially allowing the upflowing and downflowing streams of Rayleigh streaming or jet-driven streaming to experience counterflow heat exchange, so that the tube would enjoy $\dot{E}_2 \sim \dot{H}_2$ even in the presence of a $\dot{m}_{2,0}$ greater than the value estimated above. If $\dot{m}_{2,0}/\rho_m \sim 0.1$ m/s and $\Delta x \sim 0.1$ m, then 1 s is available while the stream traverses the length of the tube. If the oscillation frequency is of order 100 Hz, then 100 acoustic periods are available. During this long time, heat can diffuse roughly $20\delta_\kappa$. Only the pulse tubes of the smallest cryocoolers are this small. However, note that the weak turbulence of most pulse tubes and buffer tubes probably enhances such radial heat transport.

7.4.4 Streaming Within a Regenerator or Stack

Little is known about mass streaming *within* a regenerator or stack, except that it should be avoided for the same reasons that Rayleigh streaming should be avoided. We know it can occur, because we once suffered from harmful toroidal streaming within a regenerator when a narrow jet blew strongly on the heat exchanger at the ambient end of the regenerator, due to an abrupt transition from a small-diameter driver duct to the larger diameter of the ambient heat exchanger and regenerator. Such toroidal time-averaged flows sometimes also arise spontaneously, as has been reported when two or more refrigerator regenerators are operated in parallel [116] and as we have observed with side-wall thermometers on large-aspect-ratio regenerators in orifice pulse-tube refrigerators similar to that shown in Fig. 1.17.

7.4.5 Deliberate Streaming

In common with many other heat engines and refrigerators, most thermoacoustic systems suffer from a practical difficulty: They usually need heat exchangers to transfer heat between the thermodynamic working gas and some other gas (such as air or combustion products), here referred to as the "process gas." Heat exchangers are expensive, and contribute to system inefficiency via temperature differences and viscous effects in the working gas, in the process gas, and often in an intermediate heat-exchange fluid such as water. This difficulty is serious. For example, the historic decline of the Stirling and steam engines and the rise of the Diesel and other internal-combustion engines occurred in large part because the internal-combustion engines need no combustion-temperature heat exchangers (and also reject much of their waste heat in their exhaust instead of through heat exchangers).

For some applications, I hope that some or all of the heat exchangers can be eliminated from thermoacoustic systems eventually, by using the process gas as the thermoacoustic working gas, and superposing the steady flow needed to deliver the process gas with the oscillating flow needed for the thermoacoustic cycle. I hope that this idea will lead ultimately to thermoacoustic equipment for purposes such as drying of compressed air, combustion-powered air conditioning, combustion-powered dehumidification drying (such as for lumber), gas purification, or cryogen liquefaction.

Some of the expected features that thermoacoustic systems with deliberate steady flow might have will be illustrated by one specific case: a standing-wave thermoacoustic air-conditioning system. The ideas illustrated using this example should also be applicable to other standing-wave and traveling-wave thermoacoustic engines and refrigerators.

7.4.5.1 Parallel Flow

As a point of departure, Fig. 7.17 shows the main parts of an air-conditioning system using standing-wave thermoacoustic refrigeration. Four heat exchangers are required: two in the thermoacoustic working gas and one in each of the two air streams. Heat transfer between working-gas heat exchangers and air heat exchangers, indicated by heavy black arrows, is accomplished via heat-transfer

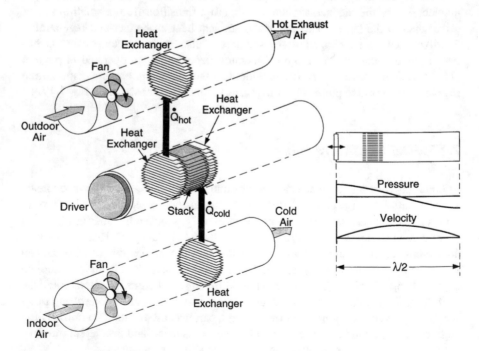

Fig. 7.17 A half-wavelength standing-wave thermoacoustic refrigerator, thermally coupled to two air ducts to form an air-conditioning system

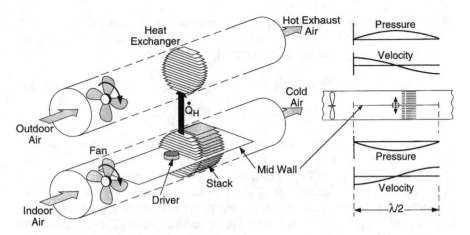

Fig. 7.18 Two side-by-side half-wavelength thermoacoustic refrigerators in an air duct allow direct thermoacoustic cooling of the air. The standing-wave refrigerator of Figs. 1.13 and 1.14 has this topology, with the midwall in this figure equivalent to the large empty space in the center of the torus in Figs. 1.13 and 1.14; it was built to study this type of deliberate steady flow [3, 4, 35]

loops (such as pumped water, heat pipes, thermosyphons, etc.). The four heat exchangers and two heat-transfer loops account for most of the capital cost of the system.

The air-conditioning system in Fig. 7.18 illustrates a simplification that is possible by using the indoor air itself as the thermoacoustic gas. A midwall in the indoor-air duct separates two acoustic resonators, driven 180° out of phase from each other by an oscillating piston in the center of the midwall. The drive frequency is chosen to make the acoustic wavelength equal to twice the midwall length, so there will be pressure nodes at the ends of the midwall and, hence, negligible acoustic power radiated to distant parts of the duct. The position of the stack relative to the nodal pattern of the standing wave is chosen so that conventional standing-wave thermoacoustic processes pump heat from right to left. Superimposed on those thermoacoustic processes, the gas drifts slowly through the stack from ambient to cold, so that it leaves the right end of the apparatus at the cold temperature.

This superposition can give the system of Fig. 7.18 a higher efficiency than that of the system of Fig. 7.17, for two reasons. First, two heat exchangers, and their internal small temperature differences, are eliminated. Second, a much more subtle improvement in efficiency arises because the system of Fig. 7.18 essentially puts the air stream sequentially in thermal contact with a large number of refrigerators in series—a sort of continuum limit of staged refrigeration. To understand this point, first imagine that thermoacoustic refrigerators are ideal, having Carnot's $COP = T_C/(T_0 - T_C)$. Then, in the case of Fig. 7.17, removing heat $\dot{M}c_p(T_0 - T_C)$ at temperature T_C with the refrigerator requires work

$$\dot{W} = \dot{M}c_p(T_0 - T_C)^2/T_C. \tag{7.87}$$

This is more than twice the minimum work required by the first and second laws of thermodynamics for this process, which (as discussed in Chap. 6) is given by the difference between the outgoing and incoming flow availabilities:

$$\dot{W} = \dot{M}\,[(h_C - h_0) - T_0\,(s_C - s_0)] \tag{7.88}$$

$$= \dot{M}c_p[T_C - T_0 + T_0\ln(T_0/T_C)]. \tag{7.89}$$

(The second expression results from using ideal-gas expressions for h and s.) The trouble with the simple, one-stage refrigerator is that it removes *all* the heat load at T_C, where every watt of cooling power requires the same amount of work. It is much more efficient to remove as much of the heat load as possible at higher temperatures T_C', where each watt of cooling power requires less work because the Carnot $COP = T_C'/(T_0 - T_C')$ is higher. The perfect embodiment of this idea would employ an infinite number of ideal refrigerators, each providing an infinitesimal cooling power $\dot{M}c_p\,dT_C'$ at T_C' and rejecting an infinitesimal amount of waste heat at T_0. The flow-through thermoacoustic refrigerator of Fig. 7.18 has some features of this perfect situation, with each length dx of the stack acting as a tiny refrigerator unto itself, lifting heat from T_C' to $T_C' + dT_C'$. The flow-through system removes each watt of heat from the flowing stream at the highest possible temperature, using an "infinite" number of tiny refrigerators, each of which must handle the waste heat of all its downstream neighbors in addition to the local heat load of the flowing stream. Unfortunately, these tiny thermoacoustic refrigerators are less than ideal, suffering from the irreversibilities discussed in Chaps. 5 and 6.

With the addition of nonzero mean velocity along x, the gas moves through the system in a repetitive, "$N + \delta$ steps forward, $N - \delta$ steps back" manner, equivalent to 2δ steps forward each cycle within a peak-to-peak oscillation of N steps, in position, in temperature, and in density and entropy. This violates one assumption on which Chaps. 4 and 5 are based—that $u = \mathrm{Re}[u_1 e^{i\omega t}]$, with no "$u_m$". The derivation of corresponding results in the presence of nonzero parallel steady flow $u_m \neq 0$ begins with the momentum, continuity, and energy equations. The variables are expressed as

$$p = p_m(x) + \mathrm{Re}\left[p_1(x)e^{i\omega t}\right], \tag{7.90}$$

$$u = u_m(x, y, z) + \mathrm{Re}\left[u_1(x, y, z)e^{i\omega t}\right] + u_{2,0}(x, y, z), \tag{7.91}$$

$$v = \mathrm{Re}\left[v_1(x, y, z)e^{i\omega t}\right], \tag{7.92}$$

$$T = T_m(x) + T_m'(x, y, z) + \mathrm{Re}\left[T_1(x, y, z)e^{i\omega t}\right], \tag{7.93}$$

$$\rho = \rho_m(x) + \rho_m'(x, y, z) + \mathrm{Re}\left[\rho_1(x, y, z)e^{i\omega t}\right], \tag{7.94}$$

$$\mu, k,\ \text{etc.} = \text{similar to } \rho \text{ (but see below).} \tag{7.95}$$

As usual, p_1/p_m and v_1/a are assumed to be small, and terms are kept to first order in smallness in the momentum and continuity equations and to second order in smallness in the energy equation as usual. The temperature gradient dT_m/dx is given—of zeroth order. Below, the energy equation will require that u_m/a be as small as a typical dimensionless second-order quantity, but this can be treated as an assumption for now.

Substituting Eqs. (7.90)–(7.95) into the momentum equation, and requiring that the time-independent terms and the terms proportional to $e^{i\omega t}$ must equate separately, yields the same first-order momentum equation as in Chap. 4, and a time-averaged equation:

$$0 = -dp_m/dx + \mu \nabla^2_{y,z} (u_m + u_{2,0}) . \tag{7.96}$$

This time-averaged equation simply expresses the fact that viscosity and steady flow cause a steady pressure gradient. It shows that dp_m/dx is of second order (the same order as u_m), so p_m can be regarded as an x-independent constant in the other equations.

The general equation of heat transfer yields here the same first-order equation for T_1 as in Chap. 4, and a time-averaged equation

$$\rho_m c_p (u_m + u_{2,0}) \, dT_m/dx = k \nabla^2_{y,z} T'_m . \tag{7.97}$$

The only use of Eq. (7.97) here is to note that it implies T'_m is of second order. Hence T'_m, and similarly ρ'_m, μ'_m, k'_m, etc., can be neglected in the other equations, because their contributions are of negligible order.

Next, the continuity equation

$$\partial \rho / \partial t + \nabla \cdot (\rho \mathbf{v}) = 0 \tag{7.98}$$

can be integrated with respect to y and z to obtain to lowest order

$$i\omega A \langle \rho_1 \rangle + \frac{d(\rho_m U_1)}{dx} = 0, \tag{7.99}$$

and time averaged to obtain

$$\frac{d\dot{M}_2}{dx} = 0, \tag{7.100}$$

where

$$\dot{M}_2 = \frac{1}{2} \mathrm{Re} \left[\int \rho_1 \tilde{u}_1 \, dA \right] + \rho_m \int (u_m + u_{2,0}) \, dA. \tag{7.101}$$

In Chap. 5, the important terms in the time-averaged, laterally integrated energy equation were

$$0 = \frac{d}{dx} \left[A \langle \overline{\rho u h} \rangle - A \langle \overline{k \, \partial T / \partial x} \rangle - A_{\text{solid}} \left\langle \overline{k_{\text{solid}} \, \partial T_{\text{solid}} / \partial x} \right\rangle_{\text{solid}} \right]. \tag{7.102}$$

In principle, each of these terms must be considered to second order here, but in practice, for situations of interest, zeroth order suffices for the last two terms, as it did in Chap. 5. The first term's expansion to second order is

$$
A \left\langle \rho_m u_m h_m + \frac{1}{2} \text{Re}\,[\rho_1 \widetilde{u_1}]\,h_m + \rho_m u_{2,0} h_m + \rho_m \frac{1}{2} \text{Re}\,[\widetilde{u_1} h_1] \right\rangle , \tag{7.103}
$$

where six terms have already been set equal to zero because the time average of sinusoidally oscillating quantities (e.g., $\overline{\rho_1}$ or $\overline{u_1}$) is zero. Hence (7.103) reduces to

$$
\dot{M}_2 h_m + \frac{1}{2} \rho_m \int \text{Re}\,[h_1 \widetilde{u_1}]\, dA . \tag{7.104}
$$

Using $dh = c_p\, dT$, and combining this with the other significant terms from Eq. (7.102), yields

$$
\dot{H}_2 = \dot{M}_2 c_p (T_m - T_0) + \frac{1}{2} \rho_m c_p \int \text{Re}\,[T_1 \widetilde{u_1}]\, dA - (Ak + A_{\text{solid}} k_{\text{solid}})\, \frac{dT_m}{dx} , \tag{7.105}
$$

which is the sum of Chap. 5's Eq. (5.25) and the steady-flow term given by $\dot{M}_2 c_p (T_m - T_0)$.

The form of this equation bounds the magnitude of the applied steady mass flow $\rho_m U_m$ (included in $\dot{M}_2$) that is allowed by the approximations made here. This mass flow must be independent of x within a stack, regenerator, etc., but the first term in Eq. (7.105) can vary with x via the x dependence of T_m. In laterally insulated stacks, regenerators, etc., $\dot{H}_2$ is independent of x. Hence, $\rho_m U_m$ must be small enough that any x dependence in the first term can be cancelled by an equal and opposite x dependence in either or both of the other terms. In useful engines and refrigerators, the $(Ak + A_{\text{solid}} k_{\text{solid}})\, dT_m/dx$ term is an undesirable nuisance, usually small, so generally the new term, $\rho_m c_p (T_m - T_0) U_m$, must be of the same size as the thermoacoustic term, $\frac{1}{2} \rho_m c_p \int \text{Re}\,[T_1 \widetilde{u_1}]\, dA$, that was obtained in Chap. 5. Hence, it is usually convenient to think of U_m as essentially of second order, like $U_{2,0}$ in the discussion of Gedeon streaming above. It may be useful to maintain some notational distinction, however: The $U_{2,0}$ discussed earlier results from the first-order acoustics, so the physics *causes* it to be of second order. In contrast, U_m discussed here is externally forced and could be of any magnitude in real hardware, although these approximations are valid only if it is no larger than a typical second-order term.

With $\dot{H}_2$ and $\dot{M}_2$ independent of x in a stack, regenerator, or other thermally insulated channel, Eq. (7.105) enables prediction of the x dependence of T_m during numerical integration along x when steady flow is superimposed on thermoacoustic oscillations. [Note that this analysis can also be applied to the Gedeon flow illustrated in Fig. 7.12a.] Generally, the steady flow causes $T_m(x)$ to bend significantly, as steady flow from hot to cold pushes the center of the stack or regenerator to warmer temperatures, or steady flow from cold to hot pushes the center of the stack or regenerator to colder temperatures [3, 4, 117]. Hence, in a stack or regenerator experiencing steady flow parallel to x, superimposed on simpler, familiar thermoacoustics, the "point of view for computations" introduced near the end of Chap. 5 must be modified: The evolution of p_1, U_1, and T_m with x can be obtained numerically from equations of the form

$$dp_1 = \mathcal{F}_{\mathrm{mom}}(T_m, U_1; \omega, \text{gas properties, geometry})\, dx \qquad (7.106)$$

$$dU_1 = \mathcal{F}_{\mathrm{cont}}(T_m, p_1, U_1; \omega, \text{gas properties, geometry})\, dx \qquad (7.107)$$

$$dT_m = \mathcal{F}_{\mathrm{pow}}(T_m, p_1, U_1, \dot{M}_2; \dot{H}_2, \omega, \text{gas properties, geometry})\, dx. \qquad (7.108)$$

The first two equations are not changed by the presence of mean flow. The third equation is changed by the addition of the new term in Eq. (7.105).

Example: Standing-Wave Refrigerator
The standing-wave refrigerator introduced in Figs. 1.13 and 1.14 was intended for quantitative testing [3, 35] of some of these issues of parallel superimposed steady and oscillating flow. Ten temperature sensors were arrayed along the centerline of the central plate of each stack to detect $T_m(x)$. The bending of T_m in response to U_m discussed above was apparent (see Fig. 9.5), and was in reasonable agreement with Eq. (7.105).

An electric resistance heater was included at the cold end of each stack. In one set of experiments, both the applied flow U_m and the heat $\dot{Q}_C$ supplied to these two electric resistance heaters placed significant loads on the refrigerator. In these measurements, U_m was increased and $\dot{Q}_C$ was reduced in such a way as to keep T_C constant. These measurements showed that the total cooling power $\dot{Q}_C + \dot{M}_2 c_p (T_0 - T_C)$ increased significantly as U_m was increased, even while the acoustic power supplied to the refrigerator remained nearly constant. If you imagine that the electric load $\dot{Q}_C$ on the cold ends of the stacks could in principle have been caused by an imaginary second stream of gas, external to the thermoacoustic system, being cooled from T_0 to T_C by crossflow heat exchangers at the cold ends of the two stacks (connected in parallel with respect to the imaginary gas flow), then these data illustrate the efficiency improvement discussed above near Eq. (7.89): As the "total" gas flow was shifted from the imaginary stream to the actual through-flow stream U_m, more total gas flow (through flow plus imaginary flow) was cooled from T_0 to T_C with no increase in consumption of acoustic power.

7.4.5.2 Perpendicular Flow

Thus far, steady flow superimposed *parallel* to the oscillating flow has been discussed. To introduce the idea of steady flow superimposed *perpendicular* to the oscillating flow, the air-conditioner motif will be revisited in Fig. 7.19, which shows an idea for the elimination of *all* heat exchangers from a thermoacoustic

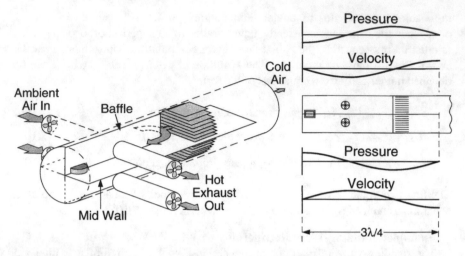

Fig. 7.19 Two side-by-side three-quarter-wavelength thermoacoustic refrigerators, each with a baffle to direct the superimposed steady flow, in an air-duct network. This arrangement allows direct thermoacoustic cooling of some of the steady air flow, rejecting waste heat to the remainder of the steady air flow, with no heat exchangers except the stack itself

system. A midwall separates the duct into two regions that are driven 180° out of phase from each other, at a resonance frequency chosen so that 3/4 of the acoustic wavelength equals the midwall length, thereby putting a pressure antinode at the hard duct closure at the left end of the midwall and a pressure node at the right, open end of the midwall. Deflector walls further divide each of these two resonators in order to direct the steady flow of the air. This steady flow is introduced into the resonator at the pressure node 1/4 of a wavelength from the hard closure, so that negligible acoustic power is radiated into the air inlet duct. A significant fraction of the steady flow passes through the stack, is thereby cooled and dehumidified, and leaves the right end of the duct. Most of the steady flow passes vigorously past the hot end of the stack, moving perpendicular to the oscillating flow, in order to remove the waste heat from the thermoacoustic system and exhaust it. This portion of the steady flow is removed from the resonator at another pressure node. The quantitative understanding of oscillating thermodynamics in the presence of such a perpendicular steady flow is a significant and exciting challenge.

The air conditioner described above will, incidentally, dehumidify the air, returning cold dry air. But for many important drying applications, it is desired to dehumidify air and return it *warm* and dry. Hence, configurations comprising coolers and heat pumps in the same resonator could also be considered, to put the waste heat from the hot end of the cooling stack back into the air stream. This is but one of many possible configurations for open thermoacoustic systems in which multiple stacks, serving multiple functions, share a resonator. As another example, industry has long dreamed of a combustion-powered heat-pump hot-water heater that would have a (first-law) efficiency greater than 100%: It would use part of the heat of combustion to drive an engine, which would drive a heat pump to draw heat from ambient air and deliver it to the hot water. Such devices can (and have) been built with existing technology, but only at a cost that is unattractive to today's consumer. Such a system might use open flows through a thermoacoustic system, with a pulse combustor [118], thermoacoustic engine, and thermoacoustic heat pump in an air stream, all delivering heat to hot water passing through a heat exchanger.

7.5 Harmonics and Shocks

Sometimes the most obvious evidence of behavior beyond Rott's thermoacoustics is seen on an oscilloscope display of $p(t)$ from one of the pressure transducers, which can be distorted far from the low-amplitude sine wave $\mathrm{Re}[p_1 e^{i\omega t}]$ by harmonics $\mathrm{Re}[p_{2,2} e^{i2\omega t}]$, $\mathrm{Re}[p_{3,3} e^{i3\omega t}]$, etc., as illustrated in Fig. 7.20. This is most dramatic for low-frequency, "valved" (also called "GM-type") pulse-tube refrigerators, in which pressure waveforms may more closely resemble square waves than sine waves.

Surprisingly, even in such extreme cases, harmonic analysis may yield reasonable accuracy for the quantities of most interest—the time-averaged powers—because, in the presence of two or more superimposed waves, the net power (whether acoustic power, total power, or exergy flux) is often just the sum of terms of second, fourth, sixth,... orders, without odd-order terms, since the time averages of products like $\sin \omega t \sin 2\omega t$ are zero. Hence, harmonic content usually contributes fourth-order corrections to the second-order powers of Chaps. 5 and 6. An important

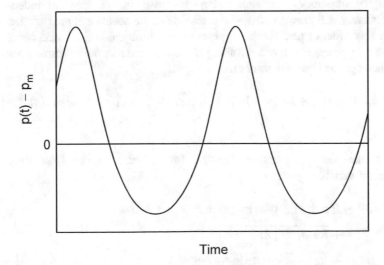

Fig. 7.20 In thermoacoustic engines and refrigerators operating at high amplitudes, harmonic content such as this may be seen when viewing $p(t)$ on an oscilloscope

exception to this rule arises from nonlinear flow resistances such as tortuous porous regenerators or minor-loss sites, which cause a second-order pressure drop with ωt time dependence. Nevertheless, harmonics that are clearly visible in an oscilloscope signal often have only a small effect on the power of a thermoacoustic engine or refrigerator, with the 2ω pressure oscillations carrying power only proportional to $|p_1|^4$.

In standing-wave thermoacoustic engines and refrigerators with acoustic or mechanical resonators, close examination often shows [98] that $|p_{2,2}| \propto |p_1|^2$, as illustrated in Fig. 7.21, suggesting that harmonic content often arises from nonlinear phenomena, usually in the hydrodynamics or thermodynamics. The fundamental acoustic wave, as represented by expressions such as $\mathrm{Re}[u_1 e^{i\omega t}]$, interacts with itself, as represented in terms such as $u\,du/dx \sim \mathrm{Re}[u_1 e^{i\omega t}]\,d\mathrm{Re}[u_1 e^{i\omega t}]/dx$ and $\rho\,du/dt \sim \mathrm{Re}[\rho_1 e^{i\omega t}]\,d\mathrm{Re}[u_1 e^{i\omega t}]/dt$ in the momentum, continuity, and temperature equations. These interactions cause oscillations at 2ω, via trigonometric identities such as $\cos^2 \omega t = \frac{1}{2}(1 + \cos 2\omega t)$, with amplitudes proportional to the square of the first-order amplitude.

In most Los Alamos thermoacoustic engines and refrigerators to date, we believe that such harmonics have not significantly affected performance, probably for a combination of two reasons. First, the powers associated with them are often no larger than fourth order; in particular, in our standing-wave systems, the third-order losses due to adiabatic–isothermal joining effects [see Eq. (7.62)] have been negligible and we have avoided significant third-order minor losses [see Eq. (7.35)]. Second, harmonics are largest when they are enhanced by resonances [98], but other design considerations (see Chap. 8) usually ensure that the higher resonant modes of a resonator occur at frequencies other than integer multiples of ω.

Nevertheless, as thermoacoustic engines and refrigerators are designed at higher and higher amplitudes, harmonic content will rise and the desire to understand it quantitatively will grow. A *complete* extension of the acoustic approximation to the next higher order of perturbation appears to be a formidable challenge, because the next step in power is 4th order. Taking $\overline{pu}$ as an example, the dynamical variables must be expanded through third order, e.g.,

$$p = p_m + p_{2,0}(x) + \mathrm{Re}\left[p_1(x)e^{i\omega t} + p_{2,2}(x)e^{2i\omega t} + p_{3,1}(x)e^{i\omega t} + p_{3,3}(x)e^{3i\omega t}\right],$$

$$(7.109)$$

because the power expressions through fourth order require some third-order dynamical variables:

$$\overline{pu} = (p_m + p_{2,0})(u_m + u_{2,0}) + p_m u_{4,0} + p_{4,0}u_m$$

$$+ \overline{\mathrm{Re}\left[p_1 e^{i\omega t}\right] \mathrm{Re}\left[u_1 e^{i\omega t}\right]}$$

$$+ \overline{\mathrm{Re}\left[p_{2,2} e^{2i\omega t}\right] \mathrm{Re}\left[u_{2,2} e^{2i\omega t}\right]}$$

$$+ \overline{\mathrm{Re}\left[p_{3,1} e^{i\omega t}\right] \mathrm{Re}\left[u_1 e^{i\omega t}\right]} + \overline{\mathrm{Re}\left[p_1 e^{i\omega t}\right] \mathrm{Re}\left[u_{3,1} e^{i\omega t}\right]}. \qquad (7.110)$$

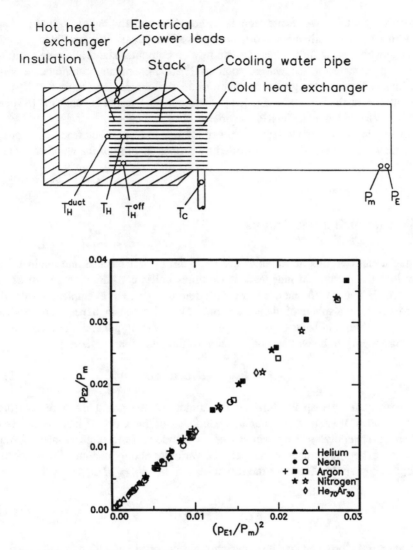

Fig. 7.21 In a standing-wave engine with a right-circular-cylinder resonator and driving no load, illustrated in the *upper part* of the figure, $p(t)$ showed strong harmonic content [98]. The *lower part* of the figure shows $|p_{2,2}|/p_m$ vs $|p_1|^2/p_m^2$ at the rightmost sensor "P_E"

The derivation of expressions for the "2,2" variables, such as $p_{2,2}$—second order in the perturbation expansion, with oscillations at frequency 2ω—is difficult enough, but the challenge of the "3,1" variables—third order in the perturbation expansion, with oscillations at ω—is even greater. These "3,1" variables arise from nonlinear interactions between the ordinary acoustic (subscript "1") variables and the second-order variables, with subscripts "2,0" and "2,2". [Fortunately, "3,3" variables are not needed in Eq. (7.110).] Rott described the effort involved to reach a valid fourth-order total-power expression as "a hopeless undertaking" [13], but perhaps modern symbolic mathematics software will someday make this possible.

Meanwhile, accurate calculations can rely on computer codes allowing arbitrary time dependences of the waves, such as [70, 71], or can accommodate large nonlinear effects by rough approximations in the simpler harmonic "$e^{i\omega t}$" time dependence, such as in DELTAEC [61] where one such effect is "built in" [63] and others can be added using its RPN segments.

At very high amplitudes, harmonics can interact to form shock fronts in standing acoustic waves, which can be controlled by proper shaping of the resonators [119, 120].

7.6 Dimensionless Groups

Dimensionless groups of variables are used throughout science and engineering, to reduce the number of independent variables under consideration and to assess similarity between different experimental circumstances. For example, consider the drag force F on a sphere of diameter D moving with a constant speed u through an incompressible fluid having viscosity μ and density ρ. In general, the force on the moving sphere can be written as a function of the other four variables:

$$F = F_{\text{general}}(D, u, \mu, \rho). \qquad (7.111)$$

To experimentally map the form of this function, one might measure the force for ten values of each of the four independent variables, or 10^4 measurements in all. However, applying the principles of similitude to form dimensionless groups shows that the true number of independent variables is only one, the dimensionless independent variable known as the Reynolds number $N_R = uD\rho/\mu$:

$$\frac{F}{\rho u^2 D^2} = \mathcal{F}_{\text{dimless}}(N_R). \qquad (7.112)$$

The functional form can now be experimentally mapped out with only ten measurements, and making these measurements on a marble moving through water would enable an accurate prediction of the drag force on an iridium sphere moving through molten plutonium. These principles are routinely used in scale-model testing of designs for airplanes and ships.

A formal approach to similitude and dimensionless groups, e.g., using the Buckingham Pi theorem [121], has been applied to Stirling machines by Organ [14, 16] and to thermoacoustic engines and refrigerators by Olson and Swift [122]. For complete thermoacoustic devices, it may be convenient to use dimensionless groups from among these:

$$\frac{T(x,y,z,t)}{T_0}, \quad \frac{p(x,y,z,t)}{p_m}, \quad \frac{\mathbf{v}(x,y,z,t)}{a_0}, \quad \frac{U(x,t)}{a_0 A_0}, \quad \frac{\dot{m}}{p_m/a_0},$$

$$\frac{\dot{Q}}{p_m a_0 A_0}, \quad \frac{\dot{W}}{p_m a_0 A_0}, \quad \frac{\dot{H}}{p_m a_0 A_0}, \quad \frac{\dot{E}_2}{p_m a_0 A_0},$$

$$\frac{x_j}{r_{h,0}}, \quad \frac{x_k}{\Delta x}, \quad \frac{f\,\Delta x}{a_0}, \quad \frac{\delta_{\kappa,0}}{r_{h,0}},$$

$$\gamma, \; \sigma_0, \; b_\mu, \; b_k, \; k_{\text{solid}}/k_0. \tag{7.113}$$

Here, the subscript "0" signifies ambient temperature, with A_0 and $r_{h,0}$ the cross-sectional area and hydraulic radius at the ambient end of the stack or regenerator. Other geometrical dimensions in the system are symbolized by x_j and x_k. Ideal-gas laws determine the gas properties at $T \neq T_0$, with the gas transport properties given by $\mu(T) = \mu_0(T/T_0)^{b_\mu}$ and similarly for $k(T)$.

The choice of variables, especially which variables are treated as independent and which are treated as dependent (see Chap. 9), depends largely on the point of view brought to a particular thermoacoustic system. For example, one possible set of dimensionless groups appropriate for a driven thermoacoustic refrigerator, such as the standing-wave refrigerator of Fig. 1.13 or the pulse-tube refrigerator of Fig. 1.17, is

$$\begin{pmatrix} T(x,y,z,t)/T_0, \\ p(x,y,z,t)/p_m, \\ \mathbf{v}(x,y,z,t)/a_0, \\ \dot{Q}_C/p_m a_0 A_0 \end{pmatrix} = \mathcal{F} \begin{pmatrix} A_0/r_{h,0}^2, \; x_j/r_{h,0}, \\ \gamma, \; \sigma_0, \; b_\mu, \; b_k, \\ k_{\text{solid}}/k_0, \\ T_C/T_0, \; \delta_{\kappa,0}/r_{h,0}, \; fr_{h,0}/a_0, \; p_0(t)/p_m \end{pmatrix}. \tag{7.114}$$

Equation (7.114) reflects this point of view: The ambient temperature T_0 is given; a particular gas and mean pressure p_m are chosen, and hence the gas properties at T_0 (subscript "0") are given; all hardware dimensions are given; the refrigerator is driven at a given frequency f in such a way that the pressure at the ambient end of the stack is $p_0(t)$; and its cold temperature T_C is given. The desired results are the temperature distribution throughout the apparatus, the acoustic variables p and $\mathbf{v}$ everywhere, and the cooling power $\dot{Q}_C$. This point of view is appropriate for an experimentalist seeking to understand existing hardware, while a different point of view might be appropriate during the design of a new apparatus.

This might look like no more than a lot of arbitrary, weird bookkeeping, but in fact it brings meaningful insight, organization, and even scale-model testing to thermoacoustics beyond Rott's approximations.

7.6.1 Insight

Analyzing data in terms of dimensionless groups sometimes gives insight into poorly understood phenomena. For example, consider again the harmonic data plotted in Fig. 7.21, where the amplitude $|p_{2,2}|$ of the first harmonic is plotted against $|p_1|^2$ for many different gases used in a single apparatus. When we first saw these data in Los Alamos, we were surprised that the slope on this plot was independent of the choice of gas used in the apparatus. Changing from a monatomic gas to a diatomic gas made no difference, so apparently the phenomena responsible for the harmonic generation did not depend on γ. This suggested that the compressibility of the gas was unimportant. Changing to a gas mixture, with a lower value of $\sigma = \delta_\nu^2/\delta_\kappa^2$ than that of the pure gases, also made no difference, and changing mean pressure of a given gas (which affects both δ_κ and δ_ν) made no difference; it seemed probable that phenomena within the stack, where δ_κ and δ_ν have profound influence, were unimportant. Thus, we eliminated many phenomena as probable causes of this $p_{2,2}$ harmonic generation. We did not rule out the $u\,du/dx$ term in the momentum equation, which could have generated 2ω phenomena through the interaction of u_1 with its own spatial derivative, so an experiment to change u_1 somewhere, while keeping other things constant, might be a good next step.

As a second example of this type of reasoning, consider the disagreement between measurements and acoustic-approximation calculation shown in Fig. 7.1. One might ask whether the disagreement in the lower graph should be interpreted as evidence of a heat-transfer mechanism acting in parallel with the thermoacoustic processes, thereby adding to the calculated heat consumed at a given pressure amplitude. One might further ask whether that extra heat consumption was due to gravity-driven convection within the stack (x was horizontal in those measurements, so gravity could have caused gas to stream from T_H to T_0 in the upper channels of the stack and from T_0 to T_H in the lower channels), or to one of the thermoacoustically generated streaming mechanisms we discussed earlier in this chapter. Repeating the measurements with the apparatus rotated to make x vertical would be a straightforward approach to this question, but for a large apparatus this might be awkward. An alternative approach is to repeat the measurements with the helium in the engine replaced with argon at a mean pressure of 0.37 times the original helium pressure. At this particular pressure, the thermophysical lengths λ, δ_κ, and δ_ν in the argon are the same as they were in the helium, so the two "systems" are thermophysically similar. One would then plot both sets of data in dimensionless form, dividing heater power by $p_m a_0 A_0$ and dividing the pressure amplitude by p_m. If the argon and helium data coincided on these plots, one would know that the source of the extra heat consumption was something included in the similitude analysis, which includes all phenomena depending on the variables listed in Eq. (7.113), such as thermoacoustically driven streaming and all the other phenomena discussed in detail in this chapter. On the other hand, if the argon and helium data did not coincide on the dimensionless plots, one would know that the source of the extra heat consumption was something not included in the similarity between the helium and argon setups, such as gravity-driven convection [the acceleration of gravity does not appear in Eq. (7.113)], structure-borne power loss, radiation of sound into the room, fluttering of flexible or loose plates in the stack, etc. (In fact, we never learned the source of the disagreement between measurements and acoustic-approximation calculation shown in Fig. 7.1, because the project was canceled. Comparable disagreements have been observed in comparable apparatus [122].)

7.6.2 Empirical Correlation

Large quantities of data can be organized and presented concisely by use of dimensionless groups of variables, so that other people can easily use the results for their own circumstances. For example, the friction-factor diagram shown in Fig. 7.3 concisely summarizes the practical knowledge about the pressure drop when an incompressible fluid flows steadily through a straight, circular pipe, and one can easily use the results summarized in this figure to predict the behavior of this type of flow with any fluid in a pipe of any dimensions, even though one's own fluid and dimensions differ significantly from those of the original experiments on which Fig. 7.3 is based.

Hence, when planning a high-amplitude thermoacoustics experiment, it is sometimes important to think about the anticipated results in dimensionless terms before performing the experiment, to strive for the greatest generality possible. An early choice of which dimensionless variable groups will serve as independent variables can guide the experimental design to cover the greatest possible range of these dimensionless variables. Furthermore, such planning can save time and effort in needless duplication of results—for example, we could have skipped one of the helium data sets, one of the neon data sets, and one of the nitrogen data sets when doing the experiments that led to Fig. 7.21, without diminishing our ability to reach valid conclusions about the possible cause of harmonic generation.

7.6.3 Scale Models

As with scale-model testing of airplanes and ships, understanding of dimensionless groups of variables in thermoacoustics can enable meaningful scale-model testing of thermoacoustic engines and refrigerators. For example, a steel model operating near room temperature with a low p_m could be used to test a design intended to operate at high temperature and high p_m, where refractory metals or ceramics would have to be used. Examination of the dimensionless groups $f \Delta x / a_0$ and $\delta_{\kappa,0} / r_{h,0}$ (and using the ideal-gas equation of state) shows the scaling requirement to be $p_m \propto T_0^{0.5+b_k}$ when all dimensions of the apparatus are kept constant and the same gas is used. Alternatively, a scale model can be created by changing dimensions, gas, and mean pressure while keeping mean temperatures constant. For example, we built a half-scale model of part of Cryenco's first large thermoacoustic natural-gas liquefier [36]. The full-size system used helium, while the half-scale model used argon whose mean pressure was 0.8 times the helium pressure, ensuring that the thermophysical length scales δ_κ, δ_ν, and λ in the argon system were half as large as they were in the helium system.

Care must be exercised in scale-model building, because imperfect similarity is typical. For example, in using the half-scale argon model as a test of full-scale helium behavior, the $k_{\text{solid}} \, dT_m/dx$ term in the $\dot{H}_2$ equation would only scale correctly if we could build the argon model's spiral parallel-plate stack with a solid material having a thermal conductivity 8 times lower than that of the stainless steel in the helium engine's spiral parallel-plate stack, because the ratio of thermal conductivities of helium and argon is 8. We avoided the need to build a ceramic stack with this geometry by realizing that this term was negligible anyway, at the high amplitudes of most interest. The heat leak from the hot end of the engine through its ceramic-fiber insulation to ambient temperature was not negligible, however, so we had to subtract it from the argon measurements, scale the "thermoacoustic" results up to helium size, and add an appropriate heat leak back in. This procedure worked because the insulation heat leak could be calculated with confidence using the insulation-manufacturer's data, and could be measured separately as a check.

7.7 Exercises

7.1 Calculate the molecular mean free path for your favorite gas in your favorite thermoacoustic apparatus, by looking up the appropriate equations in something like an introductory statistical-mechanics textbook such as Reif's *Statistical and Thermal Physics* [123]. Compare to dimensions in the apparatus and to δ_ν and δ_κ. Is the mean free path so negligible compared to δ_ν that the representation in Ani. Viscous should be accurate? Are the crevices between screen wires in your regenerator large compared to the mean free path? [80]

7.2 Calculate the peak Reynolds number and either $\mathcal{R}/\delta_\nu$ or r_h/δ_ν at many locations in your favorite thermoacoustic apparatus. Where do these points appear on Fig. 7.4?

7.3 The steady-state friction factor of a particular tortuous porous medium has a power-law dependence on steady-state Reynolds number: $f = C\,(N_R)^{c'}$ where C and c' are constants. Assume that this is applicable at each instant of time during sinusoidal flow, and take the fundamental Fourier transform of the friction-factor equation to show that

$$\frac{dp_1}{dx} = -I \frac{\mu}{8r_h^2} C \, |N_{R,1}|^{1-c'} \frac{U_1}{\phi A}, \tag{7.115}$$

where

$$I = \frac{2}{\pi} \int_0^\pi \sin^{3-c'}(z) \, dz. \tag{7.116}$$

7.4 In the section on minor losses, $U(t)$ was assumed to be sinusoidal. In some situations, it may be better to assume that $\Delta p(t)$ is sinusoidal. Repeat the derivation of Eq. (7.36) assuming that $\Delta p(t)$ is sinusoidal. Can you reconcile this answer with Eq. (7.36)?

7.5 To help assure yourself that the similitude-variable list given in Eq. (7.113) is complete, express μ_0 in terms of these variables: σ_0, k_0, a_0, p_m, T_0, and γ.

7.6 Express the Mach number $|u_1|/a$ and the Reynolds number in terms of ratios of lengths, using $|\xi_1|$, r_h, λ, and δ_ν.

7.7 Your sponsor wants you to build a thermoacoustic engine to generate electricity on a satellite. The engine will operate from a plutonium heat source at 1800 K, and dump its waste heat to a tiny black-body radiator at 900 K. Helium will be the gas, at a pressure of 20 bar. You want to use similitude to do some preliminary tests in a model that operates between 600 and 300 K—exactly 1/3 the ultimate temperatures—so that you can experiment using stainless steel instead of iridium and platinum. You want the model to have the same dimensions as the ultimate hardware. What mean pressure of helium must you use in the model? How much higher or lower will all powers be in the model? If you wanted to use a different monatomic gas instead of helium in the model, could you keep the mean pressure in the model lower/safer? Would the powers then be lower?

7.8 Figure 7.3 shows that the laminar friction factor is $f_M = 64/N_R$ for steady flow in a circular pipe. Derive this from Eq. (7.2). Begin this process by justifying the neglect of most terms in Eq. (7.2), leaving only $0 = -dp/dx + \mu \nabla^2 u$.

For a serious mathematical challenge, show that combining Eqs. (4.54) and (4.60) gives the same result, in the limit $\omega \to 0$.

7.9 Review Exercise 6.2. Repeat that exercise, assuming that all refrigerators operate at the same fraction η_{II} of Carnot's *COP*. In which direction does the comparison between Case 2 and Case 3 shift? Now extend Case 3 of the original exercise from two refrigerators to an infinite number of refrigerators, showing by comparison to Eq. (7.89) that the work required is the minimum allowed by the laws of thermodynamics. Can you repeat this continuum calculation, but assuming that all the refrigerators operate at the same fraction η_{II} of Carnot's *COP*? Which of these situations is analogous to a thermoacoustic refrigerator with deliberate streaming parallel to the direction of acoustic oscillations?

7.10 A piston and a cylinder contain N moles of a monatomic ideal gas. The piston moves so that the volume $V(t) = V_m + \mathrm{Re}[V_1 e^{i\omega t}]$. (a) Assume that the motion is slow enough that the gas temperature T_m is constant, equal to the temperature of the cylinder, and derive expressions for p_1, $p_{2,0}$, $p_{2,2}$, $p_{3,1}$, and $p_{3,3}$. (b) Assume that the motion is rapid enough that the oscillations are adiabatic, and that the temperature is T_m when the volume is V_m. Derive expressions for p_1, $p_{2,0}$, $p_{2,2}$, $p_{3,1}$, $p_{3,3}$, T_1, $T_{2,0}$, $T_{2,2}$, $T_{3,1}$, and $T_{3,3}$.

7.11 The subsection on Insight claimed that switching from helium at pressure p_m to argon at pressure $0.37p_m$ in a thermoacoustic engine will keep all thermophysical lengths constant. Verify this statement. The subsection on Scale Models claimed that halving all hardware dimensions in a thermoacoustic engine, changing from helium to argon, and setting the argon's mean pressure equal to 0.8 times the helium pressure will also maintain similarity. Verify that this change reduces λ_0, $\delta_{\kappa,0}$, and $\delta_{\nu,0}$ by a factor of 2.

7.12 Discuss appropriate dimensionless groups for your favorite thermoacoustics hardware. Adopt a useful point of view, and decide which groups should be considered independent variables and which should be considered dependent.

7.13 The section on joining conditions described a second-order discontinuity in U_1, obtained from Eqs. (7.56). Show that Eqs. (7.56) also represent a discontinuity in $U_{2,0}$. Show that this discontinuity is consistent with the discussion at the beginning of the subsection about Gedeon streaming, by considering the difference between the ideal isothermal and the ideal adiabatic expressions for $\frac{1}{2}\mathrm{Re}[\widetilde{\rho_1}U_1]$.

7.14 If $\frac{1}{2}\mathrm{Re}[\widetilde{\rho_1}U_1]$ were 3% of $\rho_m|U_1|$ in the stack, regenerator, pulse tube, or thermal buffer tube of your favorite thermoacoustics hardware, and $\rho_m U_{2,0} \equiv 0$, how much unwanted heat load would flow?

7.15 To check your understanding of the complex notation, pick one of the terms in Eq. (7.82) having a factor of $\frac{1}{2}$, and derive it from the corresponding term in Eq. (7.81).

7.16 Show that the line marking the right boundary of the weakly turbulent regime in Fig. 7.4 is described by $|\xi_1|/\delta_\nu \simeq 250$.

7.17 Show that

$$\frac{1}{2}\int \mathrm{Re}\left[\rho_1\widetilde{u_1}\right]dA = \frac{\rho_m}{p_m}\left\{\dot{E}_2 - \frac{\gamma-1}{\gamma}\left[\dot{H}_2 + (Ak + A_{\mathrm{solid}}k_{\mathrm{solid}})\frac{dT_m}{dx}\right]\right\}$$

(7.117)

is a more accurate version of Eq. (7.69).

7.18 Sketch the system described in the last paragraph of Sect. 7.4.5. Identify the standing-wave nodes and antinodes.

7.19 Figure 7.3 shows that f_M is independent of N_R in the "fully rough zone." Equation (7.25) shows that the viscous dissipation of acoustic power is proportional to $|U_1|^3$ in this case. However, thermoacoustic hardware is more likely to contain "smooth pipes," whose friction factor is often approximated by the Blasius correlation [50]

$$f_M \simeq \frac{0.3164}{N_R^{0.25}}.$$

(7.118)

(a) Show that $d\dot{E}_2/dx \propto |U_1|^{2.75}$ in this case. (b) In a smooth duct with a large enough radius $\mathcal{R}$ to allow the boundary-layer approximation, show further that

$$m \simeq 0.05 \left(\frac{|\xi_1|}{\delta_\nu}\right)^{0.75} \left(\frac{\delta_\nu}{\mathcal{R}}\right)^{0.25}, \tag{7.119}$$

so that the viscous dissipation per unit surface area is approximately

$$3\gamma p_m a \frac{\delta_\nu^{0.5} |\xi_1|^{2.75}}{\lambda^3 \mathcal{R}^{0.25}}. \tag{7.120}$$

7.20 Try to derive joining conditions for the interface between an isothermal space and a large, *well-mixed* adiabatic space. When the flow goes toward the adiabatic space, assume that it mixes instantly with all the gas in that space, so that the temperature in that space is always spatially uniform.

7.21 Beyond Rott's acoustic approximation, Eq. (7.8) is the more general expression corresponding to $\dot{H}_2$ and Eq. (7.9) is the more general expression corresponding to $\dot{X}_2$. Why didn't this author write down a more general expression corresponding to $\dot{E}_2$?

Chapter 8
Hardware

The previous chapters considered ideal-gas dynamics and thermodynamics—the hardware confining the gas was discussed only insofar as it defined the boundary conditions for the gas. This chapter discusses the hardware itself, especially its role in providing thermal and mechanical interfaces between the thermoacoustic gas and the rest of the world.

This discussion of hardware issues should serve as a beginners' outline of the practical problems that must be confronted when building thermoacoustic engines and refrigerators. The examples presented here should give beginners to thermoacoustics a starting point, but these probably do not represent the best fabrication methods—please discover better ones, and browse the literature for better ideas. Published proceedings of conferences on Stirling engines, Stirling refrigerators, cryocoolers, etc. are full of examples of more sophisticated hardware for oscillating-gas heat engines and refrigerators. Many of these use designs whose roots are decades old.

In this field, the literature is full of examples of independent re-discovery and publication of ideas that were well understood and published by earlier generations of researchers. The references I have cited here merely give a few entry points into this vast literature—I have probably not found the earliest or "best" citations for most of these topics, especially in the two-century-old Stirling literature.

8.1 Prelude: The Gas Itself

The choice of gas for a thermoacoustic engine or refrigerator involves trade-offs among many issues, including power, efficiency, and convenience. There is no easy answer; the final choice depends on the specific goals of the apparatus. Here are some of the issues:

© Acoustical Society of America 2017
G.W. Swift, *Thermoacoustics*, DOI 10.1007/978-3-319-66933-5_8

"Dimensionless groups" in Chap. 7 showed that thermoacoustic powers generally scale as $p_m a A$. Hence, for a given $|p_1|/p_m$, high mean pressure and high sound speed yield high power per unit volume of hardware. The lightest gases—H_2, ^{3}He, He, Ne—have the highest sound speeds (review Sect. 2.3 if necessary), and hence will give the highest powers. These light gases are particularly necessary for cryocoolers, because heavier gases condense or freeze at low temperatures, or at least exhibit strongly nonideal behavior (see below).

Light gases also have high thermal conductivity, leading to higher δ_κ and hence larger stack and heat-exchanger gaps and easier heat-exchanger fabrication. However, high mean pressure reduces δ_κ, so devices with high power density tend to have narrow heat-exchange gaps, which may be difficult to build with reasonable precision.

Generally, practitioners have preferred helium, and occasionally hydrogen, using the highest mean pressure that reasonable fabrication methods can accommodate. Hydrogen must be handled with extra care, because it is flammable. Diatomic, it has a higher heat capacity per mole than helium, so regenerating it requires better heat transfer in the regenerator than is necessary for helium. Nevertheless, it is appealing for some applications, because it has a very high thermal conductivity and a low third-order adiabatic-isothermal interface dissipation (see the end of "Joining conditions" in Chap. 7).

In principle, and sometimes in practice, the efficiency of a thermoacoustic engine or refrigerator can be improved by adding a small amount of a heavy gas to a light gas, thereby reducing the Prandtl number below the value 2/3 typical of pure gases. For example, 20% (by mole) argon in helium has a Prandtl number of 0.4, and 20% xenon in helium has a Prandtl number of 0.2 [55]. The reduction in Prandtl number lowers viscous dissipation, and has allowed significant improvement in the efficiency of standing-wave refrigerators [124]. However, the added mass reduces the sound speed, so power per unit volume is sacrificed.

This book describes only ideal gases, but sometimes nonideal gases are used. For example, one standing-wave engine [60] used sodium at $p_m = 9.7$ MPa, operating between 500 and 115 °C. Sodium's critical temperature is 2700 K and its critical pressure is 40 MPa. It was far from ideal-gas conditions in this engine, where $T_m \beta$ ranged from 0.22 to 0.10 (cf. the ideal gas, for which $T_m \beta \equiv 1$) and the Prandtl number ranged from 0.005 to 0.010. As another example, several research groups use helium in pulse-tube refrigerators at pressures of 2–3 MPa and temperatures from ambient down to about 2 K [125]. Helium's critical temperature is 5.2 K and its critical pressure is 0.23 MPa. For traveling-wave devices using such fluids, Eqs. (3.34) and (5.25) are no longer true, so it is no longer possible to imagine an ideal regenerator to be one in which $\dot{H}_2 = 0$. The interaction of such non-ideal-gas effects with Gedeon streaming and deliberate parallel-flow streaming should be interesting.

8.2 Stacks and Regenerators

Stacks and regenerators are the heart of standing-wave and traveling-wave systems, respectively. They provide solid heat capacity. In engines and refrigerators of high power, the stacks or regenerators have a large cross-sectional area A. In order to maintain modest (standing-wave) or good (traveling-wave) thermal contact between the working gas and the solid heat capacity across such a large cross-sectional area, stacks and regenerators are finely subdivided into many parallel channels or pores, with hydraulic radius r_h comparable to δ_κ in stacks and smaller than δ_κ in regenerators.

Stacks and especially regenerators should be located where the magnitude of the local acoustic impedance, $|Z|$, is significantly larger than $\rho a / A$, because keeping $|u_1|$ small helps reduce the viscous dissipation of acoustic power. Hence, both stacks and regenerators should be close to the pressure antinode of a standing wave. However, according to Eqs. (5.12)–(5.13) and (5.30), standing-wave systems produce acoustic power or pump enthalpy in rough proportion to $\mathrm{Im}[\widetilde{p}_1 U_1]$, so they are powerless exactly at the pressure antinode of a standing wave. A reasonable compromise between high power (requiring high velocity) and high efficiency (requiring low velocity) typically puts stacks about $\lambda/20$ from the nearest pressure antinode. In contrast, regenerators produce power or pump entropy in rough proportion to $\mathrm{Re}[\widetilde{p}_1 U_1]$, so minimization of viscous losses usually puts regenerators at a location in the wave where p_1 and U_1 are nearly in phase. If the viscous pressure drop across the regenerator could be neglected, this location would be exactly at the pressure antinode of the wave.

It is important to avoid "leaks" in parallel with stacks and regenerators, because these can provide low-impedance paths for streaming similar to the type shown in Fig. 7.12d, essentially within the stack or regenerator. As described in Chap. 7, a nonzero $\Delta p_{2,0}$ is to be expected across the stack or regenerator, and this pressure difference will drive a nonzero $\rho_m U_{2,0}$ through any low-impedance leak in parallel with the stack or regenerator, convecting heat unnecessarily. The most obvious location for such leaks is at the outside diameter of the stack or regenerator, where it should fit snugly into its case. We generally try to keep the gap there no larger than the hydraulic radius of the stack or regenerator itself.

8.2.1 Standing Wave

In standing-wave systems, the stack provides local heat capacity for the thermoacoustic gas, enabling both the bucket-brigade shuttling of entropy along the mean temperature gradient and the nonzero $\oint p \, d\rho$, by providing the temporally isothermal boundary condition $T_1 = 0$ at the solid surface. A good stack provides this heat capacity for the gas while minimizing ordinary conduction of heat along the temperature gradient and minimizing viscous dissipation of acoustic power.

Fig. 8.1 Inexpensive metal honeycomb such as this stainless-steel hexagonal honeycomb is commercially available. This piece was purchased from Kentucky Metals. Other suppliers can be found by searching the Internet for "honeycomb metal"

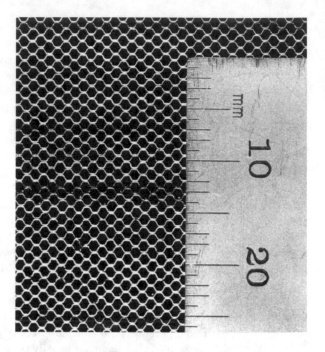

Commercially manufactured metal honeycomb (as shown in Fig. 8.1), ceramic honeycomb, and even paper or plastic honeycomb make inexpensive, easily available stacks. A few units in series can be used to reach a desired total stack length if long honeycomb is not available. There is almost always plenty of heat capacity available, so the material can be chosen with the smallest possible thermal conductance, to minimize ordinary heat leak down the temperature gradient.

However, $\mathrm{Im}[-f_\kappa]$ is larger for parallel-sided channels than for the circular, hexagonal, or square pores of honeycomb, as suggested by Fig. 4.14, so parallel-sided channels should be better. Indeed, numerical studies seem always to confirm that efficiency, power, or other figures of merit are 10–20% higher with parallel-sided channels than with honeycomb. Hence, many stacks are built to resemble the ideal of laterally infinite parallel plates.

The spiral configuration is one way to build a stack with approximately parallel-plate geometry. Stack material is wound into a well-spaced spiral, with the gaps between the layers of the spiral forming the nearly parallel-sided thermoacoustic channels. Figure 1.11 is one example of such a spiral stack. The spacing between layers is maintained by the radial ribs on each end—in between the ends there are no spacers. When the entire stack is at ambient temperature, the spacing between layers seems to be very uniform, but we do not know how thermal stresses deform it when one end is at ambient temperature and the other end is red hot.

To make the stack shown in Fig. 1.11, we wound a 7.6-cm-wide, 50-μm-thick stainless-steel sheet and a 7-cm-wide, 250-μm-thick copper strip together around the 5-mm-diameter core visible at the center of Fig. 1.11. Holding that assembly

Fig. 8.2 A stainless-steel spiral stack, larger than the one shown in Fig. 1.11, with an overall diameter of 15 cm and gaps between layers of 0.4 mm. In this stack, the ribs were welded on "by hand" with conventional arc welding

firmly, we cut 24 radial slits 240 μm wide and 1.6 mm deep into each end, filling them with stainless-steel ribs 200 μm thick and 1.2 mm wide. As is visible in Fig. 1.11, one half of the ribs reached the core. All 24 ribs extended past the circumference at this stage of the fabrication. Next, an electron-beam welder fused each intersection of rib and sheet. Finally, we left the assembly in a nitric-acid bath for many days to dissolve away the copper. This is the most intricate, delicate stack we have ever made—and the most expensive, due to the cost for the electron-beam welding and for the disposal of the copper–nitric-acid solution. We have more recently made similar stacks with sacrificial spacing materials other than copper, and joined the ribs to the spiral with less expensive welding techniques, such as is shown in Fig. 8.2. Low-cost, environmentally benign sacrificial spacing materials may include anything that can be melted away for re-use, dissolved away with water, or burned away cleanly.

Spiral stacks can also be made with long spacers parallel (or nearly parallel) with x. Hofler [73, 126] described making a spiral stack with long spacers parallel with x, by gluing tiny nylon "rods" (monofilament fishing line) onto a plastic strip before rolling up the strip into a spiral. This results in a stack with channels having essentially rectangular cross section in the y–z plane, with an aspect ratio of the order of 10:1 (and with odd corners, depending on how the glue forms fillets between each rod and the strip). Calculations of f_κ and f_ν for rectangular channels [40] incorporated into a full-system calculation [61] show that the ideal parallel-plate geometry yields higher efficiency than does such a 10:1 rectangular channel.

Spiral stacks can also be made with bumps on the stack material maintaining the spacing between layers (see regenerators below). For the wide gaps typical of stacks,

Fig. 8.3 A stack from the
refrigerator of Fig. 1.13. The
15-cm-long stack is in a white
plastic pipe of the same
length. The 0.81-mm gaps
between plates are too small
to be seen in this photo, but
are visible in Fig. 8.5.
Thermocouples penetrate the
pipe along the *left side* and
extend into the stack

the "shadow" of each bump may extend downwind of the bump along x a distance of
the order of $|\xi_1|$ (see "Entrance effects" in Chap. 7), so the parallel-plate calculations
of f_κ and f_ν may not be applicable if the bumps are closely spaced along x.

Another approximation to parallel-plate geometry is obtained by stacking flat
sheets with spacers.[1] The stack [3, 35] shown in Fig. 8.3, which is from the standing-
wave refrigerator of Fig. 1.13, is of this type. The plates are fiberglass, the spacers
are nylon (monofilament fishing line), and the assembly is held together with epoxy.
The resulting channels are essentially rectangular, with gaps $2y_0 = 0.81$ mm and
widths 12.7 mm. This stack was made in two halves, each with semicircular cross
section, to enable careful placement of many tiny thermocouples along a central
channel.

Once, we tried making a stainless-steel stack by stacking flat sheets with spacers,
similar to the fiberglass stack of Figs. 8.3, 8.4 and 8.5. The sheets were stainless
steel, and the spacers were copper-plated stainless-steel wires. Bonding between
sheets and spacers was achieved by heating the assembly in a hydrogen atmosphere
to the melting point of copper. This high temperature caused warping of the stainless
steel sheets, so that the resulting stack, shown in Fig. 8.6, looked irregular to
the naked eye. In an engine, this stack yielded performance roughly a factor of
two worse than predicted by calculations. From this one incident, we tentatively
conclude that if a stack looks bad, it will perform badly.

Other stacks have been used less routinely. The pin-array stack [62, 127]
promises good performance, but has been extremely difficult to build accurately,
especially for small gas thermal penetration depths (i.e., high pressure or high
frequency). Porous stacks made of loose-weave screens [128, 129] and reticulated
vitreous carbon (RVC) [130] work reasonably well, although calculating their
properties in this imperfect-thermal-contact regime appears challenging and the

---

[1]Yes, that's why we called it a "stack" in 1981.

Fig. 8.4 An early practice piece, made before the fabrication of the stack shown in Fig. 8.3. The gaps between plates are maintained by glued-string spacers on 13.5-mm centers. The string was wound between large pegs, forming the curved portions seen in this photo; these curved portions were cut for removal from the pegs, and would subsequently be cut off entirely

Fig. 8.5 Another practice piece, used to test fabrication methods for the stack shown in Fig. 8.3. The US penny gives a sense of scale. The stack can be built up from plates of varying widths to form the outer circular contour, or the circle can be cut out of a rectangular stack formed from plates of equal widths

Fig. 8.6 End view of a
"parallel-plate" stack that
performed poorly. This view
was exposed by a wire
electrical discharge
machining cut, after the
stainless-steel sheets and
stainless-steel spacer wires
had been bonded by furnace
brazing

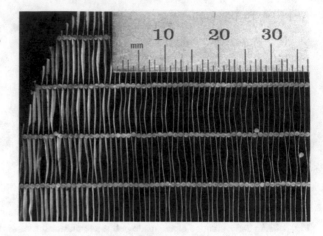

RVC is fragile. In high-pressure gas, RVC and pin arrays may have low enough heat capacity that they do not enforce a perfect $T_1 \equiv 0$ boundary condition on the gas at their surfaces. A sliced-up parallel-plate stack in a heat pump [131, 132] essentially kept the entire stack in the entrance regime to enhance heat transfer (see "Entrance effects" in Chap. 7), because every other slice was rotated 90° and the length of each slice was $\Delta x \sim |\xi_1|/3$. This increased both $\dot{H}_2$ and $\Delta \dot{E}_2$ in this heat pump, and improved the efficiency.

Even the complete elimination of the stack may be possible in some circumstances [133], when the temperature spanned by the engine or refrigerator is of the same order of magnitude as the oscillating temperature amplitude.

8.2.2 Traveling Wave

Most Stirling engines and refrigerators and pulse-tube refrigerators use plain-weave metal screens as regenerators. For small-diameter regenerators, the screens can be individually punched to the desired diameter and slipped into their case, which should fit closely. To ensure that the regenerator has a uniform number of screens per unit length along x in spite of friction between the perimeter of each screen and the case, it may be necessary to carefully press the pile of screens with a predetermined force after each screen is added. For larger diameters, irregularities at the perimeter comprise a smaller fraction of the total regenerator area, so a cruder, faster cutting method can be considered. Always turn each layer approximately 45° relative to its predecessor in the pile, for reasons explained at the end of Sect. 8.4.

The regenerator shown in Fig. 8.7 for the thermoacoustic-Stirling heat engine of Figs. 1.22 and 1.23 was made of 120-mesh screen (120 wires per inch) with a wire diameter of 65 μm, by crudely cutting the screens oversized, stacking them between aluminum plates, squeezing the plate–screen assembly to the desired screen

Fig. 8.7 The regenerator for the thermoacoustic-Stirling heat engine of Figs. 1.22 and 1.23. The screen is so fine that the individual 65-μm-diam wires cannot be seen in this photograph. The *thin case* shown here held the screens and was later slipped into the pressure vessel. *Ribs across the top* provided spacing between the regenerator and an adjacent heat exchanger, and reduced bulging of the screens at the end of the regenerator

density between a lathe's chuck and a "live" (i.e., free to rotate) tailstock support, and cutting the diameter of the screens with a conventional lathe tool. The thin case shown in the figure (not the pressure vessel) had been pre-placed on the tailstock, so it could be slid over the cut screen assembly while the assembly was still held to the desired final length in the lathe. In addition, the aluminum plates already had slots in place so that the end ribs visible in the figure could be slipped in and welded to the thin case while the assembly was still held to the desired final length. Thus, when the assembly was removed from the lathe it should have looked as shown in Fig. 8.7. However, springiness in the screens exerted enough force to cause the end screens and ribs to bow outward a few mm in the centers, so the assembly was annealed under force between flat surfaces to flatten the ends. (Close examination of the central rib in Fig. 8.7 reveals slight, residual bowing.)

It is important to know the volumetric porosity ϕ and hydraulic radius r_h of the regenerator. We use the total weight and volume, and the known density of the wire material, to determine the volumetric porosity, and we check the manufacturer's claimed wire diameter as well as possible with a good machinist's micrometer, so we can calculate an as-built value for the hydraulic radius using Eq. (7.10).

Plain-weave metal screens still provide the most predictable tortuous-regenerator behavior, because measurements throughout the world on Stirling engines, Stirling refrigerators, and pulse-tube refrigerators have produced results consistent with calculations based on the friction-factor and heat-transfer data shown in Figs. 7–8 and 7–9 in Kays and London [64]. However, some groups use less costly porous media, such as metal wool, pressed metal wool [134], sintered metal [16], and plastic screens [135]. Very-low-temperature regenerators ($\lesssim 20$ K) typically use solids with high heat capacity per unit volume, such as lead or rare-earth materials, that are either too fragile or too brittle to fabricate into fine screens. In this situation, randomly packed spheres (sometimes as small as $\sim$0.2 mm diameter) are used [64].

Rolled spiral regenerators, with bumps on the material maintaining the gaps between adjacent layers, have been used successfully in Stirling engines (metal spirals) and in small-temperature-difference Stirling refrigerators (plastic and metal spirals). However, such parallel-plate geometry has thus far not succeeded in cryogenic pulse tube refrigerators [136], although there are rumors that *several* groups have tried. Perhaps parallel-plate regenerators are more susceptible than screen regenerators to streaming within the regenerator, as illustrated in Fig. 7.12d.

Microscopically engineered regenerators have also been described [79], and are available from Mitchell Stirling (Berkeley CA).These are made of a spiral roll of very intricate, photochemically milled stainless-steel sheets, with a repetitive three-dimensional pattern intended to enhance the heat transfer between gas and solid while minimizing the conduction of heat along x through the solid and minimizing the friction factor.

8.3   Heat Exchangers

8.3.1   Common Arrangements

While stacks and regenerators only store heat temporarily, accepting it from the thermoacoustic gas during one half of the thermoacoustic cycle and returning it to the thermoacoustic gas during the other half of the cycle, the heat exchangers in a thermoacoustic engine or refrigerator have the more difficult task of transferring time-averaged heat from the gas to an external sink of heat (such as ambient-temperature water) or to the gas from an external source of heat (such as the combustion products from a burner). For small, low-power systems this is some-times easy, but at Los Alamos we are usually interested in high enough powers that heat-exchanger design is difficult. The fundamental design challenge: provide good thermal contact between two flowing streams, while causing minimal pressure drop in either stream.

Engineers must often exchange heat between two gas or liquid streams flowing unidirectionally with steady volume flow rates, so many introductory and advanced textbooks (e.g., [137]) and handbooks (e.g., [64, 138]) teach the basic principles and give tables and charts of heat-transfer coefficients and friction factors for a wide variety of geometries. The overall approach is to consider the heat passing through three temperature differences in series: first from the flow-weighted-average temperature of one stream to the temperature of the surface between that stream and the solid separating the two streams, second from the temperature of that surface through the solid itself to the temperature of the surface between the solid and the

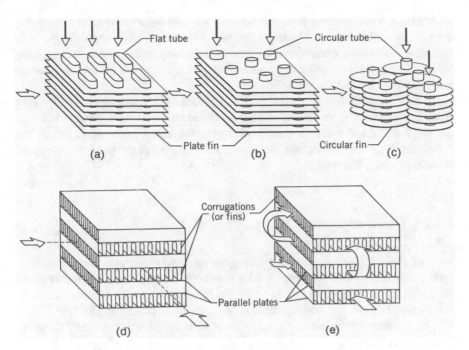

Fig. 8.8 Some examples of crossflow heat-exchanger geometry, reproduced from [137]. *Arrows* indicate flowing streams of gas or liquid. (**a**), (**b**), and (**c**) Finned-tube heat exchangers, with fins attached to individual tubes in (**c**), while (**a**) and (**b**) show shared fins. (**d**) and (**e**) Corrugated-fin heat exchangers, with headers (not shown) creating single-pass flow in (**d**) and multi-pass flow in (**e**)

second stream, and third from that surface temperature to the flow-weighted-average temperature of the second stream. An overall topology of crossflow, as shown in Fig. 8.8, is commonly used in traditional heat-transfer engineering and is well suited to the geometry of thermoacoustic systems.

Unfortunately, the handbook data are for steady flow, while thermoacoustic gas experiences oscillating flow. Often the hydraulic radius is large and the thermoacoustic displacement amplitude $|\xi_1|$ is comparable to the length (along x) of the heat exchanger, so there is little reason for confidence in the quasi-steady hypothesis (see Chap. 7) that each instant of the time-dependent flow depends only on that instant's velocity, not on the time history. Some numerical calculations of such oscillating heat transfer have been made in the context of standing-wave systems [94, 95], but no one has yet done so for the important situation with an open duct ($r_h \gg \delta_\kappa$) on one side of the heat exchanger and a stack ($r_h \gtrsim \delta_\kappa$) on the other side. Some relevant experiments and calculations have been made in the context of traveling-wave devices [139], but these have provided less easily usable information than exists for steady flow. The entire situation is complicated by the fact that entrance effects usually cause very large increases in both friction factor and heat-transfer coefficient in the short passages on the thermoacoustic side of these heat exchangers.

At Los Alamos, without good justification, we have been in the habit of bringing the *rms* Reynolds numbers to the published steady-flow correlations (e.g., [64], including end-effect corrections as appropriate), and using the resulting heat-transfer coefficients to estimate the time-averaged temperature difference between the oscillating gas and the solid surface.

To date (January 2015, Version 6.3b11.12) DELTAEC uses an *extremely* naive algorithm to generate an estimated time-averaged temperature difference between the solid surface and the thermoacoustic gas. We don't know why, but in our experience it usually agrees with the procedure described in the previous paragraph to within about a factor of 2.

8.3.2 Thermoacoustic Choices

Most of the thermoacoustic heat exchangers we have made at Los Alamos have been either finned-tube exchangers with the thermoacoustic gas flowing between the fins, similar to those shown in Fig. 8.8a, b, or shell-and-tube exchangers with the thermoacoustic gas flowing through the tubes, as shown in Fig. 1.19. These will be illustrated with some thermoacoustic-to-water heat exchangers, shown in Figs. 8.9, 8.10, and 1.24.

One of the two identical ambient-temperature heat exchangers for the standing-wave engine of Figs. 1.8 and 1.9 is shown in Fig. 8.9. In this finned-tube exchanger, water flowed around the circumference of the case and through eight small water tubes, clustered into two sets of four in order to block minimal area in the helium. Small copper spacers between each pair of fins and around each set of four tubes maintained the desired gaps between fins during assembly, gave the thermoacoustic gas a streamlined environment, and helped make thermal contact between the copper fins and all the surface area of the tubes, instead of only that fraction of tube surface

Fig. 8.9 Water-cooled ambient-temperature heat exchanger for the ambient end of each stack in the standing-wave engine of Figs. 1.8 and 1.9. The copper fins extend 13 mm along x, into the page. At this stage of the construction of the heat exchanger, the eight water tubes pass through the case and fins, but have not yet been cut to length

Fig. 8.10 Six banks of tubes (almost vertical here) were required in this water-cooled heat exchanger. Fins, almost horizontal, are 0.5-mm-thick copper with 1.3-mm gaps. The diameter of the finned zone is 13 cm

area in contact with the fins themselves. After the photo was taken, the next step in the assembly was to solder tubes to case, fins to spacers, and (we hoped) fins and spacers to tubes. The tube-to-case solder joints had to be leak tight, and the other solder helped fin-to-tube thermal contact.

At larger diameters, more banks of tubes are required, such as in the heat exchanger shown in Fig. 8.10.

The shell-and-tube exchanger used in the thermoacoustic-Stirling heat engine of Fig. 1.22, and identified as the "Main ambient heat exchanger" in Fig. 1.23, is shown in Fig. 1.24. The 299 stainless-steel tubes, each 2.0 cm long and with 2.5-mm inside diameter, were welded on each end to flat stainless-steel tubesheets. The thermoacoustic helium oscillated through these 299 tubes, while cooling water flowed perpendicular to them through the shell around them.

The design of heat exchangers such as those shown in Figs. 8.9, 8.10, and 1.24 is a process of compromising among many worthy goals: high heat-transfer coefficient on the thermoacoustic side, low acoustic power dissipation on the acoustic side, small temperature difference within the metal barrier between thermoacoustic gas and water, high heat-transfer coefficient on the water side, low pressure drop on the water side, low cost, and low risk. There is obviously no universal "best" design for a thermoacoustic heat exchanger, because each experimental or practical situation will impose different priorities among these conflicting goals. As thermoacousticians learn better and better to make these compromises, heat exchangers will be invented that are farther and farther from a naive "physics-lab" acoustics design. For example, the heat exchanger shown in Fig. 8.11, used in an automobile Stirling engine, shows evidence of careful consideration of the combustion-side heat transfer and of thermal expansion stresses.

Fig. 8.11 One of four hot heat exchangers in an automobile Stirling engine built by Mechanical Technology, Inc. (Albany, New York) [140]. The engine working gas oscillated through the tubes, which were bent to accommodate thermal expansion without plastic deformation. Combustion gases flowed past the outside of the tubes, generally *left to right* in this photograph. Fins on the downstream portion (with respect to combustion-gas flow) of each tube enhanced the heat transfer there

Sometimes (especially for research purposes) it is only necessary to electrically heat the thermoacoustic gas, instead of transferring heat to it from another gas or liquid stream. We sometimes accomplish this by inserting common, commercial tubular heaters or cartridge heaters in the resonator near the hot end of the stack. We orient these heating elements perpendicular to the x direction, and may optionally add fins (typically nickel) to improve heat transfer to the thermoacoustic gas. However, when operating such heating elements simultaneously at red-hot temperatures and near their rated powers, they burn out more quickly than you would expect based on experience with home appliances. Hence, for the most demanding situations we have made high-power electric resistance heaters with NiCr ribbons directly exposed to the thermoacoustic gas. The heat exchanger shown in Fig. 1.10, used at the hot end of each stack in the standing-wave engine of Figs. 1.8 and 1.9, featured NiCr ribbon wound zigzag around NiCr pins, with the pins (not visible in the photo) trapped in holes in two alumina (ceramic) rings. (The NiCr pins helped short-circuit the ribbon as it turned each corner of the zigzag pattern. Otherwise these locations might have run hotter than the rest of the ribbon, because these locations were partly sheltered from the vigorous helium flow in the central portion of the circle.) Thermal expansion of NiCr is greater than that of alumina, so the twelve combs visible in the photo were necessary to prevent hot, sagging NiCr from shorting to itself. Each comb had slits to hold the ribbons, and the three ribs perpendicular to the combs added stability to the combs. The two nickel wires extending toward the camera were spot welded to the ends of the ribbon, serving as electrical leads. Before assembly, the ribbon was pre-formed into the zigzag pattern by winding it around hard steel pins set in steel holes, with the hole locations a few tenths of a mm closer than in the alumina parts, to ensure zero slack in the NiCr when it was transferred from the steel pins to the NiCr pins on the alumina rings. During assembly, the alumina parts were bonded to each other as needed with ordinary superglue (cyanoacrylate) so that the heat exchanger could be handled without falling apart. However, the parts were designed so that, when the heat exchanger and stack were installed in the resonator, a step on the inside of the resonator trapped one alumina ring and all the ribs and combs, while the stack trapped the other alumina ring. The superglue could then be baked away under vacuum without loss of structural integrity to the heat exchanger. More recently, we have had the combs, ribs, and one ring of similar exchangers made (Progressive Technology, Auburn CA) from a single piece of ceramic, eliminating the superglue and yielding a much stronger assembly.

8.4 Thermal Buffer Tubes, Pulse Tubes, and Flow Straighteners

Most thermal buffer tubes and pulse tubes are simply large, round tubes. They must certainly be longer than the peak-to-peak gas displacement amplitude $2\,|\xi_1|$ in them, if thermal isolation is to be maintained by a slug of oscillating gas executing plug flow. Ray Radebaugh has recommended that the length of pulse tubes for small cryocoolers should be roughly $6\,|\xi_1|$ to make the slug of oscillating gas long enough to provide good insulation.

To ensure the validity of the mathematics of [88] for suppression of streaming (summarized in "Rayleigh streaming" in Chap. 7), the pulse tube must operate in either the laminar regime or the weakly turbulent regime shown in Fig. 7.4, and the inside surface must be polished so that the surface roughness is $\ll \delta_\nu$.

We have tapered our tubes at Los Alamos by straightforward, tedious machining, starting with a straight, thick-walled pipe and machining it to form a thinner-walled tube with both the i.d. and o.d. tapered. Other standard, less-expensive tapered-tube fabrication methods could easily be used, especially in a production setting.

Flow straighteners should also be used at one or both ends of a pulse tube or thermal buffer tube to prevent jet-driven streaming as described in "Jet-driven streaming" in Chap. 7. We have typically used several layers of plain-weave screen. Turning each layer roughly $45°$ relative to its predecessor in the pile is important for preventing close angular alignment of layers and two-dimensional nesting of the array of peaks of one layer's wires into the array of holes between wires in the adjacent layer. Such nesting at the end of a flow straightener can cause gas blowing from the flow straightener into a pulse tube or thermal buffer tube to lean toward one wall, creating jet-driven streaming.

8.5   Resonators

8.5.1   Dissipation

Resonator components provide the inertances L_i and compliances C_j needed to store acoustic energy and shift the amplitude or phase of oscillating pressure or volume flow rate, connecting a few heat-exchanger components into a complete, no-moving-parts heat engine or refrigerator. Unfortunately, as discussed in Chaps. 4 and 5, inertance is generally accompanied by viscous resistance R_v and compliance is generally accompanied by thermal-relaxation resistance R_κ, with both resistances dissipating acoustic power. A primary goal of resonator design is to provide sufficient inertance and compliance, thereby maintaining a desired resonance frequency, while simultaneously minimizing the dissipation of acoustic power.

A straight, uniform-diameter resonator is seldom the best design. Usually the portions of the resonator where inertance is vital can be somewhat shortened and reduced in diameter, maintaining the necessary length-to-area ratio while reducing the surface area on which viscous and thermal-relaxation dissipation occurs (see Exercise 8.11).

For the lowest dissipation of acoustic power, the transitions between portions of the resonator, and indeed all internal surfaces in the engine or refrigerator, should be smooth and "streamlined" like the external surfaces of an airplane. Steps, misalignments, and abrupt transitions generate turbulence unnecessarily. In the standing-wave engine shown schematically in Fig. 1.9, the internal angle of the gradual tapers joining the ambient heat exchangers and the central resonator pipe was chosen to minimize minor losses (as well as possible, based on steady-flow data [91, 92]), and represents good streamlining. On the other hand, the base of the branch to the refrigerator shown in Fig. 1.9 has no streamlining, and probably dissipated significant acoustic power. Similarly, the long-radius 540° bend in the inertance of the orifice pulse-tube refrigerator shown in Fig. 1.17 is an example of good streamlining and minimum dissipation, but the two short-radius elbows in series with it (at the lower-right end, near the compliance) suffered from severe minor-loss dissipation, and were eliminated soon after this photograph was taken.

The various parts of resonators may be joined by means of welds, solder, or rubber O-rings. These methods sometimes create small, unwanted compliances weakly connected with the thermoacoustic gas. For example, when O-ring grooves are cut according to the O-ring catalog's recommendations, the compressed O-ring will typically fill only 70% of the volume of the groove. When the system is assembled and pressurized, the mean pressure will push the O-ring to the outside extremity of the groove, so 30% of the volume of the groove forms a compliance C weakly connected to the resonator through a crevice between flanges. Let R represent the flow resistance of the crevice. If this part of the resonator is experiencing pressure oscillations with amplitude $|p_1|$, the RC network comprising the inter-flange crevice and the O-ring groove dissipates acoustic power at a rate

$$\dot{E}_2 = \frac{|p_1|^2}{2R} \frac{1}{1 + 1/\omega^2 R^2 C^2}. \tag{8.1}$$

In situations like this, estimating R can be difficult, so we often assume the worst case: With respect to variations in R, the maximum value of Eq. (8.1) is

$$\dot{E}_{2,\text{max}} = \frac{|p_1|^2 \omega C}{4}. \tag{8.2}$$

With $C = V/\gamma p_m$, a 1-cm^3 compliance connected in this way to a 30-bar helium resonator with $|p_1|/p_m = 0.1$ at 350 Hz could dissipate 10 W of acoustic power. (This is the high-amplitude operating condition of the standing-wave engine shown in Figs. 1.8 and 1.9.)

To avoid such an accidental cavity associated with a rubber O-ring, make the O-ring-groove volume only a little larger than the volume of the O-ring itself.

When a strong welded joint is made partly by welding from the inside of the resonator and partly by welding from the outside, such a cavity can also occur between the inner and outer welds, and it may be weakly connected to the resonator if the inner weld has one or more accidental pores. To prevent this, ask for a deliberate "weep hole" (standard welding terminology) through the *outside* weld, so that the cavity is deliberately connected to outside ambient air and any accidental pore in the inner weld will immediately call attention to itself as a leak.

The separation at the interfaces between heat-exchange components also demands attention. As one example, consider the interface between the regenerator and the main ambient heat exchanger in the thermoacoustic-Stirling heat engine shown in Fig. 1.23. Both had the same overall diameter, but the heat exchanger was comprised of tubes whose internal cross-sectional areas added up to less than 25% of the regenerator area. If the regenerator had pressed tightly against this heat exchanger, only 25% of the screen area in the first few layers of screen would have been available to carry the entire U_1, resulting in locally high viscous dissipation of acoustic power. On the other hand, an excessively large gap between this heat exchanger and the regenerator would essentially have provided two additional adiabatic-isothermal interfaces of the type described in "Joining conditions" in Chap. 7 and [103], which would also have dissipated acoustic power. We do not have a quantitative design procedure to address this issue; the ribs in Fig. 8.7 show the compromise we chose. As a second example, imagine the similar concerns at the interface between the cold heat exchanger and the pulse tube in the orifice pulse-tube refrigerator shown in Fig. 1.18, where the gap had to accommodate a factor-of-four area reduction and the flow straightener (a few layers of screen) had to enforce plug-like flow upward into the pulse tube. The compliance in such gaps can be appreciable, so we always include these volumes in our DELTAEC models of apparatus.

8.5.2 Size, Weight, and Pressure-Vessel Safety

The hoop stress s_{hoop} in the wall of a cylinder of radius $\mathcal{R}$ and wall thickness $t \ll \mathcal{R}$ due to internal pressure p is

$$s_{\text{hoop}} = \frac{\mathcal{R}}{t} p, \qquad (8.3)$$

which must be kept within safe limits to prevent bursting of the cylinder. The weight of such a cylinder is

$$W = \rho_{\text{solid}} g 2\pi \mathcal{R} t \, \Delta x \qquad (8.4)$$

$$= \rho_{\text{solid}} g 2\pi \mathcal{R}^2 \, \Delta x \frac{p}{s_{\text{hoop}}}, \qquad (8.5)$$

with Δx its length. If the cylinder is used as an inertance $L = \rho_m \, \Delta x / \pi \mathcal{R}^2$, then W can also be written

$$W = \frac{\rho_{\text{solid}}}{\rho_m} \frac{p}{s_{\text{hoop}}} L g 2\pi^2 \mathcal{R}^4. \qquad (8.6)$$

When the resonator size and weight are somewhat larger than those of the heat-exchange components, the $\mathcal{R}^2$ and $\mathcal{R}^4$ dependences in Eqs. (8.5) and (8.6) show the importance of keeping $\mathcal{R}$ small, especially in resonator inertances, to keep system weight and volume as low as possible. The parts of the pressure vessel enclosing stacks, regenerators, and other components having nonzero dT_m/dx must also be kept as small and as thin-walled as possible, to reduce heat leak carried by the pressure vessel.

However, it is vital to keep stresses in the solid pressure boundary of thermo-acoustic devices within safe limits, because a sudden failure of part of your pressure vessel can kill or injure you. (See Exercise 8.1.)

Simple concepts such as that described by Eq. (8.3) are at the foundation of the very advanced discipline of safe pressure-vessel design [141, 142]. A complete description of this discipline can occupy half a meter on your bookshelf [143]. The design of thermoacoustic pressure vessels is made especially challenging by the possibility of fatigue failure caused by oscillating stress at the acoustic frequency, by stress induced by thermal expansion or contraction in parts of the pressure vessel with temperature gradients (such as regenerators, stacks, pulse tubes, and thermal buffer tubes), and by creep and oxidation at the hot ends of engines.

At Los Alamos, we try our best to design and build our thermoacoustic engines and refrigerators according to the standard US safety code [143]. The design is made and documented by one person, and is always checked by a second person. Either the designer or the checker must be someone with considerable relevant experience. Fabrication is also done according to the code. This may require vendor certification of material specifications, welding by a certified welder, or X-ray inspection of welds. When an apparatus is first assembled, we pressure test it well above any anticipated working pressure according to the code requirements, either hydraulically or using gas pressure. If this test is done with gas pressure, we prepare for the worst, putting the device under test in an empty room or a pit in the ground and raising and lowering the pressure remotely so that people and valuable equipment are protected from the effects of an unexpected rupture. The design process includes other engineering controls in all cases: relief valves, temperature and cooling-water-flow interlocks as needed to shut off heat if temperatures were to become dangerously high, etc.

In spite of these precautions, we have had several pressure-vessel accidents through the years, fortunately with no injury. These accidents generally occurred after a system had been in routine use for weeks or months, and have generally been due to failure of small components: pressure transducers, windows, tubes connecting the system to the gas supply, etc. Small-component failures usually launch small, high-velocity projectiles.

In the constantly changing environment of a research laboratory, you must always be ready for a pressure-vessel failure, even after doing your absolute best to design and build safely. So, as a final line of defense: stay as far away from the apparatus as possible, as much of the time as possible, to gain personal safety according to $1/r^2t$; put polycarbonate sheets (bulletproof "Lexan") or other barriers at key locations; aim potential projectiles such as pressure transducers toward the floor or wall; and above all protect your eyes—they are valuable, fragile, and impossible to repair or replace.

8.5.3 Harmonic Suppression

Nonlinear terms in the hydrodynamic and thermodynamic equations, such as $\rho_1 \widetilde{u_1}$ arising from $\rho \, \partial \mathbf{v}/\partial t$ in the momentum equation, create oscillations at 2ω from the fundamental ω oscillations, because $\cos \omega t \times \cos \omega t = \frac{1}{2}(1 + \cos 2\omega t)$. Once these 2ω waves exist, they interact with the fundamental ω oscillations via the nonlinear terms, yielding 3ω oscillations because $\cos \omega t \times \cos 2\omega t = \frac{1}{2}(\cos \omega t + \cos 3\omega t)$. If the resonator's cross-sectional area is independent of x along its entire length Δx, its resonance frequencies are $f_n = na/2\Delta x$, where n is an integer, so all these harmonically generated oscillations coincide with resonances, and hence may rise to very high amplitudes, such as illustrated in Fig. 7.20. One could imagine a deliberate use of such harmonics to enhance some desired aspect of performance, but in our experience harmonics have only been harmful to system efficiency.

Fortunately, harmonics are easy to avoid by using a resonator with a nonuniform area $A(x)$ [120, 144, 145], which is often necessary anyway in order to minimize dissipation of acoustic power, length, and weight (see above). We generally design our resonators based on these three concerns, and check briefly for any unlucky resonance-frequency harmonicity at the end of the design process, by numerical calculation. The resonators shown in Figs. 1.8, 1.9, 1.13, 1.14, and 1.22, 1.23 were designed in this way, and had no unlucky harmonic resonances, either calculated or experimental.

Radial waves in cylinders also have nonharmonic modes, so radial flow through cylindrical heat-exchanger components has also been used to avoid resonance-frequency harmonicity [146, 147].

8.6 Electroacoustic Power Transducers

In electric-driven refrigerators or combustion-driven electric generators, electric power must be converted to acoustic power or vice versa. Several transduction mechanisms can be used, but the most common is "electrodynamic," using copper wires and permanent magnets. The ordinary loudspeaker and the linear motor–alternator represent two approaches to the implementation of the electrodynamic transduction mechanism.

For many simple experiments, ordinary audio loudspeakers are excellent, with low cost, easy availability, and versatility. Figure 8.12 shows one of the four loudspeakers used in the standing-wave refrigerator of Figs. 1.13 and 1.14, and Fig. 8.13 shows a line drawing of such a speaker. Unfortunately, the power-transduction efficiency of loudspeakers is usually poor, their paper cones are weak and fragile, their strokes are limited, and their impedances are poorly matched to gas at high p_m. Without heroic modifications, pressure amplitudes are limited.

People in the cryocooler and Stirling-engine community seldom think of using an audio loudspeaker as a driver, even for a simple experiment. Instead, they routinely design and build their own electrodynamic linear motors and linear alternators, which can have moving coils and stationary magnets or moving magnets and stationary coils. One commercially available [148] moving-magnet linear motor–alternator is shown in Fig. 8.14. All that is visible in the photo is part of the outside of the massive iron "yoke" structure that channels magnetic flux between the magnets and the coils, and the flexible straps that keep the moving piston

Fig. 8.12 One of the four ordinary loudspeakers used to drive the standing-wave refrigerator shown in Figs. 1.13 and 1.14

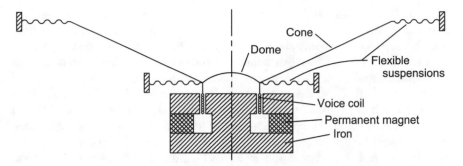

Fig. 8.13 Schematic of an electrodynamic loudspeaker. The geometry ensures that electric current, magnetic field, and motion are mutually perpendicular

Fig. 8.14 An efficient 2-kW linear motor-alternator, from CFIC Qdrive, Troy NY. Photo courtesy of John Corey. Dimensions are approximately $24 \times 24 \times 13 \, \text{cm}$

and magnets centered in the cylinder and iron yoke. Linear motor–alternators are generally bulkier, heavier, more sharply mechanically resonant, and more efficient than loudspeakers.

The fragile loudspeaker and the stout linear motor–alternator are two examples of the spectrum of possible electrodynamic devices, all described approximately by these linear, harmonic-approximation equations [5]:

$$\Delta V_1 = (R_{\text{elec}} + iX_{\text{elec}}) I_1 - \frac{\tau}{A} U_1, \tag{8.7}$$

$$\Delta p_1 = \frac{\tau}{A} I_1 + \frac{R_{\text{mech}} + iX_{\text{mech}}}{A^2} U_1. \tag{8.8}$$

These two deceptively simple equations express Ohm's law (including motional electromotive force) and Newton's law, respectively, for the transducer. The complex volume flow rate of the transducer's motion is U_1, the complex electric current through the transducer is I_1, the complex voltage difference across the electric terminals is ΔV_1, and the difference between the complex pressures on the front and back sides of the moving element is Δp_1. The basic physics of the transduction mechanism is captured in the transduction coefficient τ, which appears in both equations. In loudspeakers, τ is called the "$\mathcal{B}l$ product," because it is the product of the magnetic field and the length of the wire in the field. [Faraday's law, which relates the relative velocity between the wire and magnetic field to the induced voltage, is responsible for the presence of τ in Eq. (8.7), and the magnetic force law is responsible for its presence in Eq. (8.8).] The electric resistance R_{elec} and electric reactance X_{elec} also appear in Eq. (8.7), often with $X_{elec} = \omega L_{elec}$ due to electric inductance L_{elec}. The mechanical resistance R_{mech} and mechanical reactance

$$X_{mech} = \omega M - K/\omega \qquad (8.9)$$

appear in Eq. (8.8), where M is the moving mass and K is the spring constant.

In commercially available loudspeakers, audio fidelity over a broad range of frequency is most important. To achieve this, ωM is made larger than K/ω over most of the frequency range (so that the frequency dependence of the motion and the frequency dependence of the radiation impedance [5] conspire to give a frequency-independent acoustic power) and modest frequency-independent resistances R_{mech} and R_{elec} are accepted. In linear motors and linear alternators, high efficiency is most important, so resistances are minimized and ωM and K/ω are typically much larger than R_{mech} —the mechanical characteristics of the transducer are usually dominated by its moving mass and spring constant.

To consider the efficiency of transduction [149], focus on the situation of a thermoacoustic refrigerator driven electrically at frequency ω by an electrodynamic transducer described by Eqs. (8.7) and (8.8). Suppose that the acoustic environment in which the transducer is located requires a pressure difference Δp_1 across the transducer when the volume flow rate of the transducer is U_1. In other words, the acoustic impedance of the acoustic network, including the acoustic components on both sides of the transducer, is

$$R_{acoust} + iX_{acoust} = -\Delta p_1/U_1. \qquad (8.10)$$

The electric power into the transducer can be obtained easily from Eq. (8.7):

$$\dot{W}_{elec} = \frac{1}{2}\mathrm{Re}\left[\Delta V_1 \widetilde{I_1}\right] = -\frac{1}{2}\frac{\tau}{A}\mathrm{Re}\left[U_1\widetilde{I_1}\right] + \frac{1}{2}R_{elec}\left|I_1\right|^2. \qquad (8.11)$$

Similarly, the acoustic power out of the transducer, which is the difference between the acoustic power flowing into the transducer on one face and out on the other face, is obtained easily from Eq. (8.8):

$$\Delta\dot{E}_2 = -\frac{1}{2}\mathrm{Re}\left[\Delta p_1 \widetilde{U_1}\right] = -\frac{1}{2}\frac{\tau}{A}\mathrm{Re}\left[U_1\widetilde{I_1}\right] - \frac{1}{2}\frac{R_{mech}}{A^2}\left|U_1\right|^2. \qquad (8.12)$$

Equations (8.11) and (8.12) show that a portion $R_{elec} |I_1|^2 /2$ of the electric power $\dot{W}_{elec}$ is dissipated by the electric resistance, and that the rest of the electric power, $-(\tau/2A) \operatorname{Re}[U_1 \tilde{I}_1]$, is transduced from electric power to mechanical power from one equation to the other. A portion $(R_{mech}/2A^2) |U_1|^2$ of this mechanical power is dissipated by the mechanical resistance, and the remainder is delivered as acoustic power $\Delta \dot{E}_2$ to the acoustic network.

Taking the ratio of these two equations gives the efficiency η_{trans} of the transduction of electric to acoustic power. Dividing top and bottom by $|U_1|^2$ gives

$$\eta_{trans} = \frac{\Delta \dot{E}_2}{\dot{W}_{elec}} = \frac{-(\tau/A) \operatorname{Re}[I_1/U_1] - (R_{mech}/A^2)}{-(\tau/A) \operatorname{Re}[I_1/U_1] + R_{elec} |I_1/U_1|^2}. \tag{8.13}$$

This efficiency depends on the acoustic impedance $R_{acoust} + iX_{acoust}$ to which the transducer is connected. The ratio I_1/U_1 appearing in Eq. (8.13) can be expressed in terms of this acoustic impedance by solving Eq. (8.8) for I_1/U_1 and using Eq. (8.10):

$$\frac{I_1}{U_1} = -\frac{A}{\tau} \left[\left(R_{acoust} + \frac{R_{mech}}{A^2} \right) + i \left(X_{acoust} + \frac{X_{mech}}{A^2} \right) \right]. \tag{8.14}$$

Substituting this into Eq. (8.13) yields [149]

$$\frac{1}{\eta_{trans}} = 1 + \frac{R_{mech}}{A^2 R_{acoust}}$$
$$+ \frac{R_{elec}}{R_{acoust}} \frac{A^2}{\tau^2} \left[\left(R_{acoust} + \frac{R_{mech}}{A^2} \right)^2 + \left(X_{acoust} + \frac{X_{mech}}{A^2} \right)^2 \right]. \tag{8.15}$$

Obviously, Eq. (8.15) shows that setting

$$X_{acoust} + X_{mech}/A^2 = 0 \tag{8.16}$$

leads to higher efficiency, if all other variables are held constant. Since the reactances X_{acoust} and X_{mech} depend strongly on frequency, this condition is easily interpreted: The system should be operated "on resonance" to achieve the highest efficiency. With powerful linear motors, it is often the case that ωM is considerably larger than K/ω, so the transducer's spring is too weak to make X_{mech} small. In this situation, a gas spring contributing to X_{acoust} can be used to achieve resonance. However, it would usually be better to avoid the use of such a strong gas spring if possible, as it necessarily leads to a large pressure difference across the transducer, which in turn causes undue strain on flexing seals (e.g., bellows) or leakage through piston–cylinder gaps. Hence, the most desirable resonance condition is twofold: simultaneous acoustic resonance $X_{acoust} = 0$ and transducer self-resonance $\omega M = K/\omega$.

Even with resonance enforced so that $X_{acoust} + X_{mech}/A^2 = 0$, Eq. (8.15) is still complicated, but it can be further simplified by picking the area A_{best} that yields the highest efficiency. Using Eq. (8.16) in Eq. (8.15), differentiating with respect to A, and setting the result equal to zero gives the optimum [149]:

$$A_{\text{best}}^4 = \frac{R_{\text{mech}}^2}{R_{\text{acoust}}^2} \left(\frac{\tau^2}{R_{\text{elec}} R_{\text{mech}}} + 1 \right). \tag{8.17}$$

At this area, the efficiency is

$$\eta_{\text{best}} = \frac{\sqrt{c+1} - 1}{\sqrt{c+1} + 1}, \tag{8.18}$$

where

$$c = \frac{\tau^2}{R_{\text{elec}} R_{\text{mech}}} \tag{8.19}$$

is an easy-to-remember figure of merit for the transducer. Not surprisingly, Eq. (8.19) shows that a good electrodynamic transducer has a large transduction coefficient and small electric and mechanical resistances.

Much basic analysis of transducers can begin with these simple Eqs. (8.7) and (8.8), following different algebraic steps and reaching different conclusions for different circumstances and different assumptions. For example, consider the goal of obtaining the most power instead of the best efficiency, and assume that all parameters of the acoustic system and the transducer are given except for the area A connecting them. The transduction term in Eqs. (8.11) and (8.12) can be written

$$\frac{1}{2} \tau \omega \, \text{Im} \left[\xi_1 \widetilde{I_1} \right], \tag{8.20}$$

showing that maximum power transduction is achieved by operating at maximum $|\xi_1|$ and maximum $|I_1|$. A given transducer has a given maximum displacement amplitude $|\xi_1|_{\text{max}}$ (set by the allowable stresses in the flexible suspension) and a given maximum current amplitude $|I_1|_{\text{max}}$ (set by electric power dissipation and the allowable temperature of the transducer). Picking signs for an electrically driven refrigerator, and setting $X_{\text{acoust}} = 0$ and $X_{\text{mech}} = 0$ for good efficiency, Eq. (8.14) yields

$$A_{\text{most}}^2 = \frac{\tau \, |I_1|_{\text{max}}}{\omega R_{\text{acoust}} |\xi_1|_{\text{max}}} - \frac{R_{\text{mech}}}{R_{\text{acoust}}}. \tag{8.21}$$

For a given acoustic system *and* a given transducer, some compromise must generally be chosen between Eqs. (8.17) and (8.21), between highest efficiency and highest power.

Additionally, as with thermoacoustics itself, reality goes beyond the basics of such linear equations as soon as amplitudes reach non-negligible values. For example, the parts of the coil of a moving-coil transducer may move outside the pole faces of the magnet-yoke structure at the extremes of high-amplitude motion, where they contribute nothing to τ but still contribute to R_{elec} [150]. As a second example, the magnetic field from a large electric current can add to the field of the permanent magnet, exceeding the capability of the yoke's iron to carry magnetic flux and reducing the effective value of τ at high currents.

8.7 Exercises

8.1 (a) Calculate the energy that would be released from your favorite thermoacoustics system, if it burst and the mean pressure went down to atmospheric pressure. Compare this energy to the kinetic energy of an automobile moving at freeway speed and the energy you would acquire if you fell off a 3-m-high cliff. Is it more realistic to assume an isentropic expansion to atmospheric pressure, or an isothermal expansion? Why? Which type of expansion releases the most energy? (b) You can assure yourself of your pressure vessel's strength by filling it with liquid water and pressurizing it to 1.5 times your usual operating pressure. How much energy would be released if the vessel burst during this test? (The compressibility of liquid water is approximately 5×10^{-10} Pa^{-1}.)

8.2 Calculate the ratio of resonance frequencies for the lowest harmonic to the fundamental for the resonator of Fig. 4.23, with $\Delta x_b = 2\Delta x_c$ and $A_b = 2A_c$. Your choice: use software such as DELTAEC, or work out the trigonometric functions.

8.3 You think that your electrodynamic transducer obeys Eqs. (8.7) and (8.8), but you don't know any of the constants. (a) Which of the constants (or products or ratios of constants) can you determine by making only electrical measurements, with the transducer in vacuum? Assume that you can make dc electrical measurements, and ac measurements at a wide variety of frequencies, but you cannot touch or observe the motion of the moving piston or dome. (b) Now assume that you can also touch and observe the moving piston or dome. Design an experimental program to measure all constants, using common laboratory equipment. Use dc measurements when possible. (Example: Orient the driver so that its axis of motion is vertical; set up a machinist's dial indicator or other sensitive sensor of piston position; add weight and dc current simultaneously in whatever proportions are needed to keep the position of the piston constant; deduce τ.) (c) Do you really expect these "constants" to be perfectly independent of frequency? Give an example of a physical mechanism that might lead one of these "constants" to depend on frequency.

8.4 What is the flow rate of the water in the room-temperature heat exchanger of your favorite thermoacoustic system? What is the temperature rise? What is the Reynolds number in the water?

8.5 Make an order-of-magnitude estimate of the oscillating force exerted by the thermoacoustic gas on one end cap of the standing-wave engine example, at the "10%" operating point, where $|p_1| \sim 300$ kPa. (Obtain dimensions from the

DELTAEC file in Appendix B, or by careful examination of Fig. 1.9.) If the apparatus weighs ~ 50 kg, and hangs freely from two ropes like a pendulum, what is the resulting vibrational acceleration (compared to the acceleration of gravity) and what is the vibrational displacement amplitude?

8.6 Make an order-of-magnitude estimate of the force exerted on the stack by the oscillating gas for the standing-wave engine example, for the operating point detailed in the DELTAEC file in Appendix B. Consider two forces: viscous shear forces on the total spiral surface area of the stack (i.e., the total "wetted" surface area), and pressure forces on the thin edges of the spiral at each end of the stack. Show these forces on a phasor diagram, together with the phasors for p_1 and U_1 in the center of the stack. How does the sum of the two forces compare with the weight of the stack? Hence, should the stack be trapped rigidly between the heat exchangers, or could it simply rest there lightly?

8.7 Assume that power is proportional to $p_m a A \left(|p_1|/p_m\right)^2$ in your favorite thermoacoustic device. Consider doubling p_m and halving all areas, while keeping the same gas, the same resonator length, and the same value of $\left(|p_1|/p_m\right)^2$, in order to keep the same power while reducing the volume of the apparatus to only half of the original volume. (a) By what factor must you reduce the stack hydraulic radius, in order to keep r_h/δ_κ the same as before? (b) If the resonators in both situations are designed with the maximum hoop stress in the walls (due to $p_m + |p_1|$) equal to 1/4 of the yield stress of the resonator material, is the weight of the smaller resonator less than the original weight?

8.8 When we built the NiCr heat exchanger shown in Fig. 1.10, we worried that the thin NiCr ribbon might flutter as the helium flowed through, causing fatigue and failure of the ribbon. At a typical fully loaded operating point, the peak velocity of the 30-bar, 800 °C helium was expected to be 9 m/s. We decided to test the heat exchanger for the suspected flutter using steady flow of air at the same Reynolds number. Was this a reasonable test, or can you think of another dimensionless number that might be relevant here? What air velocity was needed to achieve this Reynolds number? (This was a two-person experiment. One person drove a car, while the other person closely inspected the heat exchanger, which was held outside at a suitable angle near the top center of the car's windshield.)

8.9 Verify Eqs. (8.1) and (8.2).

8.10 For a heat exchanger in your favorite thermoacoustics hardware, estimate $|N_{R,1}|$, r_h/δ_κ, $|\xi_1|/\Delta x$, and the average heat-flux density $\dot{Q}/S$ across the gas-to-solid "wetted area" S.

8.11 Consider a quarter-wavelength resonator with length Δx and diameter D, as illustrated in Fig. 8.15a. The pressure amplitude is p_0 at the left end and zero at the right end. (a) Integrate an expression for $d\dot{E}_2/dx$ from Chap. 5 to show that the dissipation in this resonator is

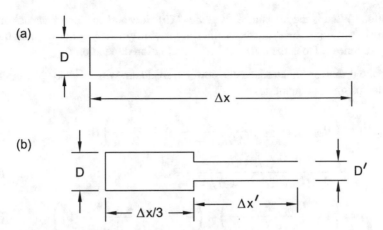

Fig. 8.15 Resonators for Exercise 8.11. (**a**) Uniform-diameter resonator. (**b**) Similar resonator with reduced diameter near the open end

$$\Delta \dot{E}_2 = \frac{\pi}{8} \frac{p_0^2}{\rho_m a^2} D \Delta x \omega \delta_v \left(1 + \frac{\gamma - 1}{\sqrt{\sigma}} \right) \tag{8.22}$$

for laminar flow in the boundary-layer approximation, neglecting thermal-relaxation losses on the closed end cap and radiation from the open end. (b) The quality factor Q of a resonator is $Q = \omega E_{\text{stored}} / \Delta \dot{E}_2$ where E_{stored} is the kinetic energy in the resonator at that time phasing when $p_1 \simeq 0$ and $|u_1|$ is a maximum in the resonator. This kinetic energy is the spatial integral of $\frac{1}{2} \rho u^2$ at that instant of time. What is the Q of this resonator? (c) Review Exercises 4.1 and 8.2. Keep the resonance frequency the same as in part (a) of this exercise, but reduce the diameter of the high-velocity part of the resonator, as shown in Fig. 8.15b. Ignore minor loss at the area change. What is $\Delta x'$? What value of D' gives the lowest $\Delta \dot{E}_2$? [73] What value gives the highest Q? (d) Review Exercise 7.19. Repeat (c), assuming that the viscous losses are turbulent instead of laminar but continuing to ignore minor loss at the transition.

8.12 A thermoacoustic cryocooler is filled to 30 bar with 1 mole of H_2 at 27 °C. It bursts. How much $\int p \, dV$ energy is released in the explosion? The H_2 then mixes stoichiometrically with air and ignites from a spark of static electricity. How much chemical energy is released in this explosive combustion? Roughly how much of that chemical energy might show up as sudden $\int p \, dV$ expansion in this second explosion? Which do you think is more dangerous, the pressure-vessel failure or the explosive combustion?

8.13 A linear alternator obeys Eqs. (8.7) and (8.8), and is connected to a combustion-powered thermoacoustic engine characterized by Eq. (8.10). The electric terminals of the alternator are connected to an electric load, whose electric impedance R_{load} is real. (a) What value of R_{load} is required to achieve steady-state operation? What is the resonance frequency? (b) Now add an electric capacitor C_{load} in series with R_{load}. Investigate whether adjustability of C_{load} allows a significant ability to tune ω. Does the efficiency of transduction depend on C_{load}?

8.14 Show that an electroacoustic power transducer with a leaky piston can be modeled using the equations

$$
\Delta V_1 = \left(R_{elec} + iX_{elec} + \frac{\tau^2}{A^2 R_{leak} + R_{mech} + iX_{mech}} \right) I_1
$$
$$
- \frac{(\tau/A)\, U_1}{1 + (R_{mech} + iX_{mech})/A^2 R_{leak}}, \tag{8.23}
$$

$$
\Delta p_1 = \left(1 + \frac{R_{mech} + iX_{mech}}{A^2 R_{leak}} \right)^{-1} \left(\frac{\tau}{A} I_1 + \frac{R_{mech} + iX_{mech}}{A^2} U_1 \right), \tag{8.24}
$$

where R_{leak} is the flow resistance of the leak. Here U_1 is the volume flow rate seen by the acoustic network in which the transducer is imbedded; this is not equal to the product of piston area and piston velocity.

Chapter 9
Measurements

This chapter introduces some measurement approaches that yield good understanding of thermoacoustics hardware.

Sophisticated measurement techniques will not be considered; they are covered in great detail elsewhere [151–153]. On the contrary, the techniques briefly described in the first half of this chapter are those that are simple enough for everyday use by everyone: primarily measurements of mean temperature and of amplitude and phase of oscillating pressure.

Nor will mathematical data-analysis techniques be discussed; they are also covered in great detail elsewhere [154]. Thermoacoustic measurements usually yield such high signal-to-noise ratios that statistical analysis is unnecessary. Attention to possible systematic errors (which can be due to sensor miscalibrations, uncertainties in hardware dimensions or surface finishes, gas impurities, etc. in thermoacoustics) is clearly required for good results, but methods of analysis of the effects of systematic errors are also covered in great detail elsewhere.

So, instead of dwelling on such *techniques* of measurement and analysis, the second half of this chapter will focus on measurement *points of view*. Sophisticated points of view can enable advanced interpretation of simple measurements. Points of view seem to receive no formal attention in the classroom or in the literature of science and engineering, so many researchers unconsciously adopt and apply only a limited number of points of view to all their work, handicapping themselves as surely as they would if they chose to use only one instrument for all measurements.

9.1 Easy Measurements

For a radically new thermoacoustics hardware project, or a lecture demonstration, a knowledge of powers at the order-of-magnitude level, pressure amplitudes at the factor-of-two level, pressure phases plus or minus tens of degrees, and temperatures

© Acoustical Society of America 2017

G.W. Swift, *Thermoacoustics*, DOI 10.1007/978-3-319-66933-5_9

plus or minus tens or hundreds of degrees is often sufficient, allowing significant savings of time and expense in hardware fabrication and in any measurement program. However, in this chapter, let us assume that one is concerned with a relatively mature project, for which one might aspire to 10% accuracy in powers, 2% accuracy in pressure phasors (2% in amplitude, 1° in phase), and one or a few degrees accuracy in temperatures. Successful measurements at this level of accuracy require some discipline and care.

As a general rule, slight skepticism about everything is wise. We never rely completely on only one sensor or one instrument for a vital measurement. We double check everything, and we use multiple independent measurements when possible. We check critical sensors when it's easy to do so: We put thermocouples in liquid nitrogen, dry ice, freezing water, boiling water, melting solder, or some other easy fixed point (we found a spool of bad thermocouple wire this way); we check the sensitivity of pressure sensors to dc pressure *in situ* (we found one that was internally strained by overtightening, so installation changed its sensitivity); we cross-calibrate pressure sensors with oscillating pressure to identify dysfunctional or miswired sensors; etc.

9.1.1   Pressures and Frequency

A pressure transducer [153] is used to sense the amplitude and phase of oscillating pressure and, incidentally, frequency. Both piezoelectric and piezoresistive transducers are satisfactory. The piezoelectric transducer, with integral integrated-circuit amplifiers, typically has a high signal-to-noise ratio. The piezoresistive transducer, being dc coupled, allows measurement of the mean pressure and easy calibration checks via deliberate changes in the mean pressure.

For historical reasons, we adopted Endevco (Endevco, San Juan Capistrano, CA, www.endevco.com) piezoresistive transducers, as shown in Fig. 9.1, for our work at Los Alamos. Each transducer simply screws into a tapped hole in the pressure vessel, sealing with a small rubber O-ring against a flat that is machined on the outside of the pressure vessel. With minimal effort, this often puts the sensing surface nearly flush with the inside of the typical pressure-vessel wall, so the transducer perturbs the flow but little. Actually, these transducers measure the pressure *difference* between this sensing surface and a cavity inside the transducer, which is accessible through a small capillary. We usually leave this capillary vented to the atmosphere, but it can also be plumbed to vacuum or to another part of the pressure vessel for differential pressure measurements.

These transducers can be used at cryogenic temperatures (Ray Radebaugh, personal communication) by replacing the rubber O-ring with a thin teflon gasket and using a raised, flat-topped annular ridge machined on the outer surface of the pressure vessel to fill most of the volume previously occupied by the rubber O-ring. For such extreme conditions, the transducer must be calibrated at its operating temperature. Alternatively, the pressure amplitude and phase can be measured at a cryogenic location or an extremely hot location by connecting a room-temperature

Fig. 9.1 An Endevco
piezoresistive pressure
transducer, screwed into a
small hole in the side of a
pipe. The US penny gives a
sense of scale. Note the small
flat machined onto the pipe,
to allow the rubber O-ring to
seal

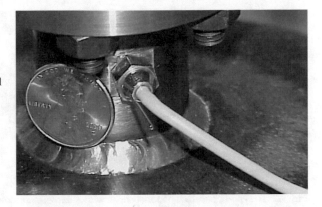

transducer to the desired location through a short, low-thermal-conductance tube.
In that situation, we first model the thermoacoustics of the connecting tube, to
assure ourselves that the pressure amplitude and phase at the sensor location will be
reasonably close to those at the other end of the connecting tube and that significant
enthalpy flux does not flow along the tube.

The actual sensing element in these pressure transducers is a flexible semi-
conductor strain gauge, with four strain-sensitive resistors connected in a bridge
configuration. A dc supply voltage must be connected to two terminals of the bridge;
the other two terminals produce the signal. If either side of the dc supply voltage
is grounded, a differential amplifier or a floating amplifier must be used on the
sensing terminals. Alternatively, if the amplifier grounds one of the signal lines, the
dc supply must float. If two or more transducers are powered by a single, floating
dc supply, only one transducer can be grounded at any one time.

At Los Alamos we usually use a lock-in amplifier (e.g., Model SR830, Stanford
Research Systems, Sunnyvale CA, www.srsys.com) to measure the magnitude and
phase of the oscillating voltages from piezoresistive pressure transducers (and from
any other oscillating signal sources). Alternatively, direct computer acquisition of
$V(t)$ can be used with piezoelectric transducers, which typically have high signal-
to-noise ratios.

We often use many pressure transducers in a new thermoacoustic system, to help
verify that the acoustic behavior of the system is correct. To understand all details
of a complicated system such as any of the four extended examples introduced
in Chap. 1, a dozen transducers might be desired, to enable separate verification
of complex pressure differences across each inertance, orifice, heat exchanger,
regenerator, etc. and to be sure that no unexpected, extra pressure differences arise
from poorly designed interfaces between any of those components. Unfortunately,
good pressure transducers are expensive. To compromise, we sometimes make
a dozen tapped holes of the proper size, but with transducers installed in only
some of them, plugging the other holes with bolts and gasket-washers. Thus,
as measurements proceed and we find ourselves wishing we knew the pressure

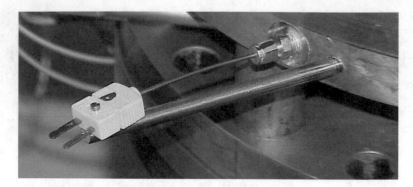

Fig. 9.2 A type-K sheathed thermocouple with subminiature connector, sealed with a metal compression fitting into a thermoacoustic pressure vessel. The sheath diameter is 1/16 in. The temperature-sensing tip of the sheath is deep inside the vessel

"everywhere," we can easily move transducers from place to place without taking the hardware to the machine shop. Once we have developed confident knowledge of our phasors and we proceed to more sophisticated investigations (such as streaming heat transport), we may even remove a few pressure transducers in order to use them in other experiments.

9.1.2 Mean Temperature

We usually use thermocouples [153] to measure T_m. Often we use commercially available (Omega Engineering, Stamford CT, www.omega.com) sheathed, ungrounded-junction thermocouple probes with integral standard electric connectors, as shown in Fig. 9.2. The sheathed type, with either stainless-steel or Inconel sheaths, is available in the same standard sizes as many metal compression fittings (e.g., Swagelok fittings) and rubber O-ring fittings, so it is easy to use a sheathed thermocouple with such fittings to seal a temperature-sensing probe tip inside the pressure vessel while the electrical connection is outside, as shown in Fig. 9.2. The ungrounded junction allows checking for shorts to ground; the absence of such shorts helps give us confidence that the thermocouple is not damaged. However, we sometimes make our own thermocouples from spools of thermocouple wire, if size constraints preclude the use of sheathed probes or if we want to spot-weld the two individual thermocouple wires to a metal surface to ensure the best possible thermal contact to the metal.

We choose the specific placement of thermocouples in a thermoacoustic system based on what we want to learn about the system. If we are concerned only about acoustic-approximation thermoacoustic performance, we might need only a small number of thermocouples in the thermal reservoirs (e.g., in water streams, heat-transfer-fluid circulation loops, metal cold tips, or combustion-gas streams). Measurement of system performance might require thermocouples on both the inlet and outlet of each heat-transfer stream. However, to diagnose the internal thermoacoustic behavior in an engine or a refrigerator when details "beyond Rott's acoustic approximation" are of interest, we sometimes install dozens of thermocouples. Acoustic network components such as inertances and orifices in orifice pulse-tube refrigerators might not be directly water cooled, so knowledge of their mean temperatures is sometimes important for accurate modeling. Measurement of the axial temperature profiles along pulse tubes, thermal buffer tubes, stacks, and regenerators helps diagnose Gedeon streaming and Rayleigh streaming. If internal regenerator or stack streaming, as illustrated in Fig. 7.12d, is suspected, many thermocouples within the regenerator or stack itself might be required. A detailed understanding of the heat transfer within thermoacoustic heat exchangers (which we have not yet attempted at Los Alamos) might require many thermocouples on the thermoacoustic side of the heat exchanger, many more attached to the metal boundary surfaces, and yet more on the non-thermoacoustic side (e.g., water-stream side).

In many cases, the details of the attachment and mounting of thermocouples are important. When knowledge of the temperature of a solid surface is desired, we try to attach a thermocouple directly to the solid. However, the pores in thermoacoustic heat exchangers, regenerators, and stacks are typically very small, so attachment of a thermocouple without significantly altering the thermoacoustic flow in the vicinity is usually challenging, if not impossible. In the standing-wave refrigerator of Figs. 1.13 and 1.14, one entire pore of the stack was a dedicated thermocouple-routing path; we deliberately blocked that pore completely.

Measurement of the average *gas* temperature with a thermocouple is, if anything, even more difficult than measuring a solid temperature, because the thermocouple itself is a solid surface fully capable of imposing the $T_1 = 0$ boundary condition on the thermoacoustic gas, thereby supporting boundary-layer thermoacoustic enthalpy flux along its surface. Hence, if a thermocouple is aligned parallel with the x direction, its tip will run hot if it points toward the nearest pressure antinode of the standing wave and will run cold if it points toward the nearest velocity antinode of the standing wave. The most meaningful results are obtained with the thermocouple tip perpendicular to the x direction.

9.2 Power Measurements

9.2.1 Acoustic Power

Acoustic power can be measured easily wherever U_1 can be inferred confidently. In two situations described below, U_1 can be inferred easily from pressure measurements. (See also Exercise 9.7.) In the third situation it is obtained from the motion of a solid object.

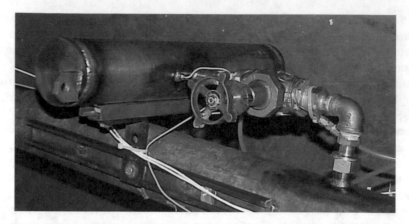

Fig. 9.3 A variable RC load connected through a 90° elbow to the thermoacoustic-Stirling heat engine of Fig. 1.22. The valve is the variable resistance; water lines have been wrapped around its two connections in order to keep it near ambient temperature. The cylindrical tank to the *left* of the valve is the compliance; a small pressure transducer is screwed into it

9.2.1.1 RC and RLC Loads

For testing thermoacoustic engines, it is useful to have a variable acoustic load with easily measurable power dissipation. Orifice pulse-tube refrigerators also need a load to absorb acoustic power at the hot end of the pulse tube; the load may incorporate phase-shifting inertance. These loads are easily constructed of lumped-impedance elements, as illustrated for an RLC load in Figs. 1.17 and 1.18, and for an RC load in Fig. 9.3. Here, the determination of the acoustic power dissipated in such loads is considered.

The volume V of the compliance can be obtained accurately from its construction details, and γ and p_m of the gas should be known accurately, so the volume flow rate into the compliance can be obtained easily and accurately from measurement of $p_{1,C}$ in the compliance, using

$$U_1 = i\omega C p_{1,C} = \frac{i\omega V}{\gamma p_m} p_{1,C}. \tag{9.1}$$

If there is negligible volume in the resistance and (if present) inertance, then U_1 is spatially uniform in them, and hence Eq. (9.1) also gives U_1 at the inlet to the acoustic network where $p_{1,in}$ can be measured. Then the acoustic power flowing at the inlet to the network is

$$\dot{E}_2 = \frac{1}{2}\mathrm{Re}\left[p_{1,in}\widetilde{U_1}\right] = \frac{\omega V}{2\gamma p_m}\mathrm{Im}\left[p_{1,in}\widetilde{p_{1,C}}\right]. \tag{9.2}$$

Equation (9.2) gives a simple, reasonably accurate result for the acoustic power that is dissipated in a lumped-element RC or RLC load. Fortunately, the mean temperature is completely irrelevant to this measurement, so there is no need to ensure that any of the components remains at a well-known temperature.

However, small inaccuracies often arise from several sources. First, the thermal-relaxation conductance on the surface area S of the compliance changes Eq. (9.1) and hence also changes Eq. (9.2). Fortunately, this effect is easy to minimize by making the compliance large and it is easy to model when $S\delta_\kappa/V$ is not negligible. Second, the unavoidable volume of gas in the inertance contributes compliance (see also Exercise 9.6), invalidating the assumption that U_1 at the inlet and U_1 into the compliance are identical (i.e., invalidating the assumption of "lumped" RL impedances), so we often use a numerical integration through the entire RLC network to obtain a more accurate estimate of $\dot{E}_2$, as illustrated for the orifice pulse-tube refrigerator in Appendix B. Finally, an effect like the volume-flow-rate discontinuity at an adiabatic–isothermal junction described near Eq. (7.58) in "Joining Conditions" in Chap. 7 can also invalidate the assumption that U_1 at the inlet and U_1 into the compliance are identical (see also Exercise 7.20).

9.2.1.2 Two-Microphone Method

The acoustic power flowing along a duct of area A can be determined from accurate measurements of the pressure amplitude and phase at two transducers A and B that are separated by a short distance Δx along the duct. The fundamental idea of the measurement is easy to understand when $\Delta x \ll \lambda$ and attenuation is neglected. In this simplified situation, the velocity is related to the pressure gradient by $i\omega\rho_m u_1 = -dp_1/dx$, so that

$$U_1 \simeq \frac{iA}{\omega\rho_m} \frac{p_{1B} - p_{1A}}{\Delta x} \tag{9.3}$$

gives an approximation to the volume flow rate midway between the two transducers in terms of p_{1A} and p_{1B}. Similarly, $p_1 \simeq (p_{1A} + p_{1B})/2$ gives an approximation to the pressure midway between the transducers in terms of p_{1A} and p_{1B}. The acoustic power is then

$$\dot{E}_2 = \frac{1}{2}\mathrm{Re}\left[p_1\widetilde{U}_1\right] \simeq \frac{A}{2\omega\rho_m \Delta x}\mathrm{Im}\left[p_{1A}\widetilde{p}_{1B}\right] = \frac{A}{2\omega\rho_m \Delta x}|p_{1A}|\,|p_{1B}|\sin\theta, \tag{9.4}$$

where θ is the phase angle by which p_{1A} leads p_{1B}.

Including attenuation in the laminar, boundary-layer approximation, and no longer requiring that Δx must be much shorter than λ, results in a similar but more accurate expression [155],

$$\dot{E}_2 = \frac{A}{2\rho_m a \sin(\omega\Delta x/a)}$$

$$\times \left(\mathrm{Im}\left[p_{1A}\widetilde{p}_{1B}\right]\left\{1 - \frac{\delta_\nu}{4r_h}\left[1 - \frac{\gamma-1}{\sqrt{\sigma}} + \left(1 + \frac{\gamma-1}{\sqrt{\sigma}}\right)\frac{\omega\Delta x}{a}\cot\left(\frac{\omega\Delta x}{a}\right)\right]\right\}\right.$$

$$\left. + \frac{\delta_\nu}{8r_h}\left(|p_{1A}|^2 - |p_{1B}|^2\right)\left[1 - \frac{\gamma-1}{\sqrt{\sigma}} + \left(1 + \frac{\gamma-1}{\sqrt{\sigma}}\right)\frac{\omega\Delta x}{a}\csc\left(\frac{\omega\Delta x}{a}\right)\right]\right), \tag{9.5}$$

which is still simple enough for routine use.

For the high standing-wave ratios typical of thermoacoustic engines and refriger-ators, Eq. (9.5) demands high accuracy in the measurement of the phase difference θ between p_{1A} and p_{1B}, except when locations A and B are chosen so that a velocity antinode is midway between them. This issue is most easily understood with reference to Eq. (9.4), where it is clear that a 1° error in the measurement of θ would have a small effect on the result for $\theta \sim 90°$ but an enormous effect on the result for $\theta \sim 0°$. Hence, we have almost always used this method very close to a velocity antinode. Elsewhere, with the $|p_{1A}|^2 - |p_{1B}|^2$ term in Eq. (9.5) significantly large, error in δ_ν (e.g., from poor knowledge of the gas temperature in the boundary layer, or from turbulent disturbance of the boundary layer, as discussed in the next paragraph) can also contribute significant error to the result.

Unfortunately, a velocity antinode is most likely to experience boundary-layer turbulence, while Eq. (9.5) was derived assuming a laminar boundary layer. Turbulence certainly invalidates the derivation of Eq. (9.5). Nevertheless, at Los Alamos we have optimistically used both Eqs. (9.4) and (9.5) in the turbulent regime near velocity antinodes and with $r_h \gg \delta_\nu$, and they have always given results that are reasonably consistent with other system-power measurements and with calculations.

A velocity antinode, where $|p_1|$ is smaller than in other parts of the system, is often a location of comparatively large $|p_{2,2}|$. Hence, the lock-in amplifier (or other instrumentation) used for two-microphone acoustic-power measurements should have good harmonic-rejection capability.

9.2.1.3 Pistons and Loudspeakers

The volume flow rate U_1 can also be determined by measurements on a moving piston or loudspeaker, whenever the moving area A is known and any variable from among the amplitude ξ_1, velocity u_1, or acceleration a_1 can be measured. Hence, if p_1 is measured in front of the piston or loudspeaker, the acoustic power delivered from the front of the piston or loudspeaker is given by any of

$$\dot{E}_2 = -\frac{\omega A}{2} \, \mathrm{Im} \left[p_1 \widetilde{\xi_1} \right] \tag{9.6}$$

$$= \frac{A}{2} \, \mathrm{Re} \left[p_1 \widetilde{u_1} \right] \tag{9.7}$$

$$= \frac{A}{2\omega} \, \mathrm{Im} \left[p_1 \widetilde{a_1} \right]. \tag{9.8}$$

Hofler [156] used a small accelerometer [153] on the bellows-sealed driving surface of a highly modified loudspeaker. The area A was a suitable average of the bellows inner and outer areas. Knowledge of A can easily be the most difficult aspect of this method when using an unmodified loudspeaker, because the moving dome, cone, and surround are so flexible that it may not be clear what area A represents the true moving area. Flexing of these parts can be due to p_1 as well as to their motion.

A rigid moving piston, such as that of a linear motor–alternator, eliminates this uncertainty. An accelerometer can be used on a moving piston, but a linear variable differential transformer [153] (search the Internet for "LVDT") can also be used, because the moving mass of pistons is typically so large that the moving mass of this type of transducer adds little.

The volume flow rate of a moving piston or loudspeaker can also be inferred from pressure measurements in a known volume that experiences spatially uniform pressure oscillations on one side of the moving element (see Exercise 9.7).

Pistons typically suffer from "blowby" leakage. If the volume flow rate through this leakage is significant, the effective volume flow rate delivered by the piston will differ from that calculated from the piston area and motion (see Exercise 8.14).

9.2.2 Heat

Figure 9.4 shows a few methods for measuring the rate at which heat is delivered to or extracted from a heat exchanger in a thermoacoustic system. In most of these methods, measurements of "the temperature" T of a flowing stream are required in order to ascertain the stream's enthalpy flux $\dot{H} = \rho c_p T U$ at two or three locations. It is important to ensure that the stream is laterally well mixed at each location of temperature measurement. [Otherwise, multiple thermometers (and velocimeters!) would be required to determine the enthalpy flux through $\dot{H} = \int \rho c_p T u \, dA$.] To ensure good mixing, one, two, or three short-radius elbows can be placed upstream of the thermometer, with no two elbows sharing the same plane, so that the centrifugally generated eddies [91] in these elbows mix the stream well.

Figure 9.4a shows the conceptually obvious method to measure the heat extracted from (or added to) a thermoacoustic system in a heat exchanger by a steady stream of a fluid such as water flowing through the heat exchanger. The volume flow rate of the stream is measured by a flowmeter, and temperatures upstream and downstream of the heat exchanger are measured by thermocouples. Everything in the vicinity of the heat exchanger and thermocouples is insulated. The heat $\dot{Q}$ is then obtained from

$$\dot{Q} = \rho c_p \left(T_{\text{out}} - T_{\text{in}} \right) U. \tag{9.9}$$

This method relies on accurate knowledge of the fluid's density ρ and specific heat c_p, which may be too uncertain for extremely accurate measurements. (Does your building's circulating chilled water contain anti-corrosion additives, algicides, or anti-freeze? Have you ever asked your building manager about their concentrations? their specific heats?) It also depends on the calibration accuracy of the flowmeter, which might be only 5 or 10% for an inexpensive model.

Sometimes greater accuracy with less effort is possible using either of the methods shown in Fig. 9.4b, c. The stream of fluid passes sequentially through the thermoacoustic heat exchanger and an electric resistance heater. The electric power

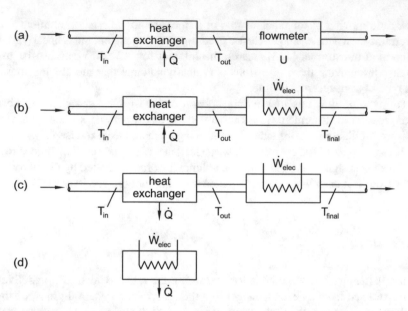

Fig. 9.4 Four methods of measuring the heat $\dot{Q}$ transferred to or from a thermoacoustic heat exchanger. For accurate measurements, all parts shown must be thermally insulated. (**a**) A heat-transfer fluid flows steadily *from left to right*, absorbing heat $\dot{Q}$ from the thermoacoustic system, and a flowmeter measures the fluid's volume flow rate U. (**b**) An electric-resistance heater takes the place of the flowmeter. (**c**) Same as (**b**), but with heat $\dot{Q}$ delivered to the thermoacoustic system. (**d**) An electric-resistance heater delivering heat $\dot{Q}$ directly to the thermoacoustic system

to the heater $\dot{W}_{elec}$ is measured accurately, as are the three temperatures T_{in}, T_{out}, and T_{final}. Assuming constant heat-capacity flux $\rho c_p U$ through the assembly, the heat delivered to the stream by the thermoacoustic heat exchanger is given by

$$\dot{Q} = \pm \dot{W}_{elec} \frac{T_{out} - T_{in}}{T_{final} - T_{out}}. \tag{9.10}$$

This method is least sensitive to thermometer errors when $\dot{W}_{elec}$ is comparable to or greater than $\dot{Q}$. The plus sign in Eq. (9.10) is appropriate for Fig. 9.4b, for which positive $\dot{Q}$ flows from the thermoacoustic system into the fluid stream, such as in a typical ambient-temperature, water-cooled thermoacoustic heat exchanger. The minus sign is needed for Fig. 9.4c, for which positive heat flows from the stream to the thermoacoustic system, such as in the hot heat exchanger of an engine or the cold heat exchanger of a refrigerator.

Adjusting $\dot{W}_{elec}$ to make $|T_{final} - T_{out}| = |T_{out} - T_{in}|$ is a special case. For Fig. 9.4b, this allows wiring four thermocouples, at T_{final}, T_{out}, T_{out}, and T_{in}, suitably in series/antiseries (using thermopiles, if desired) so that a single measurement of the total thermoelectric voltage's deviation from zero can be used to actively control the electric heater, ensuring $\dot{Q} = \dot{W}_{elec}$ always. Similarly, for Fig. 9.4c, this allows wiring two thermocouples at T_{final} and T_{in} in antiseries so that a single measurement of the total thermoelectric voltage's deviation from zero can be used to control the heater, again ensuring $\dot{Q} = \dot{W}_{elec}$. We used this second approach to measure the cooling power of the orifice pulse-tube refrigerator of Figs. 1.17 and 1.18 as a natural-gas stream (mostly methane) passed through it, without concern for the gas's ethane/propane content or for its liquid/gas ratio as it left the heat exchanger.

Figure 9.4d shows a trivial variation on this theme: For some purposes, especially in research, a heat exchanger might be directly heated by electric power, so the heat delivered to the thermoacoustic system is obviously equal to the electric power.

Electric power near 50 or 60 Hz is easily measured using commercially available "Watt meters," such as those made by Ohio Semitronics, Hilliard OH, www. ohiosemitronics.com, which produce a dc voltage or current accurately proportional to the ac electric power flowing through them. Below a few kW, such power can be easily controlled by means of a variable autotransformer ("Variac").

9.3 Difficult Measurements

It is possible to measure $\mathbf{v}(t)$ and $T(t)$ using expensive instrumentation such as laser Doppler velocimeters, hot-wire anemometers, and cold-wire thermometers [151]. With great care, these probes can even be applied over spatial dimensions that are smaller than δ_ν and δ_κ [157]. However, such difficult or expensive measurements are beyond the scope of this book.

9.4 Points of View

By "point of view" here I mean the frame of mind in which one imbeds and organizes experimental design, measurement, data analysis, and problem solving, in order to achieve a desired goal. For example, suppose my old car won't start, and my goals are to figure out why and to fix it. One point of view might be called "historical." From this point of view, I would try to figure out why the car won't start by reviewing the recent history of my car-related actions: Did I let it run low on fuel? Did I leave the lights on overnight? Did I neglect to check the oil for a few years? Another point of view might be called "functional." The problem might be electrical, mechanical, or fuel-related. What data do I have, and what further tests can I perform, to narrow it down to one of these three? If turning the key makes no sound at all, the problem is probably electrical—possibly battery, wiring, starter motor, etc. Do the horn and headlights work? If not, a dead battery is likely; if so, perhaps the starter motor or its solenoid has failed.

This simple example should make it clear that the most successful problem solvers use multiple points of view, even simultaneously [158]. Learning to apply more than one point of view, and consciously planning points of view, can enhance problem-solving skills.

In science and engineering, the point of view is often expressed finally and most clearly in the combinations of variables chosen for display in a plot. This fact can be used to help form one or more tentative points of view very early in a research project, to guide the design of the experimental hardware and, especially, its instrumentation. I once heard a mechanical-engineering professor teach this idea clearly and bluntly: He told a new student to plan every figure in his master's thesis—including the "units" and the orders of magnitude involved on the axes of all graphs—*before* building his experimental hardware.

Point of view is also expressed in the outline followed by an article or thesis describing the work. For example, the traditional "Introduction, Apparatus, Measurements, Results, Conclusions" outline usually reflects the temporal order in which research tasks were done—the historical point of view. More than one point of view is often used in technical writing. For example, a functional point of view might be used within the "Apparatus" section of a thesis whose overall organization follows the historical point of view.

In this section, four points of view will be considered: (1) "natural dependence," which corresponds to the hands-on feel of an experiment; (2) "evidence," which here means using an experiment to prove or disprove a theory; (3) "performance," which focuses on a desired behavior; and (4) "a thermoacoustic perspective," which helps diagnose unanticipated behavior. It is often natural and productive to employ more than one of these points of view. For example, a long journal publication about a new type of thermoacoustic device might go through several points of view sequentially, beginning with (1) to give the reader an overall understanding of the experiment, proceeding to (2) to compare experimental data with a theoretical prediction, and concluding with (4) to help motivate future work.

In the four examples below, I am simplifying each story to keep it illustrative of one point of view. For example, I'll neglect to describe hardware iterations that came between the original designs and the final data sets. This simplification is most extensive in the last example, where I am skipping many important details, omitting Cryenco proprietary information, and neglecting to mention such important facts as the non-negligible pressure drops in each of the three heat exchangers. These omissions and simplifications do not affect the "point of view" lessons illustrated. More complete stories about each of the four examples can be found in the cited references.

9.4.1 Natural Dependence

This very natural point of view adopts direct, experimental cause and effect as its primary focus. Some variables—"independent variables"— are easy, natural controls of experimental behavior, while other, "dependent" variables are consequences of them. This point of view arises naturally when new hardware or a new experiment is being planned, as one thinks about what it will naturally "do." Problem diagnosis and solving can be very direct with this point of view, as it focuses attention on immediate cause–effect relationships. Communication of results using this point of view is usually easy, because the audience may have enough shared or related experience to have already an intuitive appreciation for the experimental causes and effects. Sometimes it is best to stay close to actual experimental variables, e.g., converting "transducer voltage" to kPa but not going so far as to divide such a pressure by other variables to form a dimensionless group.

For example, orifice pulse-tube refrigerators usually employ an adjustable orifice—a valve. "Number of turns" of the orifice valve is a natural independent variable, because the adjustment of that variable gives direct, routine control of the gross cooling power of an orifice pulse-tube refrigerator when the driving amplitude (e.g., piston stroke or pressure amplitude) is kept fixed. Designing an experiment with this independent variable in mind is obvious and, when reporting research results, other researchers have an immediate, fundamental understanding of the role of this variable. One might consider converting "number of turns" to C_V or to R_v when reporting research results, in order to describe this variable in units that are independent of the specific valve that is used, but doing so runs the risk of losing contact with some of the audience, who might not understand MPa·s/m^3 or the even weirder units of C_V.

Some combinations of independent and dependent variables occur frequently. For example, it is common for several interesting things to happen when one natural experimental control is adjusted, so experiments are often conceived with several dependent variables and one independent variable, and results are often displayed with two or more graphs atop each other, sharing a common horizontal axis. Sometimes a second independent variable is held fixed while the easiest natural control is adjusted, and then another set of data is obtained with the second variable fixed at a new value. Such results are often plotted as a family of curves or data sets on a graph with the easiest natural control on the horizontal axis. When more than two independent variables are involved, the amount of data can be overwhelming unless the experiment is carefully planned. Sometimes it is best to select a reference (baseline) case, and to explore the vicinity of this reference case by changing one independent variable at a time while keeping the other independent variables fixed at their reference values.

Example: Standing-Wave Engine

At one time in our development of the standing-wave engine shown in Figs. 1.8 and 1.9, we had to show our Tektronix sponsors "how it worked," in preparation for delivering the hardware to them. (At that particular time, we had completed only one of the two stacks, and it was also important to show them how well one stack worked relative to DELTAE predictions, to give them confidence that the subsequent two-stack system would later behave as well as we expected.) The hot end of the engine was heated electrically, so heater power $\dot{Q}_H$ was the most obvious, natural experimental independent variable. We measured it with a commercial Watt meter (checking this measurement with voltmeter, ammeter, and multiplication). When a given heater power was selected, the engine equilibrated to a specific operating point, with a specific $p_1(x)$, $T_m(x)$, and operating frequency. (Cooling-water temperature T_0 was fixed by our building's recirculating water system.) The functions $p_1(x)$ and $T_m(x)$ could have been measured at a number of places, but we selected only a few for instrumentation. We were most concerned about overheating the hot-end pressure-vessel boundaries, so the temperature of the hot metal enclosure was measured with multiple, redundant thermocouples and became known as T_H. For convenience, we decided to measure p_1 only at ambient temperature, the most important such location being the branch to which the refrigerators would later be attached. One pressure transducer was installed at this branch, near the adjacent ambient heat exchanger, and a second transducer was initially installed at the opposite end of the resonator as a simple reality check of the half-wave resonance shape. Early measurements always showed the expected complex ratio between the p_1's at these two locations, so we soon stopped looking at the second pressure transducer.

Plotting T_H and $|p_{1,\text{branch}}|$ as functions of $\dot{Q}_H$, as shown in Fig. 7.1, gives a natural, direct explanation of how this engine behaved. We augmented these plots with a simple statement about another, less-interesting natural dependent variable, the operating frequency f: The observed operating frequency was always within 1% of DELTAE's calculated value.

9.4.2 Evidence

The evidential point of view is used when something must be proven. One might be trying to prove that a theory is correct, or that it is incorrect, or that an idea has at least some merit, or that it has no merit. One might only be trying to prove that one's understanding of something is mature enough to justify embarking on a more expensive, time-consuming development project. For the evidential point of view, it is also helpful to have in mind who must be convinced, so that one can imagine what evidence must be presented in order to convince them.

When trying to prove a theory, the theory itself often guides the design of the experiment and its instrumentation. The graphical presentation of data may also be guided by a theory one is trying to prove, as in the common case of experimental points compared to a theoretical curve on a graph. Sometimes the theory is used to calculate values of variables in units that are natural to the experiment. However, more often the experimental values are manipulated into a form that is natural to the theory—often dimensionless groups, which convey more generality and have all the advantages discussed at the end of Chap. 7.

In addition to the main plan of the experiment, it is valuable to seek opportunities to make preliminary or auxiliary measurements of quantities that are easily understood or are understood according to previously established theories, in order to build confidence in the experiment and its interpretation. Such auxiliary measurements help to prove that the experiment is sound.

Example: Standing-Wave Refrigerator

The dual purpose of the standing-wave refrigerator [3, 4, 35] shown in Figs. 1.13 and 1.14 was (1) to show that a steadily flowing stream of gas could indeed be cooled by superimposing thermoacoustic oscillations on the flowing stream, injecting and removing the steady flow at nodes in the thermoacoustic standing wave, and (2) to confirm (or perhaps disprove!) something about the theory summarized here in "Deliberate streaming: Parallel flow" in Sect. 7.4.5.

The first purpose was easily satisfied by planning the system according to the symmetric, two-node arrangement shown in Fig. 1.14, blowing steady flow from node to node, and putting thermometers in the steady streams just outside the nodes. (The left–right symmetry of the system made it easy to design, as there was no need to design the left and right acoustic networks differently.) We decided to design the hardware to work reasonably for both atmospheric air and 3-bar helium (or helium-argon mixtures). Thus, when using air, cold air could be delivered from the lower node to the room to let visitors feel the cold air directly and to show that almost no sound radiated from the open node, while higher-pressure helium could be used to give the apparatus more cooling power for careful measurements, so that unavoidable heat leaks to the room were a smaller fraction of measured powers.

In deciding what aspect of the theory to test with this hardware, we began from the fact that $p_1(x)$ and $T_m(x)$ are the easiest quantities to measure; heats are more difficult. The theory outlined in Chap. 7 shows that steady flow should have a negligible effect on p_1 and a large effect on T_m in the stack, so we focused our attention on $T_m(x)$. The steady flow was expected to cause a dramatically curved $T_m(x)$ in the stack, because this was predicted both by numerical integration of the enthalpy-flux equation and by a simple analytic approximation to that equation. So we put many thermocouples inside the stack, to measure $T_m(x)$. For confidence-building measurements, we also installed several extra pressure transducers to map out $p_1(x)$ at locations throughout the entire standing wave. We decided not to devote much attention to the loudspeaker drivers.

Choice of the primary independent variable was simple: nonzero steady flow U_m was central to the idea and the theory we were testing, and it was experimentally easy to control and to measure. As that was varied independently, we kept our well-known thermoacoustic variables f, T_0, T_C, and $|p_1|$ at a reference location constant, and measured $T_m(x)$ in one stack as the dependent variable of most interest. Typical results are displayed in Fig. 9.5, which shows $T_m(x)$ for $U_m = 0$ (a confidence-building result), $U_m > 0$ (steady flow in the cooling direction, the purpose of the hardware), and $U_m < 0$ (steady flow from the cold end toward the ambient end of the stack, which was easy to do in both the experiment and the theory). The plot compares experimental data and numerical calculations based on the theory.

Fig. 9.5 Comparison of data obtained from the standing-wave refrigerator of Figs. 1.13 and 1.14 with the theory outlined in Sect. 7.4.5. *Middle set*: $U_m = 0$; *upper set*: $U_m = 5.2$ L/s toward the cold end; *lower set*: $U_m = 5.0$ L/s toward the hot end. Details can be found in [3, 4, 35]

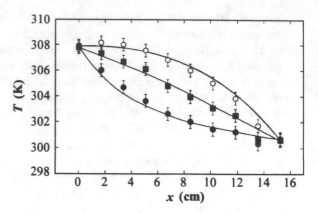

9.4.3 Performance

Sometimes the goal of a design is the optimization of performance, and the goal of a series of measurements is the demonstration of a desired level of performance. In this situation, a "figure of merit" that gives a quantitative measure of the desired performance should be identified. Some examples are cooling power, cooling power per unit weight of hardware, efficiency divided by Carnot's efficiency, and estimated mass-production unit cost for hardware of a given cooling power [159].

When the sponsor of some work identifies multiple figures of merit that cannot be optimized simultaneously, it is best to get additional information to guide the necessary compromises. For example, the sponsor of a combustion-powered natural-gas (methane) liquefier might ask for low fabrication cost and high efficiency (both at a given cooling power), which cannot be maximized simultaneously (see Fig. 1.25). In this situation one might ask for further guidance. Occasionally the sponsor can respond with a quantitative, combined figure of merit, such as "efficiency squared, divided by cost," but often the sponsor does not have such a detailed, quantitative understanding of his or her goals. Persistent questions may elicit better guidance: "At that cooling power, we might be able to give you either a fabrication cost of $100k with an efficiency of 70% liquefy/30% burn, or a $150k fabrication cost at 75%/25%. Which do you think you would prefer?" From the answers to a series of such questions, a combined figure of merit can be formulated.

Once a figure of merit has been identified, hardware can be designed to optimize it, and the instrumentation and experimental program to measure this figure of merit can be planned. As usual, many easy auxiliary measurements should be included to build confidence.

Example: Thermoacoustic-Stirling Heat Engine

When we began planning the thermoacoustic-Stirling heat engine shown in Fig. 1.22, we already had the lumped-*LCR*-impedance topology shown in Fig. 1.23 firmly in mind, and we wanted to demonstrate how well it could perform. The small floor space available in our laboratory was an important constraint, and secondary constraints included cost and time. We expected our audiences to include both reviewers of a scientific journal and program managers who might support the fabrication of a much larger version of this engine for combustion-powered natural-gas liquefaction. The sponsor of our research would be proud of our success with either of these audiences. We decided that "efficiency divided by Carnot efficiency" would be appreciated by journal reviewers, while raw "efficiency" would be more relevant to program managers. With ambient temperature fixed near 20 °C and the hot temperature limited to about 750 °C by practical materials, we believed that the same hardware design would optimize both raw efficiency and Carnot-normalized efficiency, so we adopted raw efficiency—work divided by heat—as our figure of merit. Our ultimate objective was large-scale hardware, for which resonator losses are negligible, so we decided that we could define "work" as acoustic power delivered from the toroidal *LCR* network to the resonator, without consideration of resonator losses. We also decided that both journal reviewers and program managers would not care whether the hot-end heat was delivered to this first engine from a combustion heat source or from electricity, so we chose the easily instrumented electric heater. Water cooling for the ambient-temperature heat exchanger and screens for the regenerator were obvious, practical choices. Arbitrarily, 9 cm was chosen as the diameter of the regenerator. Once these overall choices were made, the dimensions, pressures, and other operating parameters were designed by maximizing the efficiency according to DELTAE (see Appendix B).

With efficiency as our focus, the most important instrumentation was clearly identified: a Watt meter to measure $\dot{Q}_H$, and pressure sensors to measure $|p_1|$ at a reference location and for obtaining a two-microphone measurement of $\dot{E}_2$ at the entrance to the resonator. We expected the two-microphone measurement to be challenging, because this location was far from the velocity antinode, so we installed some extra pressure transducers along the resonator to build confidence. Thermocouples to measure T_0 and T_H were also obvious needs. We expected that Gedeon streaming and Rayleigh streaming would harm efficiency if they were not controlled, so we installed many more thermocouples to measure $T_m(x)$ along the regenerator and thermal buffer tube to diagnose these streamings, we installed the adjustable jet pump to control Gedeon streaming, and (after much DELTAE analysis of less promising options) we installed an adjustable resistance as part of the jet-pump assembly, essentially in series with the jet pump, to control the phase of Z in the thermal buffer tube, in an attempt to control Rayleigh streaming. Although performance demonstration was our primary goal, we also wanted to acquire as much understanding as possible about this engine's operation, to guide future developments toward higher power and efficiency. Hence, a few other pressure transducers were included for the secondary purposes of detailed understanding and confidence building.

Extensive measurements with this engine are described in detail in [38], starting with confidence-building measurements and ending with a quantitative entropy-generation tabulation of loss mechanisms throughout the engine at two particular operating points. Figures 9.6 and 9.7 are included here as examples of these results.

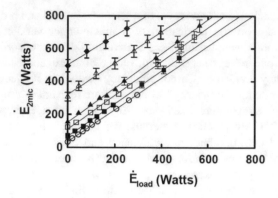

Fig. 9.6 Measured two-microphone power vs measured load power, at six different pressure amplitudes, for the thermoacoustic-Stirling heat engine of Figs. 1.22 and 1.23. Amplitudes were $|p_1|/p_m = 0.038, 0.051, 0.061, 0.069, 0.088$, and 0.10, with the larger amplitudes nearer the *top of the figure*. The *lines* are least-squares fits to the measurements

Fig. 9.7 Efficiencies vs T_H for pressure amplitudes $|p_1|/p_m = 0.051, 0.061, 0.069, 0.088$, and 0.10 (increasing *from left to right*) in the thermoacoustic-Stirling heat engine of Figs. 1.22 and 1.23. The efficiency can be defined in two ways: in terms of acoustic power delivered to the horizontal part of the resonator from the junction below the heat exchangers (*upper graph*), or in terms of acoustic power delivered to the load (*lower graph*)

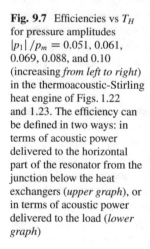

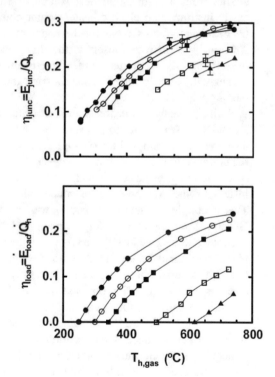

Figure 9.6 presents some confidence-building results: a comparison of two-microphone acoustic-power measurements near the entrance to the resonator and *RC*-load acoustic-power measurements at the variable *RC* load attached to the resonator. The slopes, near unity, show self consistency between the two types

of power measurement in this apparatus. The vertical offsets are a measure
of resonator dissipation as a function of pressure amplitude, which were also
checked for consistency against DELTAE calculations.

Figure 9.7 shows efficiency for several operating points, using as independent
variables T_H and the pressure amplitude at the reference location above the
regenerator. The upper plot shows our chosen figure of merit: the efficiency with
which acoustic power was delivered to the resonator, reaching nearly 30% for
several operating points. The lower plot shows the efficiency with which acoustic
power was delivered to the load.

9.4.4 A Thermoacoustic Perspective

In papers and talks describing novel heat engines and refrigerators, there is often
evidence of careful planning and design, considerable effort in building hardware,
and some time spent measuring performance, but insufficient time devoted to real
understanding, to debugging, and to iteration of the hardware. Sometimes schedules
slip, and debugging and iteration that were planned for the end of the project
are eliminated. Other times, projects are canceled before advanced activities are
reached. Consequently, many of us actually have little experience in these final
aspects of system development. At best, we know how to measure (and publish)
overall system performance, but many of us are too inexperienced in tailoring
our measurements to diagnose the cause or causes of unanticipated poor system
performance. And, in my experience, initial measured system performance is never
as good as it ought to be—never as good as the sponsors would like, never as good
as I think it ought to be—so diagnosis and iteration are always necessary.

When trying to diagnose poor performance or new phenomena in a thermoacoustic engine or refrigerator at Los Alamos, we often adopt the experimental point of
view summarized in Fig. 9.8, which parallels the thermoacoustic perspective of this
book. First of all, we measure the pressure phasors throughout the system and the
mean temperatures at the heat exchangers with great care, striving to understand
the phasors especially in the low-amplitude limit. With confidence in these, we can
be confident in the acoustic-approximation powers $\dot{E}_2$ and $\dot{H}_2$ inferred from them.

(a) Know the pressure phasors.

(b) Know what was built.

(c) Deduce $\dot{E}_2$, $\dot{H}_2$ from (a) and (b).

(d) Deduce $\dot{Q}$ from (c) and control volumes.

Fig. 9.8 Outline of a thermoacoustic point of view for the interpretation of measurements.
If calculations and measurements do not agree in (**c**) and (**d**), suspect causes described in Chap. 7,
"Beyond Rott's thermoacoustics" and go to lower amplitude

From this point of view, any deviation of measured powers from powers calculated using $\dot{E}_2$ and $\dot{H}_2$ is due to phenomena beyond Rott's acoustic approximation—the phenomena discussed in Chap. 7.

When bringing the thermoacoustic perspective to an engine or refrigerator, it is common to move far beyond the natural experimental independent and dependent variables, adopting independent variables that facilitate thermoacoustic thinking. For example, two natural control variables for an engine driving an RC load might be heater power and number of turns of the valve that comprises R, but clear thinking from the thermoacoustic perspective is probably easier if the square of the pressure amplitude at a reference location and the real part of the inverse of the RC impedance (proportional to the acoustic power dissipated) are chosen as independent variables.

Using this point of view during a program of measurement on a thermoacoustic system sometimes requires that data be taken in an apparently awkward fashion, adjusting a natural independent variable (e.g., $\dot{Q}_H$) until an unnatural, thermo-acoustic independent variable (e.g., $|p_1|$ at a reference location) is forced to take on a desired value. The time sacrificed in such unnatural data acquisition is often recovered by enabling easy diagnosis of problems and clear presentation of results.

A sophisticated, somewhat abstract point of view such as the thermoacoustic perspective is a vital guide to thinking when thermoacoustic engines or refrigerators do not perform as expected, because in these complicated systems causes and effects are extremely interdependent. The oscillating pressure gradients cause the oscillating flows, the oscillating flows cause the oscillating pressure gradients, and these simultaneously depend on and create the operating frequency. Meanwhile, these oscillating quantities are in part determined by the mean temperatures, which might also depend on the oscillations; but these temperatures are affected by streaming, which is itself a product of the oscillating quantities; etc.! Natural, experimental cause–effect relationships may be of little help in untangling such complicated relationships.

The Final Example: Orifice Pulse-Tube Refrigerator

The purpose of the orifice pulse-tube refrigerator shown in Figs. 1.17 and 1.18 was to demonstrate efficient liquefaction of methane by an orifice pulse-tube refrigerator at a cooling power roughly 30 times higher than that of any previous pulse-tube refrigerator, with hardware that could be factory-produced at low cost. Secondary purposes were to verify that the tapered pulse tube suppressed Rayleigh streaming, which at that time had been demonstrated with only one experiment [88], and to show that both the magnitude and the phase of the acoustic impedance Z at the top of the pulse tube could be varied suitably using the two adjustable valves in the RLC network at the top.

With these purposes in mind, the most important measurements were of refrigeration power $\dot{Q}_{meth}$ at the cold heat exchanger and acoustic drive power $\dot{E}_2$ below the aftercooler, with measurements of p_1 in the RLC network atop the pulse tube and of $\dot{Q}_{top}$ delivered to the water stream in the heat exchanger at the top of the pulse tube also important. The final instrumentation design is shown schematically in Fig. 9.9. Constrained by costs, we installed only the six pressure sensors shown, because we believed that the pressure drops across the various

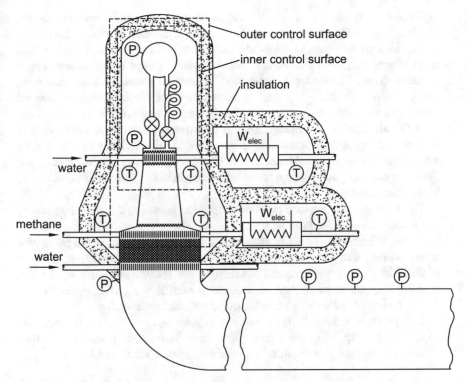

Fig. 9.9 Schematic of the instrumentation layout for the orifice pulse-tube refrigerator shown in Figs. 1.17 and 1.18. "P" indicates the location of a dynamic pressure transducer measuring p_1 and "T" indicates the location of a thermocouple measuring T_m. The parts of the refrigerator are labeled in Fig. 1.18

heat exchangers would be much smaller than those across the *RL* network and the regenerator.

I believe that using the pressure amplitude in the pulse tube as an independent variable leads to clear physical insight about experiments with orifice pulse-tube refrigerators, even though this is not as "natural" as using drive pressure at the aftercooler. There are several reasons for my belief:

First, measurement of p_1 in the compliance tank gives accurate knowledge of U_1 into the compliance tank and hence also at the hot end of the pulse tube. Adding knowledge of the pulse-tube volume and p_1 in it yields a very confident prediction for U_1 at the cold heat exchanger. The gross cooling power $\dot{E}_{2,C}$ is thus accurately known, depending only on the pressure amplitude in the pulse tube and the easily determined acoustic impedance of the *RLC* network, but not directly dependent on the aftercooler drive pressure. In contrast, trying to infer the cold heat exchanger's U_1 and $\dot{E}_2$ from measurements at the aftercooler requires knowledge of the regenerator's effective compliance, its flow resistance, and its dT_m/dx-dependent source/sink characteristics, which are usually not known with as much confidence.

Second, a consistent focus on the pulse-tube pressure amplitude as the independent variable leads to a clear conceptual separation of effects that reduce the cooling power (such as pulse-tube streaming and regenerator enthalpy flux) from effects that raise the required input acoustic power (such as regenerator pressure drop). For example, pulse-tube convection (both Rayleigh streaming and jet-driven streaming, as well as boundary-layer $\dot{H}_2$ flux) reduces cooling power and depends only on circumstances in the pulse tube, not in the regenerator or at the aftercooler. In contrast, a focus on drive pressure leads to thinking of unexpected regenerator pressure drop as something that reduces cooling power, which confuses analysis of other causes of reduced cooling power, especially if the gross cooling power is then based on the predicted p_1 instead of the actual p_1 in the pulse tube.

Third, focus on pulse-tube pressure amplitude lends itself well to lost-exergy accounting, as described in Chap. 6.

Hence, we used the pressure amplitude measured at the top of the pulse tube as our primary independent variable. Following the thermoacoustic point of view outlined in Fig. 9.8, we analyzed this refrigerator starting at the top of Fig. 9.9, focusing first on phasors, then on acoustic power, then on cooling power, and finally on enthalpy flux along x in the pulse tube and regenerator.

The pressure sensor in the compliance gave $p_{1,C}$ in the compliance. As suggested by Fig. 4.21, the volume flow rate into the compliance was then obtained from the known volume V of the compliance via Eq. (9.1):

$$U_{1,\text{in}} = \frac{i\omega V}{\gamma p_m} p_{1,C}.$$

(9.11)

If the volume of gas in the inertance and valves had been negligible, as suggested by Fig. 4.21, then the acoustic impedance Z at the location of the pressure sensor at the top of the pulse tube would have been simply

$$Z = \frac{p_{1,pt}}{U_{1,\text{in}}} = \frac{\gamma p_m}{i\omega V} \frac{p_{1,pt}}{p_{1,C}}.$$

(9.12)

However, in fact, we used the two measured pressure phasors in a simple DELTAE model to numerically integrate through the inertance, valves, and associated plumbing, as shown in Appendix B, to obtain more accurate "measured" values of Z and $U_{1,\text{top},pt}$ at the top of the pulse tube. Typically these values differed from those of the simple Eqs. (9.11) and (9.12) by roughly 10%. The circles in Fig. 9.10 show many such "measured" values of Z. (Five of these points are numbered, for later discussion.) As hoped, a broad range of Z was accessible by adjusting the valves. For these measurements, we avoided adjustments with smaller $|Z|$ in order to stay conservative in operating the thermoacoustic engine that drove this refrigerator.

Figure 9.11 shows a comparison between experimental and calculated pressure phasors below the aftercooler, with the DELTAE calculations forced to match

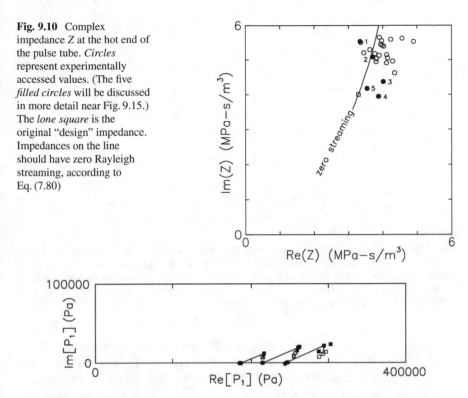

Fig. 9.10 Complex impedance Z at the hot end of the pulse tube. *Circles* represent experimentally accessed values. (The five *filled circles* will be discussed in more detail near Fig. 9.15.) The *lone square* is the original "design" impedance. Impedances on the line should have zero Rayleigh streaming, according to Eq. (7.80)

Fig. 9.11 Pressure phasors showing the pressure drop across the regenerator. *Filled squares* represent measurements below the aftercooler; *open squares* represent corresponding DELTAE calculations. The *circles* (on the horizontal axis) represent the top of the pulse tube, common to both measurements and calculations. Data at three values of pressure amplitude at the top of the pulse tube are shown. For each pulse-tube pressure amplitude, two or three aftercooler sets are shown, corresponding to different settings of the valves above the pulse tube. To guide the eye, lines link some members of each data set

the measured pressure phasors in the compliance and at the top of the pulse tube. The calculations also matched the experimental frequency, mean pressure, and heat-exchanger temperatures. Hence, this figure essentially shows a check of our understanding of the pressure drop across the regenerator, which was due to the flow resistance of the regenerator and the volume flow rate through it. The magnitude of the experimental pressure drop was typically 10% larger than the magnitude of the calculated pressure drop, and the phase of the experimental pressure drop led that of the calculated pressure drop by typically 10°. Possible systematic errors arising from the instruments were small, so these differences must represent mismatch between what we built and the DELTAE model. Candidates include error in the volume porosity of the as-built regenerator and fundamental error in DELTAE's regenerator algorithm.

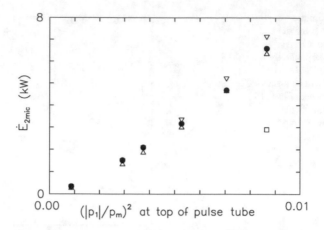

Fig. 9.12 Two-microphone power, with both *RLC* valves closed, as a function of the square of the normalized pulse-tube pressure amplitude. *Filled circles* represent measurements. *Open symbols* represent DELTAE calculations: square for laminar resonator dissipation, erect triangles for turbulent resonator dissipation with surface roughness 5×10^{-4}, and inverted triangles for turbulent resonator dissipation with surface roughness 1×10^{-3}

Thus far, we had established some confidence in our understanding of the phasors at roughly the 10% level. Hence, as we proceeded to analysis of acoustic power, we would not have been surprised at 10–20% disagreements between measurements and calculations.

The presence of three sensors near the pressure node of the resonator, as shown in Fig. 9.9, allowed two-microphone measurements to be made three different ways, using any pair of sensors selected from the three. These were always in good agreement, showing the expected, systematic differences due to the dissipation of acoustic power between the sensors themselves. Henceforth, the average of these three measurements will be called $\dot{E}_{2\text{mic}}$, the acoustic power consumed by the refrigerator and that portion of the resonator between the refrigerator and the two-microphone location.

As a further check of our understanding, we made some measurements of $\dot{E}_{2\text{mic}}$ with the valves in the refrigerator's *RLC* network closed so that the refrigerator did not cool and the resonator dissipation consumed most of the acoustic power flowing past the two-microphone location. Figure 9.12 displays these measurements as a function of the square of the normalized pressure amplitude at the top of the pulse tube. With this choice of axes, the data would have fallen on a straight line if phenomena "beyond Rott's thermoacoustics" had been irrelevant. Also shown are two sets of DELTAE calculations, modeling the resonator plus the refrigerator, and using the turbulence algorithm described in "Turbulence" in Chap. 7 for the resonator dissipation, with surface roughness factors ε of 5×10^{-4} and 1×10^{-3}. It is apparent that either value of surface roughness yielded calculations with acceptable accuracy, but that a laminar calculation was dramatically inadequate. This figure helped build confidence in the two-

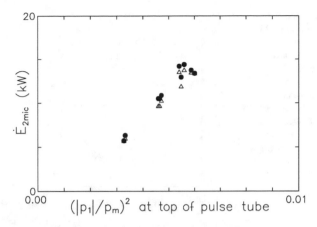

Fig. 9.13 Two-microphone power $\dot{E}_{2\mathrm{mic}}$, with *RLC* valves adjusted for normal operation and with $T_{\mathrm{meth}} = 130\,\mathrm{K}$, as a function of the square of the normalized pulse-tube pressure amplitude. *Filled circles* represent measurements. *Open triangles* represent DELTAE calculations, which used the as-built geometry and the experimental values of f, p_m, $p_{1,C}$, $p_{1,pt}$, T_0, and T_{meth} to predict, among other things, the flow resistances of the two valves, the cooling power, and $\dot{E}_{2\mathrm{mic}}$

microphone implementation and in our DELTAE model of the complete system. We chose a surface roughness factor of 5×10^{-4} for subsequent calculations.

Figure 9.13 shows more measurements of $\dot{E}_{2\mathrm{mic}}$, now with the refrigerator valves open and adjusted to produce significant cooling power at $T_{\mathrm{meth}} = 130\,\mathrm{K}$. Under these circumstances, the acoustic power consumed by the refrigerator was a few times greater than that consumed by the resonator. The good agreement between measurements and calculations shows that we understood the "acoustics" of this system well. We could proceed with our analysis of the data, trying to understand the cooling power next.

The acoustic power dissipated in the *RLC* network would have been given by Eq. (9.2),

$$\dot{E}_{2,RLC} = \frac{\omega V}{2\gamma p_m} \mathrm{Im}\left[\widetilde{p_{1,C}} \, p_{1,pt} \right], \tag{9.13}$$

if the volume of gas in the inertance and valves had been negligible. This was a reasonably accurate approximation for the present refrigerator, but we actually used the simple DELTAE model reproduced in Appendix B to obtain a more accurate estimate of $\dot{E}_{2,RLC}$ at each experimental operating point, accounting for details such as the volumes of the inertance and connecting plumbing near the valves. Such "measured" values of $\dot{E}_{2,RLC}$ are shown in Fig. 9.14. (We call this "measured" because Eq. (9.13) and the corresponding DELTAE model are very simple, direct manipulations of the experimental values of $p_{1,C}$ and $p_{1,pt}$.) This power is important, because (neglecting pressure drop in the heat exchanger at the top of the pulse tube) it is the grossest gross cooling power:

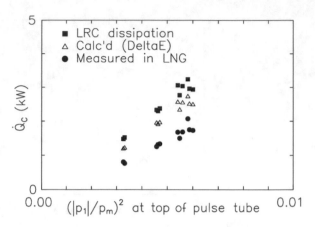

Fig. 9.14 Three cooling powers as a function of the square of the normalized pressure amplitude in the pulse tube. The *filled squares* represent measured values of $\dot{E}_{2,RLC}$, the *filled circles* represent measured values of actual cooling power $\dot{Q}_{\mathrm{meth}}$ "delivered" to the methane stream, and the *open triangles* are the corresponding results $\dot{Q}_{\mathrm{DeltaE}}$ of DELTAE calculations

In a theoretically ideal orifice pulse-tube refrigerator, with no energy flux $\dot{H}_{\mathrm{reg}}$ through the regenerator, no pulse-tube losses, and no external heat leak, the entire $\dot{E}_{2,RLC}$ would be available as cooling power. This is the ideal goal to which we should aspire, as we learn in future years to reduce those losses. In fact, we are far short of that goal: The experimental values of $\dot{Q}_{\mathrm{meth}}$ shown in Fig. 9.14, measured using the technique shown in Fig. 9.4c, were only about half of $\dot{E}_{2,RLC}$.

To get $\dot{Q}_{\mathrm{meth}}$ closer to $\dot{E}_{2,RLC}$ in the future, we wanted to understand how the difference $\dot{E}_{2,RLC} - \dot{Q}_{\mathrm{meth}}$ was distributed among regenerator losses, pulse tube losses, and heat leak. Toward that goal, the triangles shown in Fig. 9.14 are $\dot{Q}_{\mathrm{DeltaE}}$, the results of DELTAE calculations, which used the geometry of the refrigerator and the properties of stainless steel and helium, with the complex pressures in the compliance and pulse tube matched to their experimental values. (Frequency, mean pressure, and ambient and cold temperatures were also matched.) DELTAE included $\dot{H}_{2,\mathrm{reg}}$ in the regenerator (as described in "Tortuous porous media" in Chap. 7), conduction in the stainless-steel pulse-tube wall, and the (tiny) pulse-tube helium conduction and small pulse-tube boundary-layer heat transport. Most of the difference $\dot{E}_{2,RLC} - \dot{Q}_{\mathrm{DeltaE}}$ between squares and triangles in Fig. 9.14 was DELTAE's calculated value of $\dot{H}_{2,\mathrm{reg}}$. DELTAE did not include streaming in the pulse tube [i.e., Fig. 7.12b, c], regenerator internal streaming [i.e., Fig. 7.12d] or heat leak through the insulation. We concluded that most of the difference between triangles and circles in Fig. 9.14 was probably due to the consumption of cooling power by one or more of these effects.

The heat leak through the insulation around the refrigerator was easily estimated as only about 50 W for these measurements, so next we tried to learn whether the low observed cooling power is caused by regenerator internal streaming, pulse-tube streaming, or both.

To search for the presence of unwanted streaming in the pulse tube, we measured the total power $\dot{H}_{pt}$ flowing up the pulse tube, comparing it to the acoustic power $\dot{E}_{2,RLC}$ flowing up the pulse tube. Ideally these are equal, so the deviation of $\dot{H}_{pt}/\dot{E}_{2,RLC}$ from unity was a measure of pulse-tube imperfection. To

Fig. 9.15 Figure of merit $\dot{H}_{pt}/\dot{E}_{2,RLC}$ for the pulse tube, displayed against a horizontal axis showing how far the taper of the pulse tube was from the predicted zero-streaming taper. The five *filled circles* represent measurements, corresponding to the *filled circles* in Fig. 9.10. The *lines* are discussed in the text

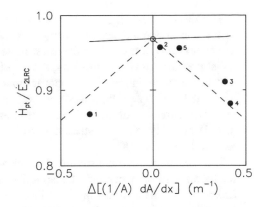

learn $\dot{H}_{pt}$, we considered the inner control surface shown in Fig. 9.9, enclosing the top of the refrigerator, cutting through the refrigerator itself in the middle of the pulse tube but otherwise lying in the thermal insulation. The control volume enclosed by this surface was in steady state, so the time rate of change of the energy in it was zero. Only three powers crossed the control surface: $\dot{H}_{pt}$ and the enthalpy fluxes carried in and out by the water stream passing through the ambient heat exchanger at the top of the pulse tube. Hence, $\dot{H}_{pt}$ equaled the heat $\dot{Q}_{top}$ carried away from the refrigerator by this water stream, which we measured using the technique shown in Fig. 9.4b.

Figure 9.15 shows the experimental values [89] of $\dot{H}_{pt}/\dot{E}_{2,RLC}$ for the experimental operating points numbered 1 through 5 in Fig. 9.10. The horizontal axis shows how far those operating points were from operating points of predicted zero streaming-driven convection, i.e. how far those operating points were from the "zero streaming" line in Fig. 9.10, which was calculated based on Eq. (7.80) and the as-built dimensions of the refrigerator. To provide a quantitative measure of the "distance" from that line, the horizontal axis of Fig. 9.15 is the difference between the "as-built" value of $(1/A)\ dA/dx$, where $A(x)$ is the pulse-tube cross-sectional area, and the value that Eq. (7.80) yields for $(1/A)\ dA/dx$ at the selected operating point. Operating points 2 and 5, which were closest to the "as built" optimal condition, had the highest value of $\dot{H}_{pt}/\dot{E}_{2,RLC}$, approximately 0.96. Slight deviations in operating point away from the "zero streaming" line in Fig. 9.10 reduced $\dot{H}_{pt}/\dot{E}_{2,RLC}$ by 10%, representing a loss of cooling power equal to 10% of $\dot{E}_{2,RLC}$.

The solid line in Fig. 9.15 near $\dot{H}_{pt}/\dot{E}_{2,RLC} = 0.97$ shows the values of $\dot{H}_{pt}/\dot{E}_{2,RLC}$ calculated by DELTAE, which included only thermal conduction in the stainless-steel pulse tube wall and thermoacoustic boundary-layer heat transport, but did not include streaming-driven convection. (The design operating point is indicated by the open circle, corresponding to the open square in Fig. 9.10.) Hence, within our experimental uncertainties of a few percent in $\dot{H}_{pt}/\dot{E}_{2,RLC}$, experimental operating points 2 and 5 exhibited no streaming-driven convection. We concluded that our flow straighteners were adequate to prevent jet-driven streaming [Fig. 7.12d] in the pulse tube.

The dashed lines in Fig. 9.15 are estimates of $\dot{H}_{pt}/\dot{E}_{2,RLC}$ taking streaming-driven convection into account, using Eqs. (10) and (11) of [88] evaluated at the center (axially) of the pulse tube to estimate the streaming mass-flux density vs radial position, and then estimating the heat load with the assumption that all the downward-streaming gas is at room temperature and all the upward-streaming gas is at 125 K. These lines are in reasonable agreement with the data, but they must be regarded as only rough estimates because many of the assumptions used in [88] are only valid for streaming so weak that its effect on time-averaged temperatures is small.

Unfortunately, time constraints, bad weather (these experiments were done outdoors), and poor planning conspired to make this part of the story a little awkward: The data shown in Fig. 9.14 did not include any of the points shown in Fig. 9.15, so we did not know exactly how much of the missing cooling power shown for the data in Fig. 9.14 was due to Rayleigh streaming in the pulse tube. We estimated that Rayleigh streaming was responsible for lost cooling power equal to 10% of $\dot{E}_{2,RLC}$ in Fig. 9.14, which was roughly a third of the difference between the measured cooling power $\dot{Q}_{meth}$ and the calculated cooling power $\dot{Q}_{DeltaE}$.

Next we wanted to obtain an experimental value for the power flux $\dot{H}_{reg}$ through the regenerator, to learn whether it was significantly larger than the value calculated by DELTAE. Consider the outer control surface shown in Fig. 9.9, enclosing the top of the refrigerator, cutting through the refrigerator itself in the middle of the regenerator but otherwise lying in the thermal insulation. The control volume enclosed by this surface was in steady state, so the time rate of change of the energy in it was zero. Only five powers crossed the control surface: $\dot{H}_{reg}$ and the four enthalpy fluxes carried in and out by the water stream passing through the ambient heat exchanger at the top of the pulse tube and by the methane stream passing through the cold heat exchanger. Hence, $\dot{H}_{reg}$ equaled the difference between the heat $\dot{Q}_{top}$ carried away from the refrigerator by the water stream and the heat $\dot{Q}_{meth}$ delivered to the refrigerator by the methane stream. For the six highest-amplitude data points in Fig. 9.14, the average value of $\dot{Q}_{meth}$ was 1700 W and the average value of $\dot{Q}_{top}$ was 2700 W, so the average measured value of $\dot{H}_{reg}$ was 1000 W. (A proper discussion would describe systematic uncertainties in these quantities, but we ignore this issue here.) DELTAE's calculation of $\dot{H}_{reg}$ for these six operating points was 450 W. Hence, the difference between the calculated and measured $\dot{H}_{reg}$, 550 W, represents something unknown to DELTAE—probably regenerator internal streaming.

So we concluded finally that the missing cooling power seen as the difference between $\dot{Q}_{meth}$ and $\dot{Q}_{DeltaE}$ in Fig. 9.14 was probably due partly to Rayleigh streaming in the pulse tube and partly to regenerator internal streaming.

9.5 Exercises

9.1 When the measurement of heat delivered to a fluid stream was discussed in the "Heat" section above, viscous pressure drops and the entropy generated by viscous shear were not mentioned. Construct a more exact theory of the measurement, including viscous effects. Consider at least two cases: (a) a fluid whose density and specific heat are independent of temperature and pressure, and (b) an ideal gas. Assume that viscosity is independent of temperature in both cases.

9.2 A particular pressure transducer is advertised to work from one Hz to one MHz. You plan to use it near 100 Hz. The manufacturer's representative explains to you that the low-frequency cutoff is essentially due to a capacitor in series with the transducer's "ideal" internal signal source and a resistor to ground downstream of the capacitor. She says that the 1-Hz cutoff means that $RC = 1$ s. What error will occur in the phase of your pressure measurements at 100 Hz due to this electric RC filter?

9.3 Draw a complete electric circuit diagram with one floating dc power supply driving two piezoresistive pressure transducers of the "bridge" type described in the text, with amplifiers connected to the signal leads of each transducer. Convince yourself that connecting the negative inputs of both amplifiers to ground simultaneously is likely to ruin the measurements.

9.4 You want to measure the oscillating pressure at a 600 °C location in an engine, using a pressure transducer at 27 °C connected to the hot location through a 2-mm-diam, 10-cm long tube of low thermal conductance. The working gas is helium at 3 MPa, and the frequency is 200 Hz. Assume laminar flow in the tube and a linear temperature profile. What errors in amplitude and phase of p_1 are introduced by the tube? Now suppose that the transition fitting between the ambient-temperature end of the tube and the transducer itself has unavoidable dead volume of 0.2 cm^3, amounting to a lumped compliance. What errors are introduced by the tube and this compliance? Should these estimates be made using the complex Bessel-function solutions for thermoacoustics in a circular duct, or is boundary-layer approximation good enough? Should numerical integration be used, or is a lumped-impedance estimate good enough?

9.5 Derive an equation for the measurement of acoustic power flowing into a lumped RC load if the compliance C has significant surface area S. How large must a spherical compliance be in order to make the difference between your result and Eq. (9.2) only 1%?

9.6 Derive an equation for the measurement of acoustic power flowing into a lumped-element load consisting of the usual flow resistance R, lumped inertance L, and a principal compliance C, with the addition of a small compliance C' between R and L which represents some unavoidable void volume in the vicinity of R and L. Does the measurement now require knowledge of the temperature of one of the four components?

9.7 (a) The front side of a loudspeaker drives a duct, and the back side is enclosed by a known volume V that is small enough to be considered lumped. Derive an equation for the acoustic power delivered to the duct by the front of the loudspeaker, in terms of gas properties and of the measured pressure amplitudes and phases in front of and behind the speaker [35]. (b) Show that this method should work even if the effective area of the moving surface of the loudspeaker is unknown because it is flexible. (c) Suppose that the back-side volume is actually divided into two portions having known volumes V_a and V_b, separated by a lumped impedance of unknown magnitude $R + i\omega L$, with V_a immediately adjacent to the moving surface of the loudspeaker. (For example, V_a might be the volume immediately adjacent to the moving dome, and the lumped impedance might be the gap between the magnet and the voice coil.) Can the acoustic power delivered by the front of the loudspeaker into the duct be determined through measurements of oscillating pressure? Are pressure measurements needed in both V_a and V_b, in addition to the pressure measurement in front of the loudspeaker? Do you need to assume that you know the effective area of the loudspeaker to derive an answer?

9.8 Two points of view—historical and functional—were described in "Points of view" in the context of trying to learn why your car won't start, and four others were described in the context of thermoacoustic examples. These are not the only possibilities. Returning to the uncooperative-automobile theme: "guilt–punishment" is another point of view, in which I ask "What did I do to deserve this?" (I yelled at my mother ... I haven't washed the car all year ...). Another is "periodic table," in which I ask "Which of the elements of the periodic table are preventing the car from starting—hydrogen? helium? lithium? ...Describe two additional points of view, which might (or might not, depending on your sense of humor) be useful for learning why a car won't start.

9.9 Describe the thermoacoustic-Stirling heat engine from the point of view of "natural dependence." Just look at Figs. 1.22 and 1.23, and the short descriptions in Chap. 1 and in this chapter, and use your imagination. What do you think are the most natural independent and dependent variables? If someone gave the engine to you, how would you organize a set of experiments to learn how it works?

9.10 Describe possible or actual experiments with your favorite thermoacoustics hardware using two of the four points of view described at length in this chapter. For each, describe what you want to learn or demonstrate, what simple instrumentation would be required in the experiment, what you would actually do in the course of the experiment, and how you would organize and graph the results for publication or for your own interpretation and understanding.

9.11 Describe a possible experiment with your favorite thermoacoustics hardware that could be approached usefully from one of the points of view described here, but could *not* be approached usefully from one of the other points of view.

9.12 Pick one of the four extended examples, and describe a possible experiment with it using one of the points of view that I did not apply to it. Feel free to use your imagination (e.g., suggesting additional sensors if necessary).

9.13 You have just assembled your standing-wave engine for the third time. It has never started oscillating at all, in spite of the fact that you've heated the hot end 150 °C hotter than the design called for. You have already double-checked every fabricated dimension against your design numerical model (see Appendix A). Using the thermoacoustic perspective, describe the first few steps of your analysis and action plan toward diagnosis of this unfortunate situation.

9.14 You have just assembled your standing-wave engine. It has oscillated, but not at high enough amplitude. You have taken some data and made some corresponding calculations, shown in this table:

	expt	expt	expt	calc	calc	calc
$\dot{Q}_H$	T_H	f	$|p_1|$	T_H	f	$|p_1|$
kW	°C	Hz	kPa	°C	Hz	kPa
1	300	no osc	no osc	600	200	28
2	630	200	14	620	200	42
3	660	200	35	640	200	53
4	700	200	47	660	200	62

Using the thermoacoustic perspective, describe the first few steps of your analysis and action plan toward diagnosis of this unfortunate situation.

9.15 You have just assembled your standing-wave engine for the third time, and for the third time it has oscillated, but not at high enough amplitude. You have already double-checked every fabricated dimension (see Appendix A). You have taken some data and made some corresponding calculations, shown in this table:

	expt	expt	expt	calc	calc	calc
$\dot{Q}_H$	T_H	f	$|p_1|$	T_H	f	$|p_1|$
kW	°C	Hz	kPa	°C	Hz	kPa
1	300	200	20	600	200	28
2	630	200	30	620	200	42
3	660	200	37	640	200	53
4	700	200	44	660	200	62

Using the thermoacoustic perspective, describe the first few steps of your analysis and action plan toward diagnosis of this unfortunate situation.

9.16 Plan a simple experiment to prove or disprove something simple in one of the earlier chapters. (The experimental techniques described in this chapter are acceptable.) Describe the axes of the graphs you would use to present the results of this experiment.

9.17 Do you think that an aesthetic point of view can play a valuable role in thermoacoustics research? Why or why not?

9.18 Based on Figs. 9.10, 9.11, 9.12, 9.13, 9.14, and 9.15 and the related discussion in the text, suggest three interesting questions regarding these measurements, beginning with "Why …," and suggest corresponding additional measurements that might have been made to answer these questions. You can suggest adding transducers as needed, but do not propose any "difficult" measurements as defined in Sect. 9.3.

Appendix A
Common Pitfalls

Having built many thermoacoustic engines and refrigerators, and helped others build even more, I've made and seen a few human errors more than once. Try to avoid these, in thermoacoustics and in all experimental R&D.

Computer Intoxication Don't spend too much time computer simulating the hardware. You must leave enough time for design of realistic hardware, fabrication and assembly, measurement, and meaningful debugging and iteration.

Multi-vendor Gridlock You'll have to rely on many machine shops, sensor suppliers, etc. Each of these vendors in turn relies on many other vendors such as material suppliers. The probability of all these being on time for you is low if they are numerous, so try to get as many as possible to deliver well ahead of your true scheduled requirement. For key components, order duplicates from independent vendors to insure yourself against delay.

Unrealistic Optimism When you've spent many days on your computer designing, and many nights imagining how glorious the results will be, you'll probably believe that the hardware will work right away. No way! It is far more likely that at first nothing will work at all, or that things will at first work qualitatively as you expect but quantitatively very poorly, especially at high amplitude. Plan for this depressing situation from the beginning. Do not try too many new or unfamiliar things simultaneously. Test subsystems whenever possible before assembling the full system. Design diagnostic features into the hardware before you need them. Put in spare holes for transducers—even if you don't own enough transducers to fill all the holes, you can swap a few around. Plan to make some measurements at low enough amplitude that you are certain that Rott's acoustic approximation is valid. Above all, leave slack in the schedule and in the budget for these unpredictable problems.

Designer–Builder Disconnect Assemble it yourself, or watch when it is assembled, to be sure you know exactly what you're building. Measure and record

© Acoustical Society of America 2017

G.W. Swift, *Thermoacoustics*, DOI 10.1007/978-3-319-66933-5

dimensions during assembly. Watch out for details that your assembly technician might not know are important: gaps and volumes between parts, unintended turbulence generators, alignment, etc. Take photographs.

Dilution of Responsibility Remember the words of George Washington,[1] who said, "... whenever one person is found adequate to the discharge of a duty by close application thereto, it is worse executed by two persons, and scarcely done at all if three or more are employed therein."

Blunders A typical thermoacoustic system may have 100 key dimensions or other parameters. If one of those turns out to be dramatically different from what you intend, the operation of the device will be a mystery to you. So, know your natural error rate, and do whatever is necessary to get it down below 1%: double check, enlist help, take a rest and check it all again, etc.

Poor Choice of Boss As leader of a large, complicated technology-development project, choose someone whose life and career depend on the success of the project. As leader of a research team, choose someone who loves learning and teaching.

Poor Choice of Workers Some people are well suited to collaboration on a large, complex project; other people are well suited to working in isolation. Match people's styles to their work.

Conflicting Goals Do not confuse technology development with research or with an educational experience. To prevent urgent deadlines from interfering with learning, and vice versa, do not assign a student to the critical path of a technology-development project.

Avoiding the Real Challenge Don't procrastinate. Attack the hard problems early, whether they are technical or personnel, and address them immediately. Problems seldom go away by themselves, and it only takes one serious problem to ruin the project.

---

[1] First president of the United States.

Appendix B
DELTAEC Files

Hoping that the usefulness of this book will greatly outlive the current version of DELTAEC [61], I've relegated four typical DELTAEC i/o files to this Appendix, for the four examples that are introduced in Chap. 1 and re-examined throughout the book. For interested readers, these show how to configure a simple file for DELTAEC for each of these examples, and may occasionally be consulted to learn a specific dimension of a piece of apparatus.[1]

We find that DELTAEC and experiment agree best when considerable detail is included in the DELTAEC file: Often, we have over 50 segments in a file to account for all the little details in hardware as built. To avoid intimidating the newcomer, the first two files here are stripped of such details; they use only the key segments necessary to describe the hardware. The third file, for the orifice pulse-tube refrigerator, includes only about 25% of the segments actually used to model this refrigerator, in order to hide corporate proprietary information. The fourth file, for the thermoacoustic-Stirling heat engine, is fully encumbered with ugly details, typical of what we actually work with at Los Alamos—if you are new to DELTAEC or to thermoacoustics, skip it!

DELTAEC is available at www.lanl.gov/thermoacoustics/. (This is currently redirected to www.lanl.gov/org/padste/adeps/materials-physics-applications/condensed-matter-magnet-science/thermoacoustics.)

---

[1]The 2002 printing of this book used DELTAE v. 4.5 examples [61]. The upgrade here to DELTAEC is mostly a matter of formatting, with no change in results. The addition of the MINOR segment and built-in calculation of $p_{2,0}$ makes some things much more convenient in the DELTAEC model of the thermoacoustic-Stirling heat engine. Small changes in heat-exchanger heats result from using the default INSULATE mode in DELTAEC instead of the earlier ISODUCTs.

© Acoustical Society of America 2017
G.W. Swift, *Thermoacoustics*, DOI 10.1007/978-3-319-66933-5

B.1   Standing-Wave Engine

This model of part of the standing-wave engine starts at the left end of Fig. 1.9 and integrates to the right, past the branch up to the refrigerators, ending at the pressure "node" at the center of the resonator. The point of view adopted in this particular file is this: In order to have the as-built hardware run with $|p_1| = 297\,\text{kPa}$ at the branch to the refrigerators and supply $1000\,\text{W}$ of acoustic power to the refrigerators, what hot temperature and heater power are required at the hot heat exchanger at the hot end of the stack, and at what frequency will the system resonate? (In this simplified model, it is assumed that the right engine can supply $500\,\text{W}$ of acoustic power flowing leftward at the pressure "node," but at Los Alamos we would typically fully include the right engine in the file.)

```
TITLE      One engine, branch to OPTRs; stop calc at midplane. As-built dims.
!->C:\Users\092710\Documents\book2015\deltxmpl\2015revisions\standeng.out
!Created@16:58:24  15-Jan-2015 with DeltaEC version 6.3b11.12!under win32,
using Win 6.1.7601 (Service Pack 1) under Python DeltaEC.
!------------------------------ 0 ------------------------------
BEGIN       the setup
 2.9660E+06 a Mean P Pa
   390.84   b Freq   Hz         G
   760.19   c TBeg   K          G
 3.1963E+05 d |p|    Pa         G
    0.0000  e Ph(p)  deg
    0.0000  f |U|    m^3/s
    0.0000  g Ph(U)  deg
helium              Gas type
!------------------------------ 1 ------------------------------
SURFACE     hot end
sameas   2a a Area   m^2              3.1963E+05 A |p|    Pa
                                        0.0000  B Ph(p)  deg
                                      2.3806E-05 C |U|    m^3/s
                                        180.00  D Ph(U)  deg
                                        0.0000  E Htot   W
ideal            Solid type          -3.8046   F Edot   W
!------------------------------ 2 ------------------------------
DUCT        hot duct
 2.8580E-03 a Area   m^2              3.1886E+05 A |p|    Pa
   0.1900   b Perim  m               1.3974E-03 B Ph(p)  deg
 4.5900E-02 c Length m               2.0886E-02 C |U|    m^3/s
                                       -90.264  D Ph(U)  deg
                                        0.0000  E Htot   W
ideal            Solid type          -15.418    F Edot   W
!------------------------------ 3 ------------------------------
HX          hot heat exchanger
sameas   6a a Area   m^2              3.1844E+05 A |p|    Pa
   0.8800   b GasA/A                  1.7912E-02 B Ph(p)  deg
 9.5000E-03 c Length m               2.5653E-02 C |U|    m^3/s
 3.8000E-04 d y0     m                 -91.522  D Ph(U)  deg
 2380.1     e HeatIn W         G      2380.1    E Htot   W
   0.0000   f SolidT K              -109.73     F Edot   W
                                      760.19    G GasT   K
copper           Solid type           785.51    H SolidT K
```

```
!------------------------------- 4 -------------------------------
STKSLAB    stack
sameas    6a a Area    m^2           3.0688E+05 A |p|      Pa
     0.8100  b GasA/A                    0.91041 B Ph(p)   deg
7.6200E-02 c Length m                5.3208E-02 C |U|      m^3/s
1.3000E-04 d y0      m                  -84.939 D Ph(U)    deg
2.5400E-05 e Lplate m                    2380.1 E Htot     W
                                         590.97 F Edot     W
                                         760.19 G TBeg     K
stainless           Solid type          319.15 H TEnd     K
!------------------------------- 5 -------------------------------
HX         ambient heat exchanger
sameas    6a a Area    m^2           3.0349E+05 A |p|      Pa
     0.7500  b GasA/A                    0.89157 B Ph(p)   deg
1.2700E-02 c Length m                5.7899E-02 C |U|      m^3/s
8.2600E-04 d y0      m                  -85.413 D Ph(U)    deg
-1880.2    e HeatIn W        G           499.88 E Htot     W
 295.00    f SolidT K        =5H         566.31 F Edot     W
                                         319.15 G GasT     K
copper             Solid type           295.00 H SolidT K
!------------------------------- 6 -------------------------------
CONE       adapter
3.1670E-03 a AreaI   m^2             2.9660E+05 A |p|      Pa
     0.1990  b PerimI m                   0.8175 B Ph(p)   deg
2.9000E-02 c Length m                7.0839E-02 C |U|      m^3/s
sameas    9a d AreaF   m^2              -86.112 D Ph(U)    deg
sameas    9b e PerimF m                  499.88 E Htot     W
ideal           Solid type              562.74 F Edot     W
!------------------------------- 7 -------------------------------
RPN        pressure amplitude target
 2.9660E+05 a G or T           =7A   2.9660E+05 A Pa
p1 mag
!------------------------------- 8 -------------------------------
BRANCH     branch to OPTRs: partly resistive, partly compliant
 3.9590E+07 a Re(Zb) Pa-s/m^3        2.9660E+05 A |p|      Pa
-1.3200E+07 b Im(Zb) Pa-s/m^3             0.8175 B Ph(p)   deg
sameas    8G c HtotBr W              7.3045E-02 C |U|      m^3/s
                                         -91.495 D Ph(U)   deg
                                         -500.0 E Htot     W
                                         -437.14 F Edot    W
                                          999.88 G EdotBr W
!------------------------------- 9 -------------------------------
CONE       a little more of the adapter
2.8200E-03 a AreaI   m^2             2.6735E+05 A |p|      Pa
     0.1880  b PerimI m                   1.0447 B Ph(p)   deg
7.3000E-02 c Length m                9.7946E-02 C |U|      m^3/s
sameas    10a d AreaF   m^2             -90.909 D Ph(U)    deg
sameas    10b e PerimF m                 -500.0 E Htot     W
ideal           Solid type              -446.38 F Edot     W
!------------------------------- 10 -------------------------------
DUCT       long thin duct at ambient temperature
2.0270E-03 a Area    m^2                 6613.7 A |p|      Pa
     0.1600  b Perim   m                  90.233 B Ph(p)   deg
     0.3700  c Length m                   0.1512 C |U|     m^3/s
                                         -89.767 D Ph(U)   deg
                                         -500.0 E Htot     W
                                         -500.0 F Edot     W
ideal           Solid type
!------------------------------- 11 -------------------------------
SOFTEND    pressure "node", with 500 W coming from other engine
     0.0000 a Re(z)                      6613.7 A |p|      Pa
```

```
      0.0000  b Im(z)              =11H           90.233   B Ph(p)   deg
      0.0000  c Htot    W                          0.1512  C |U|      m^3/s
                                                  -89.767   D Ph(U)   deg
                                                  -500.0    E Htot    W
                                                  -500.0    F Edot    W
                                                  -1.8854E-02 G Re(z)
                                                  1.8219E-14 H Im(z)
                                                   319.15   I T        K
!------------------------------ 12 ------------------------------
RPN          acoustic power from other engine
 -500.0       a G or T            =12A      -500.0       A W
Edot
!------------------------------ 13 ------------------------------
RPN          total power from other engine
 -500.0       a G or T            =13A      -500.0       A W
Htot
! The restart information below was generated by a previous run
! and will be used by DeltaEC the next time it opens this file.
guessz   0b   0c   0d   3e   5e
targs    5f   7a  11b  12a  13a
```

B.2 Standing-Wave Refrigerator

This model of part of the standing-wave refrigerator starts near the drivers at one side of the refrigerator shown in Fig. 1.14 and integrates downward, through a stack and its neighboring heat exchangers, ending at the pressure node at the center of the bottom leg of the resonator. The point of view adopted in this particular file is this: In order to have the as-built hardware run with $|p_1| = 6467$ Pa near the drivers and to reach an experimentally maintained cold temperature of 285.6 K, what U_1 and heat load on the cold heat exchanger should occur?

```
TITLE       segments 1-7 < Bob 9/97; segs 8-14 < Hiller ntbk, pg one, 10/97
!->C:\Users\092710\Documents\book2015\deltxmpl\2015revisions\standfri.out
!Created@17:06:02  15-Jan-2015 with DeltaEC version 6.3b11.12!under win32,
using Win 6.1.7601 (Service Pack 1) under Python DeltaEC.
!------------------------------ 0 ------------------------------
BEGIN       Initialize in main duct where drivers are attached
 3.2388E+05 a Mean P Pa
   91.650   b Freq    Hz
   307.80   c TBeg     K
 6467.0     d |p|      Pa
   90.000   e Ph(p)   deg
-5.2633E-02 f |U|      m^3/s      G
   187.09   g Ph(U)   deg         G
    0.0000  i Ndot    mol/s
    0.9200  j nL
HeAr                Gas type
!------------------------------ 1 ------------------------------
DUCT        Pre-stack Duct
 1.8430E-02 a Area    m^2           6215.5      A |p|      Pa
    0.4813  b Perim   m             89.766      B Ph(p)   deg
    0.1494  c Length  m          7.1230E-02 C |U|      m^3/s
                                     5.1494   D Ph(U)    deg
                                    21.013    E Htot     W
ideal               Solid type      20.768    F Edot     W
```

```
!-------------------------------- 2 --------------------------------
HX          Ambient Heat exchanger
 1.7211E-02 a Area    m^2                    6154.2     A |p|    Pa
    0.5227  b GasA/A                         89.884     B Ph(p)  deg
 1.2700E-02 c Length m                    7.2147E-02    C |U|    m^3/s
 6.3500E-04 d y0      m                      4.9381     D Ph(U)  deg
   -32.855  e HeatIn W          G           -11.843     E Htot   W
    0.0000  f SolidT K                       19.556     F Edot   W
       0.0000 g FracQN                       307.80     G GasT   K
                                             307.22     H SolidT K
copper              Solid type             -11.843     I H2k    W
!-------------------------------- 3 --------------------------------
RPN         first stack thermocouple location
  307.80    a G or T                         307.80     A Kelvin
Tm
!-------------------------------- 4 --------------------------------
STKRECT     rectangular-pore stack
 1.8824E-02 a Area    m^2                    5558.1     A |p|    Pa
    0.7050  b GasA/A                         93.242     B Ph(p)  deg
    0.1524  c Length  m                   8.6744E-02    C |U|    m^3/s
 4.0640E-04 d aa                             3.8489     D Ph(U)  deg
 1.1811E-04 e Lplate  m                     -11.843     E Htot   W
 6.3500E-03 f bb      m                      2.5539     F Edot   W
                                             307.80     G TBeg   K
                                             285.60     H TEnd   K
kapton              Solid type             -11.843     I H2k    W
!-------------------------------- 5 --------------------------------
RPN         last stack thermocouple location
  285.60    a G or T           =5A           285.60     A Kelvin
Tm
!-------------------------------- 6 --------------------------------
HX          Cold heat exchanger
 1.7211E-02 a Area    m^2                    5537.1     A |p|    Pa
    0.9300  b GasA/A                         93.272     B Ph(p)  deg
 6.3500E-03 c Length m                    8.7419E-02    C |U|    m^3/s
 1.0414E-03 d y0      m                      3.7948     D Ph(U)  deg
    11.843  e HeatIn W          G         5.5067E-14    E Htot   W
    0.0000  f SolidT K                       2.2066     F Edot   W
       0.0000 g FracQN                       285.60     G GasT   K
                                             285.60     H SolidT K
copper              Solid type             -11.843     I H2k    W
!-------------------------------- 7 --------------------------------
DUCT        From cold hx to bolt flange, + a half inch
 2.0180E-02 a Area    m^2                    5229.9     A |p|    Pa
    0.5036  b Perim   m                      93.256     B Ph(p)  deg
    0.1210  c Length  m                      0.1015     C |U|    m^3/s
                                             3.6954     D Ph(U)  deg
                                          5.5067E-14    E Htot   W
ideal               Solid type              2.0338     F Edot   W
!-------------------------------- 8 --------------------------------
DUCT        Duct, beginning 1/2 in below big bolt flange
 1.8485E-02 a Area    m^2                    4698.7     A |p|    Pa
    0.4820  b Perim   m                      93.235     B Ph(p)  deg
    0.1650  c Length  m                      0.11772    C |U|    m^3/s
                                             3.6066     D Ph(U)  deg
                                          5.5067E-14    E Htot   W
ideal               Solid type              1.7926     F Edot   W
!-------------------------------- 9 --------------------------------
DUCT        Elbow, 6 inch diam, 5 inch radius of curvature on centerline
 1.8240E-02 a Area    m^2                    3947.1     A |p|    Pa
```

```
        0.4788  b Perim   m                   93.211     B Ph(p)   deg
        0.2000  c Length  m                   0.13464    C |U|     m^3/s
                                              3.5339     D Ph(U)   deg
                                              5.5067E-14 E Htot    W
ideal               Solid type               1.4962     F Edot    W
!------------------------------ 10 -------------------------------
DUCT        straight part of big black cone
 1.8240E-02 a Area    m^2                     3582.2     A |p|     Pa
    0.4788  b Perim   m                       93.201     B Ph(p)   deg
 8.9000E-02 c Length  m                       0.14118    C |U|     m^3/s
                                              3.5097     D Ph(U)   deg
                                              5.5067E-14 E Htot    W
ideal               Solid type               1.3636     F Edot    W
!------------------------------ 11 -------------------------------
CONE        The long black plastic cone
sameas   10a a AreaI  m^2                     1139.0     A |p|     Pa
sameas   10b b PerimI m                       93.164     B Ph(p)   deg
    0.3635  c Length  m                       0.15467    C |U|     m^3/s
sameas   12a d AreaF  m^2                     3.4617     D Ph(U)   deg
sameas   12b e PerimF m                       5.5067E-14 E Htot    W
ideal               Solid type               0.45707    F Edot    W
!------------------------------ 12 -------------------------------
DUCT        little straight section of black cone (1st part of the "tee"
 8.1070E-03 a Area    m^2                     545.41     A |p|     Pa
    0.3192  b Perim   m                       93.179     B Ph(p)   deg
 5.7200E-02 c Length  m                       0.15509    C |U|     m^3/s
                                              3.4601     D Ph(U)   deg
                                              5.5067E-14 E Htot    W
ideal               Solid type               0.20783    F Edot    W
!------------------------------ 13 -------------------------------
DUCT        half the white part of the "tee", up to the symmetry midpt
 9.8100E-03 a Area    m^2                     4.4283E-12 A |p|     Pa
    0.3511  b Perim   m                       80.126     B Ph(p)   deg
 6.3500E-02 c Length  m                       0.15527    C |U|     m^3/s
 1.0000E-03 d Srough                          3.4595     D Ph(U)   deg
                                              5.8620E-14 E Htot    W
ideal               Solid type               7.9286E-14 F Edot    W
!------------------------------ 14 -------------------------------
RPN         Insist that Htot = Edot + Ndot*m*enth in SOFTEND
    0.0000  a G or T                          7.9286E-14           A W
Edot Ndot m * enth * +
!------------------------------ 15 -------------------------------
SOFTEND     pressure node
    0.0000  a Re(z)           =15G            4.4283E-12 A |p|     Pa
    0.0000  b Im(z)           =15H            80.126     B Ph(p)   deg
sameas   14A c Htot    W      =15E            0.15527    C |U|     m^3/s
                                              3.4595     D Ph(U)   deg
                                              5.8620E-14 E Htot    W
                                              7.9286E-14 F Edot    W
                                              9.0672E-17 G Re(z)
                                              3.8256E-16 H Im(z)
                                              285.60     I T       K
                                              0.2381     J p20HL   Pa
                                              0.9200     K nL
! The restart information below was generated by a previous run
! and will be used by DeltaEC the next time it opens this file.
guessz  0f   0g   2e   6e
xprecn  1.2793E-07  2.2681E-03  -4.9925E-04  -2.4683E-04
targs   5a  15a  15b  15c
```

B.3   Orifice Pulse-Tube Refrigerator

At the time of the first printing, Cryenco (later Chart Inc.) was unwilling to share construction details of the orifice pulse-tube refrigerator of Fig. 1.17 publicly, so this model only covers the *RLC* network at the top of Fig. 1.18. The model starts in the compliance and ends just above the hot heat exchanger at the top of the pulse tube. The point of view adopted in this particular file is this: In order to achieve the values of p_1 (both magnitude and phase) in the compliance and at the top of the pulse tube that were observed in a certain experiment, what must the flow resistances of the two valves be? One of those resistances has turned out to be slightly negative, indicating an error in the model, so perhaps DELTAEC's turbulence estimate in the inertance is inaccurate or volume in valve bodies should be included.

```
TITLE       RLC network only
!->C:\Users\092710\Documents\book2015\deltxmpl\2015revisions\optrLRC.out
!Created@17:09:59  15-Jan-2015 with DeltaEC version 6.3b11.12!under win32,
using Win 6.1.7601 (Service Pack 1) under Python DeltaEC.
!-----------------------------  0  -----------------------------
BEGIN       at compliance!
 3.1114E+06 a Mean P  Pa
    42.000  b Freq    Hz
   305.20   c TBeg    K
 9.3205E+04 d |p|     Pa
  -153.3    e Ph(p)   deg
    0.0000  f |U|     m^3/s
    0.0000  g Ph(U)   deg
helium            Gas type
!-----------------------------  1  -----------------------------
COMPLIANCE reserv. vol.
    0.3000   a SurfAr m^2               9.3205E+04 A |p|     Pa
 9.8100E-03 b Volume m^3               -153.3      B Ph(p)   deg
                                       4.6530E-02 C |U|     m^3/s
                                        116.57     D Ph(U)   deg
                                         0.0000    E Htot    W
ideal             Solid type           -4.7351     F Edot    W
!-----------------------------  2  -----------------------------
TBRANCH     to bypass path
 6.4296E+06 a Re(Zb) Pa-s/m^3 G         9.3205E+04 A |p|     Pa
 2.8416E+06 b Im(Zb) Pa-s/m^3 G        -153.3      B Ph(p)   deg
                                       1.3259E-02 C |U|     m^3/s
                                       -177.14     D Ph(U)   deg
                                        565.16     E HtotBr  W
                                        565.16     F EdotBr  W
                                       -569.9      G EdotTr  W
!-----------------------------  3  -----------------------------
DUCT        nuisance inertance
 3.4000E-04 a Area   m^2               8.0596E+04 A |p|     Pa
 6.5400E-02 b Perim  m                  173.62     B Ph(p)   deg
    1.0000  c Length m                1.3507E-02 C |U|     m^3/s
 3.0000E-04 d Srough                    176.46     D Ph(U)   deg
                                        565.16     E Htot    W
ideal             Solid type            543.64     F Edot    W
```

```
!-------------------------------- 4 --------------------------------
IMPEDANCE   bypass valve
  2.2707E+07 a Re(Zs) Pa-s/m^3 G          2.2624E+05 A |p|     Pa
      0.0000 b Im(Zs) Pa-s/m^3              -2.528   B Ph(p)  deg
                                          1.3507E-02 C |U|     m^3/s
                                            176.46   D Ph(U)  deg
                                            565.16   E Htot    W
                                           -1527.6   F Edot    W
!-------------------------------- 5 --------------------------------
COMPLIANCE little header
  1.0000E-03 a SurfAr m^2                  2.2624E+05 A |p|     Pa
  3.0000E-05 b Volume m^3                    -2.528   B Ph(p)  deg
                                          1.3506E-02 C |U|     m^3/s
                                            177.93   D Ph(U)  deg
                                            565.16   E Htot    W
ideal              Solid type            -1527.7   F Edot    W
!-------------------------------- 6 --------------------------------
DUCT        short tube
sameas    3a a Area   m^2                 2.2650E+05 A |p|     Pa
sameas    3b b Perim  m                  -0.97342   B Ph(p)  deg
    0.1200   c Length m                 1.3532E-02 C |U|     m^3/s
sameas    3d d Srough                      -180.0   D Ph(U)  deg
                                            565.16   E Htot    W
ideal              Solid type            -1532.3   F Edot    W
!-------------------------------- 7 --------------------------------
SOFTEND    connect bypass to trunk
    0.0000 a Re(z)                        2.2650E+05 A |p|     Pa
    0.0000 b Im(z)                       -0.97342   B Ph(p)  deg
    0.0000 c Htot    W                   1.3532E-02 C |U|     m^3/s
                                           -180.0   D Ph(U)  deg
                                            565.16   E Htot    W
                                           -1532.3   F Edot    W
                                            -1.128   G Re(z)
                                          1.9224E-02 H Im(z)
                                            305.20   I T       K
!-------------------------------- 8 --------------------------------
DUCT        the inertance
  4.6400E-04 a Area   m^2                 2.1066E+05 A |p|     Pa
  7.6360E-02 b Perim  m                    -3.0503   B Ph(p)  deg
    2.4900   c Length m                 3.9431E-02 C |U|     m^3/s
sameas    3d d Srough                      104.14   D Ph(U)  deg
                                           -565.16   E Htot    W
ideal              Solid type            -1227.8   F Edot    W
!-------------------------------- 9 --------------------------------
IMPEDANCE  valve-inertance set
 -1.4196E+05 a Re(Zs) Pa-s/m^3 G          2.0907E+05 A |p|     Pa
      0.0000 b Im(Zs) Pa-s/m^3             -1.5847   B Ph(p)  deg
                                          3.9431E-02 C |U|     m^3/s
                                            104.14   D Ph(U)  deg
                                           -565.16   E Htot    W
                                           -1117.4   F Edot    W
!-------------------------------- 10 --------------------------------
COMPLIANCE little header
  2.0000E-03 a SurfAr m^2                 2.0907E+05 A |p|     Pa
  6.0000E-05 b Volume m^3                  -1.5847   B Ph(p)  deg
                                          3.8817E-02 C |U|     m^3/s
```

```
                                              104.40     D Ph(U)  deg
                                             -565.16     E Htot   W
ideal             Solid type               -1117.6      F Edot   W
!----------------------------- 11 --------------------------------
DUCT        short tube
sameas    6a a Area    m^2                  2.2650E+05 A |p|     Pa
sameas    6b b Perim  m                    -0.97342    B Ph(p)   deg
sameas    6c c Length m                     3.8346E-02 C |U|     m^3/s
sameas    6d d Srough                       104.67     D Ph(U)   deg
                                           -565.16     E Htot   W
ideal             Solid type               -1171.1      F Edot   W
!----------------------------- 12 --------------------------------
UNION       connect bypass here
   7        a SegNum                        2.2650E+05 A |p|     Pa
 2.2650E+05 b |p|Sft Pa       =12A         -0.97342    B Ph(p)   deg
   -0.9734  c Ph(p)S deg       =12B         4.3776E-02 C |U|     m^3/s
    0.0000 d TSoft   K                      122.07     D Ph(U)   deg
                                             0.0000 E Htot   W
                                           -2703.4      F Edot   W
                                            305.20     G T       K
!----------------------------- 13 --------------------------------
DUCT        short tube
sameas    8a a Area    m^2                  2.3837E+05 A |p|     Pa
sameas    8b b Perim  m                      0.5000     B Ph(p)   deg
   0.1100   c Length m                      4.3251E-02 C |U|     m^3/s
 3.0000E-04 d Srough                        122.58     D Ph(U)   deg
                                             0.0000 E Htot   W
ideal             Solid type               -2738.0      F Edot   W
!----------------------------- 14 --------------------------------
RPN         pulsetube magnitud
 2.3837E+05 a G or T           =14A         2.3837E+05 A Pa
p1 mag
!----------------------------- 15 --------------------------------
RPN         pulsetube phase
   0.5000   a G or T           =15A         0.5000      A deg
p1 arg
!----------------------------- 16 --------------------------------
COMPLIANCE volume under head
 1.0000E-02 a SurfAr m^2                    2.3837E+05 A |p|     Pa
 7.5000E-05 b Volume m^3                      0.5000     B Ph(p)   deg
                                            4.2488E-02 C |U|     m^3/s
                                            123.24     D Ph(U)   deg
                                             0.0000 E Htot   W
ideal             Solid type               -2739.1      F Edot   W
! The restart information below was generated by a previous run
! and will be used by DeltaEC the next time it opens this file.
guessz   2a   2b   4a   9a
xprecn   36.547      40.384    -493.44      -2.1258
targs  12b  12c  14a  15a
```

B.4   Thermoacoustic-Stirling Heat Engine

This model of the entire thermoacoustic-Stirling heat engine seems dauntingly complicated, but it is in fact typical of how our models at Los Alamos evolve after many months of interaction between experiments and computations. The model starts near the top of Fig. 1.23, just above the ambient heat exchanger. It integrates up, left, and down, through the compliance and inertance to the junction. It then starts again just above the ambient heat exchanger, going down through the regenerator and thermal buffer tube to reach the junction a second time, where complex p_1 must match up. The integration then proceeded through the resonator (not included here).

Minor losses at the UNION are modeled here as impedances whose values are calculated based on guidance in Idelchik's book. Our understanding of these phenomena is still very incomplete, so such details in this file are not trustworthy.

The model requires nonzero streaming around the loop to satisfy the $p_{2,0}$ target at the UNION. For accurate results under these conditions, set Nint = 50.

(Page and book numbers in the comment fields refer to Scott Backhaus's lab notebooks.)

```
TITLE       TASHE at 10%
!->C:\Users\092710\Documents\book2015\deltxmpl\2015revisions\tasheShowingP20.out
!Created@14:06:45  16-Jan-2015 with DeltaEC version 6.3b11.12!under win32,
using Win 6.1.7601 (Service Pack 1) under Python DeltaEC.
!------------------------------- 0 -------------------------------
BEGIN       the setup
 3.1030E+06 a Mean P Pa
   84.120   b Freq   Hz
   325.00   c TBeg   K
 3.1120E+05 d |p|     Pa
    0.0000  e Ph(p)  deg
    0.0000  f |U|     m^3/s
    0.0000  g Ph(U)  deg
    0.0000  i Ndot   mol/s
helium               Gas type
!------------------------------- 1 -------------------------------
TBRANCH     Split
-3.5726E+07 a Re(Zb) Pa-s/m^3 G       3.1120E+05 A |p|     Pa
 1.5069E+07 b Im(Zb) Pa-s/m^3 G          0.0000 B Ph(p)  deg
 6.3632E-02 d NdotBr mol/s     G       8.0260E-03 C |U|     m^3/s
                                        -157.13 D Ph(U)  deg
                                        -1117.6 E HtotBr W
                                        -1150.7 F EdotBr W
                                         1150.7 G EdotTr W
!------------------------------- 2 -------------------------------
DUCT        Jetting space (pg 61 book 5)
 7.1200E-03 a Area    m^2             3.1117E+05 A |p|     Pa
    0.2990  b Perim   m              8.9253E-03 B Ph(p)  deg
 1.9100E-02 c Length  m              1.0506E-02 C |U|     m^3/s
 3.0000E-04 d Srough                  -134.81 D Ph(U)  deg
                                      -1117.6 E Htot    W
stainless            Solid type       -1152.2 F Edot    W
```

```
!------------------------------- 3 ---------------------------------
MINOR       jet pump (pg 26 book 4)
sameas   4a a Area   m^2              3.2293E+05 A |p|    Pa
   0.5000  b K+                          2.0722 B Ph(p)  deg
   0.7500  c K-                      1.0506E-02 C |U|    m^3/s
                                       -134.81  D Ph(U)  deg
                                       -1117.6  E Htot   W
                                       -1238.3  F Edot   W
!------------------------------- 4 ---------------------------------
CONE       Jet pump (pg 25 and 97 bk 4)
 9.0500E-05 a AreaI  m^2              3.2170E+05 A |p|    Pa
   0.2030  b PerimI m                    3.0483 B Ph(p)  deg
 3.3000E-02 c Length m                1.0697E-02 C |U|   m^3/s
 2.9400E-04 d AreaF  m^2               -133.94  D Ph(U)  deg
   0.2030  e PerimF m                  -1117.6  E Htot   W
 3.0000E-04 f Srough                   -1258.2  F Edot   W
stainless            Solid type
!------------------------------- 5 ---------------------------------
DUCT       180 bend plus brass connecting flange (pg 27 book 4)
 8.1500E-03 a Area   m^2              3.2137E+05 A |p|    Pa
   0.3200  b Perim  m                    3.0765 B Ph(p)  deg
 6.7000E-02 c Length m        7c      2.6472E-02 C |U|   m^3/s
 1.7500E-03 d Srough                   -104.21  D Ph(U)  deg
                                       -1117.6  E Htot   W
stainless            Solid type        -1264.2  F Edot   W
!------------------------------- 6 ---------------------------------
RPN        Capture p1 here
   0.0000  a G or T               (3.2091E+05, 1.7248E+04)
p1
!------------------------------- 7 ---------------------------------
DUCT       180 bend plus brass connecting flange (pg 27 book 4)
 8.1500E-03 a Area   m^2              3.1610E+05 A |p|    Pa
   0.3200  b Perim  m                    3.1995 B Ph(p)  deg
   0.2810  c Length m        Mstr       0.10047 C |U|    m^3/s
sameas   5d d Srough                   -91.458  D Ph(U)  deg
                                       -1117.6  E Htot   W
stainless            Solid type        -1289.5  F Edot   W
!------------------------------- 8 ---------------------------------
MINOR       minor loss for 180 degree bend
sameas   7a a Area   m^2              3.1611E+05 A |p|    Pa
   0.2500  b K+                          3.2129 B Ph(p)  deg
sameas   8b c K-                        0.10047 C |U|    m^3/s
                                       -91.458  D Ph(U)  deg
                                       -1117.6  E Htot   W
                                       -1293.2  F Edot   W
!------------------------------- 9 ---------------------------------
CONE       4" to 3" Concentric reducer (pg 36 book 4)
 8.1070E-03 a AreaI  m^2              3.1152E+05 A |p|    Pa
   0.3190  b PerimI m                    3.2782 B Ph(p)  deg
   0.1020  c Length m                   0.12095 C |U|    m^3/s
sameas  11a d AreaF  m^2      Mstr     -90.691  D Ph(U)  deg
   0.2390  e PerimF m         9d      -1117.6  E Htot   W
sameas   5d f Srough                   -1304.1  F Edot   W
stainless            Solid type
!------------------------------ 10 ---------------------------------
MINOR       minor loss for expansion
sameas   9d a Area   m^2              3.1154E+05 A |p|    Pa
   0.0000  b K+                          3.3109 B Ph(p)  deg
   0.2600  c K-                         0.12095 C |U|    m^3/s
                                       -90.691  D Ph(U)  deg
```

```
                                               -1117.6      E Htot     W
                                               -1314.9      F Edot     W
!-------------------------------- 11 --------------------------------
DUCT        3" FB Duct - Length given in concept.skf
   4.5600E-03 a Area    m^2        Mstr       2.9226E+05 A |p|       Pa
      0.2390  b Perim   m          11a           3.5705 B Ph(p)     deg
      0.2600  c Length  m                       0.15774 C |U|       m^3/s
sameas    5d d Srough                          -89.807  D Ph(U)     deg
                                               -1117.6      E Htot     W
stainless             Solid type               -1357.9      F Edot     W
!-------------------------------- 12 --------------------------------
MINOR        Minor loss R, elbow (pg 37,52 book 4), first half
sameas   13a a Area    m^2                     2.9227E+05 A |p|       Pa
   8.5000E-02 b K+                                3.6094 B Ph(p)     deg
sameas   12b c K-                              0.15774 C |U|       m^3/s
                                               -89.807  D Ph(U)     deg
                                               -1117.6      E Htot     W
                                               -1373.5      F Edot     W
!-------------------------------- 13 --------------------------------
CONE        3.5" to 3" Long radius reducing elbow (pg 36 book 4)
sameas   11a a AreaI   m^2         Mstr       2.7585E+05 A |p|       Pa
      0.2390  b PerimI  m          13a           3.8220 B Ph(p)     deg
      0.2090  c Length  m                       0.19036 C |U|       m^3/s
   6.2070E-03 d AreaF   m^2                     -89.258  D Ph(U)     deg
      0.2790  e PerimF  m                       -1117.6      E Htot     W
sameas    5d f Srough                          -1410.6      F Edot     W
stainless             Solid type
!-------------------------------- 14 --------------------------------
MINOR        Minor loss R, elbow (pg 37,52 book 4), second half
sameas   13d a Area    m^2                     2.7586E+05 A |p|       Pa
   8.5000E-02 b K+                                3.8543 B Ph(p)     deg
sameas   14b c K-                              0.19036 C |U|       m^3/s
                                               -89.258  D Ph(U)     deg
                                               -1117.6      E Htot     W
                                               -1425.4      F Edot     W
!-------------------------------- 15 --------------------------------
DUCT        FB connector/part of tee (Pg 55 book 4 concept.skf)
   5.3500E-03 a Area    m^2        Mstr       2.6965E+05 A |p|       Pa
      0.2590  b Perim   m          15a           3.9351 B Ph(p)     deg
   7.0000E-02 c Length  m                       0.20085 C |U|       m^3/s
sameas    5d d Srough                          -89.112  D Ph(U)     deg
                                               -1117.6      E Htot     W
stainless             Solid type               -1439.7      F Edot     W
!-------------------------------- 16 --------------------------------
RPN         Idelchik minor loss at 3-way junction needs U1 from later
      0.28083 a G or T            G           0.28083      A m3/s
inp
!-------------------------------- 17 --------------------------------
RPN          Minor loss resistor, FB-Res (pg 53-55 book 4)
      0.4220  a G or T                        5640.8       A Pa-s/m3
0.212 rho * 16a sqrd * 15C / 5.35E-3 sqrd / 17a *
!-------------------------------- 18 --------------------------------
IMPEDANCE  Minor loss resistor for FB to resonator junction
sameas   17A a Re(Zs) Pa-s/m^3                 2.6972E+05 A |p|       Pa
      0.0000  b Im(Zs) Pa-s/m^3                   4.1754 B Ph(p)     deg
                                               0.20085 C |U|       m^3/s
                                               -89.112  D Ph(U)     deg
                                               -1117.6      E Htot     W
                                               -1553.4      F Edot     W
!-------------------------------- 19 --------------------------------
```

```
SOFTEND     End of feedback branch
    0.0000 a Re(z)                        2.6972E+05 A |p|     Pa
    0.0000 b Im(z)                            4.1754 B Ph(p)   deg
    0.0000 c Htot    W                        0.20085 C |U|    m^3/s
                                            -89.112 D Ph(U)   deg
                                            -1117.6 E Htot    W
                                            -1553.4 F Edot    W
                                       -8.4511E-02 G Re(z)
                                             1.4712 H Im(z)
                                             325.00 I T       K
                                             1699.8 J p20HL   Pa
!------------------------------- 20 -------------------------------
DUCT        Dummy duct used to get local values
    1.0000 a Area    m^2                   3.1120E+05 A |p|     Pa
    1.0000 b Perim  m                        0.0000 B Ph(p)   deg
    0.0000 c Length m                     8.0260E-03 C |U|    m^3/s
    0.0000 d Srough                           22.869 D Ph(U)   deg
                                             1117.6 E Htot    W
ideal                Solid type             1150.7 F Edot    W
!------------------------------- 21 -------------------------------
RPN         Estim unmitigated streaming mass flux in grams/sec
    0.0000 a G or T                          1.7043 A g/s
rho Edot * pm / 1000 *
!------------------------------- 22 -------------------------------
RPN         reset Tm at hx input
   319.30    a G or T                         319.30 A Kelvin
inp =Tm
!------------------------------- 23 -------------------------------
TX          Main room temp water HX (pg 90 book 3)
  6.6580E-03 a Area    m^2                  3.1127E+05 A |p|     Pa
     0.2275 b GasA/A                     -5.1538E-02 B Ph(p)   deg
  2.0400E-02 c Length m                   7.6023E-03 C |U|    m^3/s
  1.2700E-03 d radius m                       15.687 D Ph(U)   deg
    -1962.1   e HeatIn W         G          -836.9 E Htot    W
     0.0000 f SolidT K                       1138.1 F Edot    W
     0.0000 g FracQN                         319.30 G GasT    K
                                             277.65 H SolidT K
stainless             Solid type           -811.37 I H2k     W
!------------------------------- 24 -------------------------------
STKDUCT     Regen cold end dead space due to ribs (pg 91 book 3)
  4.9700E-03 a Area    m^2                  3.1127E+05 A |p|     Pa
     0.7400 b Perim  m                    -5.3642E-02 B Ph(p)   deg
  3.1750E-03 c Length m                   7.4794E-03 C |U|    m^3/s
  2.7000E-02 d WallA  m^2                     11.944 D Ph(U)   deg
                                            -836.9 E Htot    W
                                             1138.6 F Edot    W
                                             319.30 G TBeg    K
                                             334.51 H TEnd    K
stainless             Solid type           -791.26 I H2k     W
!------------------------------- 25 -------------------------------
STKSCREEN   Regenerator (pg 92 book 3) (Ks frac est:pg 20 book 4)
  6.2070E-03 a Area    m^2                  2.7879E+05 A |p|     Pa
     0.7190 b VolPor                          3.5345 B Ph(p)   deg
  7.3000E-02 c Length m                   2.5611E-02 C |U|    m^3/s
  4.2200E-05 d rh      m                     -42.802 D Ph(U)   deg
     0.3000 e ksFrac                         -836.9 E Htot    W
                                             2464.8 F Edot    W
                                             334.51 G TBeg    K
                                             883.43 H TEnd    K
stainless             Solid type           -65.222 I H2k     W
```

```
!------------------------------ 26 ------------------------------
RPN        Estimated heat leak through FiberFrax (pg 22 book 4)
    0.0000 a G or T                    222.39     A W
25H 273 - 1.89 * 25H 273 - 1.11E-4 * 0.125 + *
!------------------------------ 27 ------------------------------
DUCT       All regen hot end dead space (pg 92 book 3)(area is avg)
 4.6200E-03 a Area    m^2              2.7875E+05 A |p|    Pa
    2.0000 b Perim  m                   3.5284    B Ph(p)  deg
 8.7000E-03 c Length m                 2.6458E-02 C |U|    m^3/s
 6.0000E-04 d Srough                   -44.70     D Ph(U)  deg
                                       -836.9     E Htot   W
stainless          Solid type          2456.5    F Edot   W
!------------------------------ 28 ------------------------------
HX         HHX (pg 93 book 4) heat xfer area used/not acoustic area
 5.6970E-03 a Area    m^2              2.7872E+05 A |p|    Pa
    0.9867 b GasA/A                     3.5254    B Ph(p)  deg
 6.3500E-03 c Length m                 2.7238E-02 C |U|    m^3/s
 7.9400E-04 d y0      m                 -46.609   D Ph(U)  deg
    3223.0   e HeatIn W       G         2386.1    E Htot   W
    0.0000 f SolidT K                   2433.2    F Edot   W
       0.0000 g FracQN                   883.43    G GasT   K
                                         965.32    H SolidT K
stainless          Solid type          3157.8    I H2k    W
!------------------------------ 29 ------------------------------
STKDUCT    hhx dead space (pg 94 book 3) stainless used for Qdot
 5.4400E-03 a Area    m^2              2.7871E+05 A |p|    Pa
    0.2620 b Perim  m                   3.5232    B Ph(p)  deg
 3.6830E-03 c Length m                 2.7675E-02 C |U|    m^3/s
 4.0540E-03 d WallA  m^2               -47.367    D Ph(U)  deg
                                        2386.1    E Htot   W
                                        2432.8    F Edot   W
                                        883.43    G TBeg   K
                                        856.18    H TEnd   K
stainless          Solid type          3121.7    I H2k    W
!------------------------------ 30 ------------------------------
STKDUCT    Straight section of pulse tube (pg 101 bk 4)
 6.2070E-03 a Area    m^2      Mstr    2.7803E+05 A |p|    Pa
    0.2790 b Perim  m          30a      3.4412    B Ph(p)  deg
 8.0000E-02 c Length m                 3.9570E-02 C |U|    m^3/s
 1.1600E-03 d WallA  m^2                -60.35    D Ph(U)  deg
                                        2386.1    E Htot   W
                                        2429.5    F Edot   W
                                        856.18    G TBeg   K
                                        339.31    H TEnd   K
stainless          Solid type          2438.1    I H2k    W
!------------------------------ 31 ------------------------------
STKCONE    Tapered section of pulse tube (pg 101 bk 4)
sameas  30a a AreaI  m^2               2.7481E+05 A |p|    Pa
sameas  30b b PerimI m                  3.2080    B Ph(p)  deg
    0.1600   c Length m                6.5953E-02 C |U|    m^3/s
 6.2070E-03 d AreaF  m^2      Mstr     -71.297    D Ph(U)  deg
    0.2797 e PerimF m         31d       2386.1    E Htot   W
 1.1400E-02 f f_wall                    2421.0    F Edot   W
                                        339.31    G TBeg   K
                                        305.43    H TEnd   K
stainless          Solid type          2393.3    I H2k    W
!------------------------------ 32 ------------------------------
MINOR      minor loss for flow straightener
sameas  31d a Area    m^2              2.7477E+05 A |p|    Pa
    0.6500  b K+                         3.2386    B Ph(p)  deg
```

```
sameas   32b c K-                     6.5953E-02 C |U|    m^3/s
                                       -71.297   D Ph(U)  deg
                                        2386.1   E Htot   W
                                        2415.9   F Edot   W
!-------------------------------- 33 --------------------------------
TX          Small water Xger
 6.6580E-03 a Area    m^2             2.7377E+05 A |p|    Pa
    0.2690  b GasA/A                    3.1933   B Ph(p)  deg
 1.0160E-02 c Length m                6.6461E-02 C |U|    m^3/s
 2.2860E-03 d radius m                 -71.438   D Ph(U)  deg
 -250.0     e HeatIn W                  2136.1   E Htot   W
  290.00    f SolidT K       =33H       2411.0   F Edot   W
    0.0000 g FracQN                     305.43   G GasT   K
                                        290.00   H SolidT K
ideal           Solid type             2143.3   I H2k    W
!-------------------------------- 34 --------------------------------
RPN         Area calc for minor loss
    0.0000 a G or T                   1.7910E-03 A m2
33a 33b *
!-------------------------------- 35 --------------------------------
MINOR       Minor loss, 2nd CHX pg 99 5.
sameas   34A a Area    m^2            2.7302E+05 A |p|    Pa
    1.0000  b K+                        3.7717   B Ph(p)  deg
sameas   35b c K-                     6.6461E-02 C |U|    m^3/s
                                       -71.438   D Ph(U)  deg
                                        2136.1   E Htot   W
                                        2316.1   F Edot   W
!-------------------------------- 36 --------------------------------
DUCT        PT connector (see pg 55 book 4 and concept.skf)
 6.2070E-03 a Area    m^2             2.7025E+05 A |p|    Pa
    0.2790  b Perim  m                  3.6353   B Ph(p)  deg
 9.2000E-02 c Length m                8.1915E-02 C |U|    m^3/s
 6.0000E-04 d Srough                   -74.314   D Ph(U)  deg
                                        2136.1   E Htot   W
stainless       Solid type             2310.9   F Edot   W
!-------------------------------- 37 --------------------------------
RPN         Minor loss, PT--resonator (pg 45 book 4) (a) = K in + K out
    0.9100  a G or T                  3.1737E+04 A Pa-s/m3
0.212 rho * 16a sqrd * 36C / 5.35E-3 sqrd / 37a *
!-------------------------------- 38 --------------------------------
IMPEDANCE Minor loss resistor for PT-Res (pg 45, 53-55 book 4)
sameas   37A a Re(Zs) Pa-s/m^3        2.6972E+05 A |p|    Pa
    0.0000  b Im(Zs) Pa-s/m^3           4.1754   B Ph(p)  deg
                                      8.1915E-02 C |U|    m^3/s
                                       -74.314   D Ph(U)  deg
                                        2136.1   E Htot   W
                                        2204.4   F Edot   W  .
!-------------------------------- 39 --------------------------------
UNION       Rejoin
 19          a SegNum                 2.6972E+05 A |p|    Pa
 2.6972E+05 b |p|Sft Pa       =39A      4.1754   B Ph(p)  deg
    4.1754  c Ph(p)S deg      =39B      0.28083  C |U|    m^3/s
    0.0000  d TSoft   K                 -84.84   D Ph(U)  deg
    1699.8  e p20HLS Pa       =39H      1018.5   E Htot   W
                                        651.00   F Edot   W
                                        305.43   G T      K
                                        1699.8   H p20HL  Pa
!-------------------------------- 40 --------------------------------
RPN         allow minor-loss calc far upstream to "know" | U1 | here
sameas   16a a G or T         =40A     0.28083   A m3/s
```

```
U1 mag
!------------------------------- 41 -------------------------------
RPN         Power output
  651.00    a G or T        =41A     651.00      A W
Edot
!------------------------------- 42 -------------------------------
RPN         Regenerator pressure drop
    0.0000 a G or T                  4.2199E+04  A Pa
                                    -1.9638E+04  B Pa
39A 39B sin * ~ 20A 39A 39B cos * -
!------------------------------- 43 -------------------------------
RPN         Jet pump pressure drop
    0.0000 a G or T                  9710.8      A Pa
                                     1.7248E+04  B Pa
5A 5B sin * 5A 5B cos * 1A -
!------------------------------- 44 -------------------------------
RPN         Normalized jp+regen pressure drop
    0.0000 a G or T              (0.16089, -1.6097E-02)  A Pa
42A 42B i * + + 43A 43B i * + + 6A /
!------------------------------- 45 -------------------------------
RPN         tau - 1
    0.0000 a G or T                  1.7668
25H 24G / 1 -
!------------------------------- 46 -------------------------------
RPN         Diss, not incl hhx-regen-chx
    0.0000 a G or T                  631.48      A W
19F ~ 1F ~ - - 28F 38F - +
! The restart information below was generated by a previous run
! and will be used by DeltaEC the next time it opens this file.
guessz   1a   1b   1d  16a  23e  28e
xprecn  -285.28    -442.42    8.7229E-08  2.2709E-06  4.6104E-02  5.1442E-03
targs   33f  39b  39c  39e  40a  41a
mstr-slave 7 7 5 9 -3 11 -2 13 -2 15 -2 30 -2 31 -3
```

References

1. P.S. Spoor, G.W. Swift, Thermoacoustic separation of a He–Ar mixture. Phys. Rev. Lett. **85**, 1646–1649 (2000)
2. G.W. Swift, P.S. Spoor, Thermal diffusion and mixture separation in the acoustic boundary layer. J. Acoust. Soc. Am. **106**, 1794–1800 (1999). Errata J. Acoust. Soc. Am. **107**:2299, 2000; **109**:1261, 2001
3. R.S. Reid, W.C. Ward, G.W. Swift, Cyclic thermodynamics with open flow. Phys. Rev. Lett. **80**, 4617–4620 (1998)
4. R.S. Reid, G.W. Swift, Experiments with a flow-through thermoacoustic refrigerator. J. Acoust. Soc. Am. **108**, 2835–2842 (2000)
5. L.E. Kinsler, A.R. Frey, A.B. Coppens, J.V. Sanders, *Fundamentals of Acoustics*, 4th edn. (Wiley, New York, 1999)
6. G.J. Van Wylen, R.E. Sonntag, *Fundamentals of Classical Thermodynamics* (Wiley, New York, 1986)
7. E. Kreyszig, *Advanced Engineering Mathematics* (Wiley, New York, 1979)
8. N. Rott, Thermoacoustics. Adv. Appl. Mech. **20**, 135–175 (1980)
9. A. Tominaga, *Fundamental Thermoacoustics* (1998). ISBN 4-7536-5079-0 (In Japanese)
10. G.W. Swift, Thermoacoustic engines and refrigerators, in *Encyclopedia of Applied Physics*, vol. 21. Wiley for American Institute of Physics (Wiley, New York, 1997), pp. 245–264
11. G.W. Swift, Thermoacoustic engines. J. Acoust. Soc. Am. **84**, 1145–1180 (1988)
12. N. Rott, Damped and thermally driven acoustic oscillations in wide and narrow tubes. Z. Angew. Math. Phys. **20**, 230–243 (1969)
13. N. Rott, Thermally driven acoustic oscillations, part III: second-order heat flux. Z. Angew. Math. Phys. **26**, 43–49 (1975)
14. A.J. Organ, *The Regenerator and the Stirling Engine* (Mechanical Engineering Publications, Ltd., London, 1997)
15. I. Urieli, D.M. Berchowitz, *Stirling Cycle Engine Analysis* (Adam Hilger, Bristol, 1984)
16. A.J. Organ, *Thermodynamics and Gas Dynamics of the Stirling Cycle Machine* (Cambridge University Press, Cambridge, 1992)
17. G. Walker, *Stirling Engines* (Clarendon, Oxford, 1960)
18. J.R. Senft, *Ringbom Stirling Engines* (Oxford University Press, Oxford, 1993)
19. R. Radebaugh, A review of pulse tube refrigeration. Adv. Cryog. Eng. **35**, 1191–1205 (1990)
20. G. Walker, *Cryocoolers* (Plenum, New York, 1983)
21. P.P. Steijaert, Thermodynamical aspects of pulse-tube refrigerators, Ph.D. thesis, Technische Universiteit Eindhoven (Netherlands), 1999
22. C.M. de Blok, Thermoacoustic system, 1998. Dutch Patent: International Application Number PCT/NL98/00515, US Patent 6,314,740, 13 Nov 2001

23. P.H. Ceperley, A pistonless Stirling engine—the traveling wave heat engine. J. Acoust. Soc. Am. **66**, 1508–1513 (1979)
24. P.H. Ceperley, Gain and efficiency of a short traveling wave heat engine. J. Acoust. Soc. Am. **77**, 1239–1244 (1985)
25. J.A. Corey, Two piston V-type Stirling engine, 1987, US Patent No. 4,633,668
26. I. Kolin, *The Evolution of the Heat Engine* (Longman, London, 1972)
27. C.M. Hargreaves, *The Philips Stirling Engine* (Elsevier Science, Amsterdam, 1991)
28. E.H. Cooke-Yarborough, E. Franklin, J. Geisow, R. Howlett, C.D. West, Thermo-mechanical generator: an efficient means of converting heat to electricity at low power levels. Proc. IEE **121**, 749–751 (1974) and references therein
29. C.D. West, *Liquid Piston Stirling Engines* (Van Nostrand Reinhold, New York, 1983)
30. W.E. Gifford, R.C. Longsworth, Pulse tube refrigeration progress. Adv. Cryog. Eng. **10B**, 69–79 (1965)
31. E.L. Mikulin, A.A. Tarasov, M.P. Shkrebyonock, Low-temperature expansion pulse tubes. Adv. Cryog. Eng. **29**, 629–637 (1984)
32. P.H. Ceperley, Traveling wave heat engine, 1978, US Patent No. 4,114,380
33. P.H. Ceperley, Resonant traveling wave heat engine, 1982, US Patent No. 4,355,517.
34. K.M. Godshalk, C. Jin, Y.K. Kwong, E.L. Hershberg, G.W. Swift, R. Radebaugh, Characterization of 350 Hz thermoacoustic driven orifice pulse tube refrigerator with measurements of the phase of the mass flow and pressure. Adv. Cryog. Eng. **41**, 1411–1418 (1996)
35. R.S. Reid, *Open cycle thermoacoustics*, Ph.D. thesis, Georgia Institute of Technology, School of Mechanical Engineering, Atlanta GA, 1999
36. G.W. Swift, J.J. Wollan, Thermoacoustics for liquefaction of natural gas. NETL GasTIPS **8**(4), 21–26 (2002). Also available at www.lanl.gov/thermoacoustics/Pubs/GasTIPS.pdf
37. S. Backhaus, G.W. Swift, A thermoacoustic-Stirling heat engine. Nature **399**, 335–338 (1999)
38. S. Backhaus, G.W. Swift, A thermoacoustic-Stirling heat engine: detailed study. J. Acoust. Soc. Am. **107**, 3148–3166 (2000)
39. C.M. Christensen, *The Innovator's Dilemma: When New Technologies Cause Great Firms to Fail* (Harper Collins, New York, 2000)
40. W.P. Arnott, H.E. Bass, R. Raspet, General formulation of thermoacoustics for stacks having arbitrarily shaped pore cross sections. J. Acoust. Soc. Am. **90**, 3228–3237 (1991)
41. E.S. Jeong, J.L. Smith, An analytic model of heat transfer with oscillating pressure, in *General Papers in Heat Transfer*, ed. by M. Jensen, R. Mahajan, E. McAssey, M.F. Modest, D. Pepper, M. Sohal, A. Lavine. ASME-HTD, vol. 204 (American Society of Mechanical Engineers, New York, 1992), pp. 97–104
42. F.Z. Guo, Y.M. Chou, S.Z. Lee, Z.S. Wang, W. Mao, Flow characteristics of a cyclic flow regenerator. Cryogenics **27**, 152–155 (1987)
43. B.J. Huang, C.W. Lu, Dynamic response of regenerator in cyclic flow system. Cryogenics **33**, 1046–1052 (1993)
44. B.J. Huang, C.W. Lu, Linear network analysis of regenerator in a cyclic-flow system. Cryogenics **35**, 203–207 (1995)
45. A. Bejan, *Advanced Engineering Thermodynamics*, 2nd edn. (Wiley, New York, 1997)
46. L.D. Landau, E.M. Lifshitz, *Fluid Mechanics* (Pergamon, Oxford, 1982)
47. H.B. Callen, *Thermodynamics and an Introduction to Thermostatistics* (Wiley, New York, 1985)
48. P.M. Morse, *Thermal Physics* (Benjamin, Reading, 1969)
49. A. Bejan, *Entropy Generation Minimization: The Method of Thermodynamic Optimization of Finite-Size Systems and Finite-Time Processes* (CRC Press, New York, 1995)
50. R.W. Fox, A.T. McDonald, *Introduction to Fluid Mechanics* (Wiley, New York, 1985)
51. R.B. Bird, W.E. Stewart, E.N. Lightfoot, *Transport Phenomena*, 2nd edn. (Wiley, New York, 2001)
52. H. Schlichting, *Boundary Layer Theory* (McGraw-Hill, New York, 1979)
53. G.K. Batchelor, *An Introduction to Fluid Dynamics* (Cambridge University Press, 1967)
54. M. Guillen, *Five Equations that Changed the World* (Hyperion, New York, 1995)

55. F.W. Giacobbe, Estimation of Prandtl numbers in binary mixtures of helium and other noble gases. J. Acoust. Soc. Am. **96**, 3568–3580 (1994)
56. P.W. Bridgman, A complete collection of thermodynamic formulas. Phys. Rev. **3**, 273–281 (1914)
57. D.T. Blackstock, *Fundamentals of Physical Acoustics* (Wiley, New York, 2000)
58. A.D. Pierce, *Acoustics: An Introduction to Its Physical Principles and Applications* (Acoustical Society of America, Woodbury/NY, 1989)
59. P.M. Morse, K.U. Ingard, *Theoretical Acoustics* (McGraw-Hill, New York, 1968)
60. A. Migliori, G.W. Swift, Liquid sodium thermoacoustic engine. Appl. Phys. Lett. **53**, 355–357 (1988)
61. W.C. Ward, G.W. Swift, Design environment for low amplitude thermoacoustic engines (DELTAE). J. Acoust. Soc. Am. **95**, 3671–3672 (1994). This historic software and user's guide are available from the Energy Science and Technology Software Center, US Department of Energy, Oak Ridge TN. In 2008, DELTAE was replaced by Design Environment for Low-amplitude Thermoacoustic Energy Conversion (DELTAEC), freely available at www.lanl.gov/thermoacoustics/.
62. G.W. Swift, R.M. Keolian, Thermoacoustics in pin-array stacks. J. Acoust. Soc. Am. **94**, 941–943 (1993)
63. G.W. Swift, W.C. Ward, Simple harmonic analysis of regenerators. J. Thermophys. Heat Tran. **10**, 652–662 (1996)
64. W.M. Kays, A.L. London, *Compact Heat Exchangers* (McGraw-Hill, New York, 1964)
65. M. Greenspan, F.N. Wimenitz, An acoustic viscometer for gases—I. Technical report, US National Bureau of Standards (now known as National Institute of Standards and Technology), 1953. NBS Report 2658
66. G. Huelsz, E. Ramos, On the phase difference of the temperature and pressure waves in the thermoacoustic effect. Int. Commun. Heat Mass Transfer **22**, 71–80 (1995)
67. P. Merkli, H. Thomann, Thermoacoustic effects in a resonance tube. J. Fluid Mech. **70**, 161–177 (1975)
68. N. Rott, The influence of heat conduction on acoustic streaming. Z. Angew. Math. Phys. **25**, 417–421 (1974)
69. H. Thomann, Acoustical streaming and thermal effects in pipe flow with high viscosity. Z. Angew. Math. Phys. **27**, 709–715 (1976)
70. D. Gedeon, A globally implicit Stirling cycle simulation, in *Proceedings of the 21st Intersociety Energy Conversion Engineering Conference*, pp. 550–554 (1986). Software available from Gedeon Associates, Athens OH
71. J. Gary, A. O'Gallagher, R. Radebaugh, A numerical model for regenerator performance. Technical report, NIST-Boulder, 1994. Also known as REGEN 3.1 Users Guide
72. A.A. Atchley, Standing wave analysis of a thermoacoustic prime mover below onset of self-oscillation. J. Acoust. Soc. Am. **92**, 2907–2914 (1992)
73. T.J. Hofler, Thermoacoustic refrigerator design and performance, Ph.D. thesis, Physics department, University of California, San Diego CA (1986)
74. R.C. Tolman, P.C. Fine, On the irreversible production of entropy. Rev. Mod. Phys. **20**, 51–77 (1948)
75. J.C. Wheatley, T. Hofler, G.W. Swift, A. Migliori, An intrinsically irreversible thermoacoustic heat engine. J. Acoust. Soc. Am. **74**, 153–170 (1983)
76. G.W. Swift, D.L. Gardner, S. Backhaus, Acoustic recovery of lost power in pulse tube refrigerators. J. Acoust. Soc. Am. **105**, 711–724 (1999)
77. A.T.A.M. de Waele, P.P. Steijaert, J. Gijzen, Thermodynamical aspects of pulse tubes. Cryogenics **37**, 313–324 (1997)
78. A.T.A.M. de Waele, P.P. Steijaert, J.J. Koning, Thermodynamical aspects of pulse tubes II. Cryogenics **38**, 329–335 (1998)
79. R. Yaron, S. Shokralla, J. Yuan, P.E. Bradley, R. Radebaugh, Etched foil regenerator. Adv. Cryog. Eng. **41**, 1339–1346 (1996)

80. M.A. Lewis, T. Kuriyama, F. Kuriyama, R. Radebaugh, Measurement of heat conduction through stacked screens. Adv. Cryog. Eng. **43**, 1611–1618 (1998)

81. J.R. Womersley, Method for the calculation of velocity, rate of flow, and viscous drag in arteries when the pressure gradient is known. J. Physiol. **127**, 553–563(1955)

82. M. Iguchi, M. Ohmi, K. Maegawa, Analysis of free oscillating flow in a U-shaped tube. Bull. JSME **25**, 1398–1405 (1982)

83. U.H. Kurzweg, E.R. Lindgren, B. Lothrop, Onset of turbulence in oscillating flow at low Womersley number. Phys. Fluids A **1**, 1972–1975 (1989) and references therein

84. S.M. Hino, M. Sawamoto, S. Takasu, Experiments on transition to turbulence in an oscillatory pipe flow. J. Fluid Mech. **75**, 193–207 (1976)

85. M. Ohmi, M. Iguchi, Critical Reynolds number in an oscillating pipe flow. Bull. JSME **25**, 165–172 (1982)

86. M. Ohmi, M. Iguchi, K. Kakehashi, M. Tetsuya, Transition to turbulence and velocity distribution in an oscillating pipe flow. Bull. JSME **25**, 365–371 (1982)

87. M. Ohmi, M. Iguchi, I. Urahata, Flow patterns and frictional losses in an oscillating pipe flow. Bull. JSME **25**, 536–543 (1982)

88. J.R. Olson, G.W. Swift, Acoustic streaming in pulse tube refrigerators: Tapered pulse tubes. Cryogenics **37**, 769–776 (1997)

89. G.W. Swift, M.S. Allen, J.J. Wollan, Performance of a tapered pulse tube, in *Cryocoolers 10*, ed. by R.G. Ross Jr. (Plenum, New York, 1999), pp. 315–320

90. J.R. Olson, G.W. Swift, Energy dissipation in oscillating flow through straight and coiled pipes. J. Acoust. Soc. Am. **100**, 2123–2131 (1996)

91. V.L. Streeter, *Handbook of Fluid Dynamics* (McGraw-Hill, New York, 1961)

92. I.E. Idelchik, *Handbook of Hydraulic Resistance*, 3rd edn. (Begell House, New York, 1994)

93. M. Hino, M. Kashiwayanagi, A. Nakayama, T. Hara, Experiments on the turbulence statistics and the structure of reciprocating oscillatory flow. J. Fluid Mech. **131**, 363–400 (1983)

94. A.S. Worlikar, O.M. Knio, Numerical simulation of a thermoacoustic refrigerator I: unsteady adiabatic flow around the stack. J. Comput. Phys. **127**, 424–451 (1996)

95. A.S. Worlikar, O.M. Knio, R. Klein, Numerical simulation of a thermoacoustic refrigerator II: Stratified flow around the stack. J. Comput. Phys. **144**, 299–324 (1998)

96. N. Cao, J.R. Olson, G.W. Swift, S. Chen, Energy flux density in a thermoacoustic couple. J. Acoust. Soc. Am. **99**, 3456–3464 (1996)

97. P.J. Storch, R. Radebaugh, J.E. Zimmerman, Analytical model for the refrigeration power of the orifice pulse tube refrigerator. Technical note 1343, National Institute of Standards and Technology, 1990

98. G.W. Swift, Analysis and performance of a large thermoacoustic engine. J. Acoust. Soc. Am. **92**, 1551–1563 (1992)

99. N. Rott, G. Zouzoulas, Thermally driven acoustic oscillations, part IV: tubes with variable cross section. Z. Angew. Math. Phys. **27**, 197–224 (1976)

100. J.L. Smith and M. Romm, Thermodynamic loss at component interfaces in Stirling cycles, in *Proceedings of the 27th Intersociety Energy Conversion Engineering Conference* (Society of Automotive Engineers, 1992), pp. 5.529–5.532

101. P. Kittel, Temperature profile within pulse tubes. Adv. Cryog. Eng. **43**, 1927–1932 (1998)

102. L. Bauwens, Interface loss in the small amplitude orifice pulse tube model. Adv. Cryog. Eng. **43**, 1933–1940 (1998)

103. M.J. Romm, J.L. Smith, Stirling engine loss in dead volumes between components, in *Proceedings of the 28th Intersociety Energy Conversion Engineering Conference*, pp. 2.751–2.758 (1993)

104. G.W. Swift, A Stirling engine with a liquid working substance. J. Appl. Phys. **65**, 4157–4172 (1989)

105. W.L.M. Nyborg, Acoustic streaming, in *Physical Acoustics, Volume IIB*, ed. by W.P. Mason (Academic Press, New York, 1965), pp. 265–331

106. D. Gedeon, DC gas flows in Stirling and pulse-tube cryocoolers, in *Cryocoolers 9*, ed. by R.G. Ross Jr. (Plenum, New York, 1997), pp. 385–392

107. C. Wang, G. Thummes, C. Heiden, Effects of DC gas flow on performance of two-stage 4 K pulse tube coolers. Cryogenics **38**, 689–695 (1998)

108. J.R. Olson, G.W. Swift, Suppression of acoustic streaming in tapered pulse tubes, in *Cryocoolers 10*, ed. by R.G. Ross Jr. (Plenum, New York, 1999), pp. 307–313

109. J.M. Lee, P. Kittel, K.D. Timmerhaus, R. Radebaugh, Flow patterns intrinsic to the pulse tube refrigerator, in *Proceedings of the 7th International Cryocooler Conference*, Phillips Laboratory, Kirtland Air Force Base NM, pp. 125–139 (1993)

110. J.W. Strutt (Baron Rayleigh), *The Theory of Sound* (Dover, New York, 1945). Section 352

111. S. Zhu, P. Wu, Z. Chen, Double inlet pulse tube refrigerators: an important improvement. Cryogenics **30**, 514–520 (1990)

112. T.G. Wang, C.P. Lee, Radiation pressure and acoustic levitation, in *Nonlinear Acoustics*, ed. by M.F. Hamilton, D.T. Blackstock (Academic Press, New York, 1998), pp. 177–205

113. C.E. Bradley, Acoustic streaming field structure: the influence of the radiator. J. Acoust. Soc. Am. **100**, 1399–1408 (1996)

114. X. Zhao, Z. Zhu, G. Du, A note about acoustic streaming: comparison of C. E. Bradley's and W. L. Nyborg's theories. J. Acoust. Soc. Am. **104**, 583–584 (1998)

115. B.L. Smith, A. Glezer, The formation and evolution of synthetic jets. Phys. Fluids **10**, 2281–2297 (1998)

116. C.S. Kirkconnell, Experimental investigation of a unique pulse tube expander design, in *Cryocoolers 10*, ed. by R.G. Ross Jr. (Plenum, New York, 1999), pp. 239–247

117. V. Kotsubo, P. Huang, T.C. Nast, Observation of DC flows in a double inlet pulse tube, in *Cryocoolers 10*, ed. by R.G. Ross Jr. (Plenum, New York, 1999), pp. 299–305

118. B. Zinn, Pulsating combustion, in *Advanced Combustion Methods*, ed. by F.J. Weinberg (Academic, London, 1986), pp. 113–181

119. C.C. Lawrenson, B. Lipkens, T.S. Lucas, D.K. Perkins, T.W. Van Doren, Measurements of macrosonic standing waves in oscillating closed cavities. J. Acoust. Soc. Am. **104**, 623–636 (1998)

120. Y. Il'inskii, B. Lipkens, T. Lucas, T.W. Van Doren, E. Zabolotskaya, Nonlinear standing waves in an acoustical resonator. J. Acoust. Soc. Am. **104**, 2664–2674 (1998)

121. E. Buckingham, On physically similar systems: illustrations of the use of dimensional equations. Phys. Rev. **4**, 345–376 (1914). See also any introductory fluid-mechanics textbook.

122. J.R. Olson, G.W. Swift, Similitude in thermoacoustics. J. Acoust. Soc. Am. **95**, 1405–1412 (1994)

123. F. Reif, *Fundamentals of Statistical and Thermal Physics* (McGraw-Hill, New York, 1965)

124. S.L. Garrett, D.K. Perkins, A. Gopinath, Thermoacoustic refrigerator heat exchangers: design, analysis, and fabrication, in *Heat Transfer 1994: Proceedings of the 10th International Heat Transfer Conference*, ed. by G.F. Hewitt (Institution of Chemical Engineers, Rugby, 1994), pp. 375–380

125. C. Wang, G. Thummes, C. Heiden, Experimental study of staging method for two-stage pulse tube refrigerators for liquid ^{4}He temperatures. Cryogenics **37**, 857–863 (1997)

126. T.J. Hofler, Concepts for thermoacoustic refrigeration and a practical device, in *Proceedings of the 5th International Cryocoolers Conference*, ed. by Chairman P. Lindquist. Wright-Patterson Air Force Base OH, August 1988. (available at cryocooler.org/past-proceedings/) (1988), pp. 93–101

127. R.J. Gibson, F.S. Nessler, R.M. Keolian, A thermoacoustic pin stack for improved efficiency, in *Heat Transfer-Baltimore 1997*. American Institute of Chemical Engineers Symposium Series, Number 314, vol. 93 (1997), pp. 258–264

128. M.S. Reed, Measurements with wire mesh stacks in thermoacoustic prime movers, Master's thesis, US Naval Postgraduate School, Monterey CA, 1996

129. T.J. Hofler, M.S. Reed, Measurements with wire mesh stacks in thermoacoustic prime movers. J. Acoust. Soc. Am. **99**, 2559 (1996). Abstract only

130. J.A. Adeff, T.J. Hofler, A.A. Atchley, W.C. Moss, Measurements with reticulated vitreous carbon stacks in thermoacoustic prime movers and refrigerators. J. Acoust. Soc. Am. **104**, 32–38 (1998)

131. J. Bösel, Ch. Trepp, J.G. Fourie, An alternative stack arrangement for thermoacoustic heat pumps and refrigerators. J. Acoust. Soc. Am. **106**, 707–715 (1999)

132. J. Bösel, Untersuchungen zur Erhöhung der Leistungsdichte thermoakustischer Maschinen, Ph.D. thesis, Swiss Federal Institute of Technology, 1998. Dissertation ETH 12530

133. W.A. Marrison, Heat-controlled acoustic wave system, US Patent No. 2,836,033 (1958)

134. D. Gedeon, J.G. Wood, Oscillating-flow regenerator test rig: hardware and theory with derived correlations for screens and felts. Technical Report 198442, NASA Lewis Research Center, 1996

135. J. Gerster, J. Krause, S. Wunderlich, L. Doerrer, F. Schmidl, R. Weidl, M. Thürk, P. Seidel, Operation of a high-T_c SQUID gradiometer in a nonmetallic one-stage pulse tube cryocooler, in *Book of Abstracts of the Low Power Cryocooler III Workshop*, 31 May–2 June 1999, Venlo, The Netherlands, p. 27 (1999)

136. W. Rawlings, K.D. Timmerhaus, R. Radebaugh, D.E. Daney, Measurement of performance of a spiral wound polyamide regenerator in a pulse tube refrigerator. Adv. Cryog. Eng. **37B**, 947–953 (1992)

137. F.P. Incropera, D.P. DeWitt, *Introduction to Heat Transfer*, 2nd edn. (Wiley, New York, 1990)

138. W.M. Rohsenow, J.P. Hartnett, Y.I. Cho, *Handbook of Heat Transfer* (McGraw-Hill, New York, 1998)

139. T.W. Simon, J.R. Seume, A survey of oscillating flow in Stirling engine heat exchangers. Technical Report 182108, NASA Lewis Research Center, 1988

140. Mechanical Technology Incorporated, The Stirling engine, Mod II design report. Technical Report NASA CR-175106, Latham NY, 1968

141. E.F. Megyesy. *Pressure Vessel Handbook*, 9th edn. (Pressure Vessel Handbook Publishing Inc., Tulsa OK, 1992)

142. H.H. Bednar, *Pressure Vessel Design Handbook*, 2nd edn. (Van Nostrand Reinhold, New York, 1986)

143. Boiler and Pressure Vessel Committee of the American Society of Mechanical Engineers, *ASME Boiler and Pressure Vessel Code* (American Society of Mechanical Engineers, New York, 2001)

144. D.F. Gaitan, A.A. Atchley, Finite-amplitude standing waves in harmonic and anharmonic tubes. J. Acoust. Soc. Am. **93**, 2489–2495 (1993)

145. A.B. Coppens, J.V. Sanders, Finite-amplitude standing waves within real cavities. J. Acoust. Soc. Am. **58**, 1133–1140 (1975)

146. W.P. Arnott, J.A. Lightfoot, R. Raspet, H. Moosmüller, Radial wave thermoacoustic engines: theory and examples for refrigerators and high-gain narrow-bandwidth photoacoustic spectrometers. J. Acoust. Soc. Am. **99**, 734–745 (1996)

147. J.A. Lightfoot, W.P. Arnott, H.E. Bass, R. Raspet, Experimental study of a radial mode thermoacoustic prime mover. J. Acoust. Soc. Am. **105**, 2652–2662 (1999)

148. J.A. Corey, G.A. Yarr, HOTS to WATTS: The LPSE linear alternator system re-invented, in *Proceedings of the 27th Intersociety Energy Conversion Engineering Conference*, pp. 5.289–5.294, Warrendale PA, 1992. Society of Automotive Engineers. No editor; general chairman Buryl McFadden

149. R.S. Wakeland, Use of electrodynamic drivers in thermoacoustic refrigerators. J. Acoust. Soc. Am. **107**, 827–832 (2000)

150. A.K. de Jonge, A. Sereny, Analysis and optimization of a linear motor for the compressor of a cryogenic refrigerator. Adv. Cryog. Eng. **27**, 631–640 (1982)

151. L. Marton, C. Marton, Editors in Chief *Methods of Experimental Physics: Fluid Dynamics*, vol. 18 (Academic Press, New York, 1981)

152. L.L. Beranek, *Acoustic Measurements* (Wiley, New York, 1949)

153. J. Fraden, *AIP Handbook of Modern Sensors: Physics, Designs, and Applications* (American Institute of Physics, New York, 1993)

154. P.R. Bevington, *Data Reduction and Error Analysis for the Physical Sciences* (McGraw-Hill, New York, 1969)

155. A.M. Fusco, W.C. Ward, G.W. Swift, Two-sensor power measurements in lossy ducts. J. Acoust. Soc. Am. **91**, 2229–2235 (1992)
156. T.J. Hofler, Accurate acoustic power measurements with a high-intensity driver. J. Acoust. Soc. Am. **83**, 777–786 (1988)
157. G. Huelsz, E. Ramos, Temperature measurements inside the oscillatory boundary layer produced by acoustic waves. J. Acoust. Soc. Am. **103**, 1532–1537 (1998)
158. T.B. Ward, R.A. Finke, S.M. Smith, *Creativity and the Mind: Discovering the Genius Within* (Plenum, New York, 1995)
159. J.L. Martin, J.A. Corey, C.M. Martin, A pulse tube cryocooler for telecommunications applications, in *Cryocoolers 10*, ed. by R.G. Ross Jr. (Plenum, New York, 1999), pp. 181–189

Author index

A
Adeff, J. A., **236**
Allen, M. S., 181, 205, 208, 285
Arnott, W. P., **27**, **98–100**, **235**, **250**, 250
Atchley, A. A., **139**, 236, 249

B
Bösel, J., **238**
Backhaus, S. N., **23**, **149**, 163, **202**, 202, **275**
Bass, H. E., 27, 98–100, 235, 250
Batchelor, G. K., **31**, **40**, **42**, 43, **66**
Bauwens, L., **191**
Bednar, H. H., **248**
Bejan, A., **31**, **54**, **56**, **130**, **150**, **153**, **157**, **173**
Beranek, L. L., **259**
Berchowitz, D. M., 1, 4, 196
Bevington, P. R., **259**
Bird, R. B., **31**, **50**
Blackstock, D. T., **59**
Bradley, C. E., **201**
Bradley, P. E., 175, 240
Bridgman, P. W., **54**
Buckingham, E., **222**

C
Callen, H. B., **31**
Cao, N., **189**
Ceperley, P. H., **1**, **5**, **105**, 106
Chen, S., 189
Chen, Z., 200
Cho, Y. I., 240
Chou, Y. M., 27, 105
Christensen, C. M., **26**
Cooke-Yarborough, E. H., **4**

Coppens, A. B., viii, 27, 59, 76, 80, **249**, 251, 252
Corey, J. A., **2**, **250**, 274

D
Daney, D. E., 240
de Blok, C. M., **1**
de Jonge, A. K., **255**
de Waele, A. T. A. M., **168**, **191**
DeWitt, D. P., 240
Doerrer, L., 240
Du, G., 201

F
Fine, P. C., 150, 153, 154
Finke, R. A., 269
Fourie, J. G., 238
Fox, R. W., **31**, **177**, 178, **182**, **189**, **228**
Fraden, J., **259**, 260, **262**, **266**, 267
Franklin, E., 4
Frey, A. R., viii, 27, 59, 76, 80, 251, 252
Fusco, A. M., **265**

G
Gaitan, D. F., **249**
Gardner, D. L., 163, 202
Garrett, S. L., **232**
Gary, J., **138**, **143**, **222**
Gedeon, D., **138**, **143**, **197**, 198, **201**, **222**, **240**
Geisow, J., 4
Gerster, J., **240**
Giacobbe, F. W., **50**, **232**
Gibson, R. J., **236**

Gifford, W. E., **5**
Gijzen, J., 168, 191
Glezer, A., 204, 209
Godshalk, K. M., **11**
Gopinath, A., 232
Greenspan, M., **116**
Guillen, M., **39**
Guo, F. Z., **27, 105**

H
Hara, T., 184
Hargreaves, C. M., **3**
Hartnett, J. P., 240
Heiden, C., 198, 232
Hershberg, E. L., 11
Hino, M., **184**
Hino, S. M., **180**
Hofler, T. J., **138, 141**, 155, **235**, 236, **257, 266**
Howlett, R., 4
Huang, B. J., **27**
Huang, P., 217
Huelsz, G., **116, 269**

I
Idelchik, I. E., **184, 186, 188, 246**
Iguchi, M., **180**, 180, **181**, 182
Il'inskii, Y., **222, 249**
Incropera, F. P., **240**
Ingard, K. U., 59, 115

J
Jeong, E. S., **27**
Jin, C., 11

K
Kakehashi, K., 180
Kashiwayanagi, M., 184
Kays, W. M., **101, 175**, 176, **202, 239**, 240, **242**
Keolian, R. M., 99, 236
Kinsler, L. E., **viii**, **27, 59, 76, 80, 251**, 252
Kirkconnell, C. S., **211**
Kittel, P., **191**, 198, 204
Klein, R., 189, 197, 241
Knio, O. M., 189, 197, 241
Kolin, I., **3**
Koning, J. J., 168
Kotsubo, V., **217**
Krause, J., 240
Kreysig, E., **viii**
Kuriyama, T. and F., 177, 226

Kurzweg, U. H., **180**
Kwong, Y. K., 11

L
Landau, L. D., **31, 40, 42–45, 50, 66, 90, 132**, **173, 201, 207**
Lawrenson, C. C., **222**
Lee, C. P., 201
Lee, J. M., **198, 204**
Lee, S. Z., 27, 105
Lewis, M. A., **177, 226**
Lifshitz, E. M., 31, 40, 42–45, 50, 66, 90, 132, 173, 201, 207
Lightfoot, E. N., 31, 50
Lightfoot, J. A., **250**, 250
Lindgren, E. R., 180
Lipkins, B., 222, 249
London, A. L., 101, 175, 176, 202, 239, 240, 242
Longsworth, R. C., 5
Lothrop, B., 180
Lu, C. W., 27
Lucas, T. S., 222, 249

M
Maegawa, K., 180–182
Mao, W., 27, 105
Marrison, W. A., **238**
Martin, J. L. and C. M., **274**
Marton, L. and C., **259, 269**
McDonald, A. T., 31, 177, 178, 182, 189, 228
Megyesy, E. F., **248**
Merkli, P., **131, 134**
Migliori, A., **65, 155**, 155, **232**
Mikulin, E. L., **5**
Moosmüller, H., 250
Morse, P. M., **31, 59, 115**
Moss, W. C., 236

N
Nakayama, A., 184
Nast, T. C., 217
Nessler, F. S., 236
Nyborg, W. L. M., **197**, 198

O
O'Gallagher, A., 138, 143, 222
Ohmi, M., **180**, 180–182
Olson, J. R., **181, 184**, 189, **198, 205, 207, 222**, **225, 245, 278, 286**
Organ, A. J., **1, 174, 222, 240**

P

Perkins, D. K., 222, 232
Pierce, A. D., **59**, **77**, **102**, **115**

R

Radebaugh, R., **1**, 11, 138, 143, 175, 177, 190,
    194, 198, 204, 222, 226, 240
Ramos, E., 116, 269
Raspet, R., 27, 98–100, 235, 250
Rawlings, W., **240**
Lord Rayleigh, **198**
Reed, M. S., **236**
Reid, R. S., **vii**, **16**, **217**, **236**, **273**, **288**
Reif, F., **226**
Rohsenow, W. M., **240**
Romm, M. J., **191**, 191, 196, **247**
Rott, N., **1**, **3**, **6**, **65**, **102**, **134**, 135, **191**, **205**,
    222

S

Sanders, J. V., viii, 27, 59, 76, 80, 249, 251, 252
Sawamoto, M., 180
Schlichting, H., **31**, **66**
Schmidl, F., 240
Seidel, P., 240
Senft, J. R., **1**
Sereny, A., 255
Seume, J. R., 241
Shkrebyonock, M. P., 5
Shokralla, S., 175, 240
Simon, T. W., **241**
Smith, B. L., **204**, **209**
Smith, J. L., 27, **191**, 191, **196**, 247
Smith, S. M., 269
Sonntag, R. E., viii, 31
Spoor, P. S., **vii**, **50**
Steijaert, P. P., **1**, **168**, 168, **191**, 191
Stewart, W. E., 31, 50
Storch, P. J., **190**, **194**
Streeter, V. L., **184**, **210**, **246**, **267**
Strutt, J. W., 198
Swift, G. W., **vii**, vii, 11, 16, **19**, 23, **50**, 65, **99**,
    101, 140, 149, 155, **163**, **174**, 175, **181**,
    181, 184, 189, **190**, **194**, **196**, 198, **202**,
    202, **205**, 205, 207, **208**, 217, **220**, **222**,
    222, **225**, 225, 232, **236**, 236, 245, 265,
    273, 275, 278, **285**, 286, 293
Swift, G. W. (review articles), **1**, **97**, **134**, **136**,
    139, **147**

T

Takasu, S., 180
Tarasov, A. A., 5

Tetsuya, M., 180
Thürk, M., 240
Thomann, H., 131, **134**, 134
Thummes, G., 198, 232
Timmerhaus, K. D., 198, 204, 240
Tolman, R. C., **150**, **153**, 154
Tominaga, A., **1**, **138**
Trepp, Ch., 238

U

Urahata, I., 180
Urieli, I., **1**, **4**, **196**

V

Van Doren, T. W., 222, 249
Van Wylen, G. J., **viii**, **31**

W

Wakeland, R. S., **252–254**
Walker, G., **1**, **3**, 4
Wang, C., **198**, **232**
Wang, T. G., **201**
Wang, Z. S., 27, 105
Ward, T. B., **269**
Ward, W. C., vii, 16, 101, **140**, 174, 175, 217,
    222, 236, 265, 273, **293**
Weidl, R., 240
West, C. D., **4**, 4
Wheatley, J. C., **155**
Wimenitz, F. N., 116
Wollan, J. J., 19, 181, 205, 208, 285
Womersley, J. R., **179**
Wood, J. G., 240
Worlikar, A. S., **189**, **197**, **241**
Wu, P., 200
Wunderlich, S., 240

Y

Yaron, R., **175**, **240**
Yarr, G. A., 250
Yuan, J., 175, 240

Z

Zabolotskaya, E., 222, 249
Zhao, X., **201**
Zhu, S., **200**
Zhu, Z., 201
Zimmerman, J. E., 190, 194
Zinn, B., **219**
Zouzoulas, G., 191

Subject index

A

Acoustic approximation, 65, *see also* Rott's
    acoustic approximation
Acoustic dynamometer, 264
Acoustic impedance, 80, 87, 92
    combining components, 116
    specific, 115
Acoustic intensity, 117
Acoustic power
    and critical temperature gradient, 128
    defined, 117
    and exergy, 158
    and joining conditions, 196
    measurement of, 263, 276, 282, 288
    power transducers, 250
    in regenerator, 125, 167
    source/sink term, 124
Acoustic power dissipation
    in accidental cavity, 246
    in boundary layer, 122
    by minor losses, 186
    in resonator, 246
    by thermal relaxation, 121
    by turbulence, 181
    in valve, 186
    by viscosity, 121
Adiabatic, 46
Amplitude, 60
Aneurysm, 246
Angular frequency, 7
Animations
    critical temperature gradient, 96, 128
    directions for running, 9
    how to obtain, 9
    Oscwall, 114

Ptr, 17, 126, 134, 136
    regenerator, 17, 22, 125, 126, 136
    Standing, 9, 14, 96, 128, 137
    Tashe, 20, 22, 110, 125
    Thermal, 90, 95, 123, 147
    Viscous, 86, 114, 122
    Wave, 72, 119
Attenuation constant, bulk, 76
Availability, 157
Average, spatial, 85, 87

B

Boundary layer, *see* Penetration depth
Boundary-layer approximation
    acoustic-power dissipation, 122
    animations, 86, 90, 114
    integrals, 85, 87
    pressure independent of y and z, 66
    source/sink term, 106
    spatial average, 87
    temperature, 95
    thermal-relaxation effects, 91
    viscous effects, 86
Bulk modulus, 46

C

Calculation methods, 138
Carnot efficiency and COP, 50, 150
Cavity, accidental, 246
Circuit diagrams, *see* Impedance diagrams
Coefficient of performance, 51
    Carnot, 51, 149
Complex notation, 60

Compliance
   accidental, 246
   defined, 78
   examples, 88, 90, 93, 107
   impedance of, 81
   and thermal-relaxation resistance, 92
   thermoacoustic, 104
Compressibility, 46, 78, 101
Computation, 140, 293
Conductance, thermal-relaxation, see
      Resistance, thermal-relaxation
Conservation of energy, see First law
Conservation of mass, see Continuity equation
Constants, gases, 49
Continuity equation
   derived, 40
   lossless acoustic, 78
   Rott's acoustic approximation to, 67
   source/sink term, 106
   with streaming superimposed, 215
   with thermal relaxation, 92
   thermoacoustic, 100, 103
Control volume, 34, 130
Convection, see Streaming
Critical temperature gradient
   animation of, 96, 128
   defined, 96
   and exergy, 161
   sign convention for, 128
   and total power, 137

D

dc flow, see Streaming, Gedeon
DELTAE and DELTAEC, 140, 293
Dimensionless groups, 178, 222, 272
   friction factor, 176, 177, 179
   Mach number, 227
   Prandtl number, 8, 155
   Reynolds number, 175, 178
Discontinuities, see Joining conditions
Displacement amplitude, 7
Dissipation of acoustic power, see Acoustic
     power dissipation
Double Helmholtz resonator, 81

E

Efficiency, 51, 149, 155
   Carnot, 51, 149
   electroacoustic power transducer, 252
Electric circuit analogy, 79
Electroacoustic power transducer, 288

Energy equation, see also Total power
   derived, 43
Energy flux density, 44, 132
Engine, 50, see also specific type
Enthalpy, 35, 48, 68, 132
Entrance effects, 189, see also Minor losses
Entropy, 38, 68, 135, 150, 196, see also Second
   law
   irreversible generation of, 39, 45, 153
Entropy equation, 55
   derived, 44
Entropy flux density, 55
Equation of motion, see Momentum equation
Equation of state
   ideal gas, 46
   Rott's acoustic approximation to, 67
Eulerian and Lagrangian viewpoints, 95
Exergy, 156, 173
   thermoacoustic, 158

F

First law of thermodynamics, 32, 43, 50
Flow availability, 157
Flow coefficient, 186
Flow impedance, for laminar steady
    flow, 55
Flow straighteners, 210, 245
Fluidyne engine, 4
Frequency, angular, 7
Friction factor, 176, 177, 179

G

Gas
   choice of, 231
   constants and identities, 49
   properties of, 45
   table of properties, 49
Gas diode, 202
Gas spring, 70
Gouy–Stodola theorem, 153
   Rott's acoustic approximation to, 157

H

Harmonic oscillator, 59, 62
Harmonics, 219, 249
Heat, 32, 50, 130, 131
   and exergy flux, 158
   measurement of, 267
Heat capacity, 8
Heat engine, 9, 50, see also specific type

Heat exchanger, 130, 240, 247
  examples, 13, 21, 25, 242
  joining conditions for, 190
Heat pump, 56
"Heat pumping", 135
Heat-transfer equation, 45, 90
  Rott's acoustic approximation to, 68, 94,
    100
Helmholtz resonator, 81
Hoop stress, 248
Hot wire anemometry, 259
Hydraulic diameter, 85
Hydraulic radius, 85
  of screen, 174
Hydrodynamic entropy flux, 137

I
Ideal gas, properties, 45
Iguchi hypothesis, 181, *see also* Quasi-steady
    hypothesis
Impedance, *see* Acoustic impedance, Flow
    impedance, Mechanical impedance, *or*
    Specific acoustic impedance
Impedance diagram
  for continuity equation, 79, 93
  defined, 79
  of double Helmholtz resonator, 81
  general thermoacoustic, 103
  for momentum equation, 79, 87
  of orifice pulse-tube refrigerator, 111
  of standing-wave engine, 107
  of standing-wave refrigerator, 111
  of thermoacoustic-Stirling heat engine, 110
Inertance
  associated compliance, 112
  defined, 79
  examples, 88, 90, 107
  impedance of, 81
  thermoacoustic, 102
  and viscous resistance, 87
Intensity, 117
Interface conditions, 190
Internal energy, 34
Isentropic, 46

J
Jet, 203, 209
Joining conditions, 190, 247

L
Lagrangian and Eulerian viewpoints, 95
Laser Doppler velocimetry, 259

Length scales, 7, 28
Linear motor–alternator, 250, 266
Liquid-piston engine, 4
Load, adjustable acoustic, 264
Lost work, 150
Loudspeaker, 250, 266

M
Mach number, 74, 227
Mass flow, *see* Streaming
Mass-flux density, 199, 204
Maxwell relations, 52
Measurements, 259
Mechanical impedance, 251
Minor losses, 183, 202
Mixtures, 50
Momentum equation
  derived, 41
  inviscid acoustic, 79
  Rott's acoustic approximation to, 68, 86, 87
  to second order, 207
  with streaming superimposed, 215
  thermoacoustic, 98, 102, 103

N
Navier-Stokes, *see* Momentum equation
Nonlinear acoustics, 171, 249
Notation and symbols, xv
  complex, 60, 65
  power, 62
  Rott's acoustic approximation, extensions
    beyond, 199
  sign conventions, 80, 184
  time average of a product, 62
  time derivative, 168
Numerical integration, 140, 293

O
Open system, 34, 39
Orifice pulse-tube refrigerator
  acoustic power, 127, 144
  acoustic-impedance network, 88, 94, 122,
    124, 246
  animation, 17, 127
  exergy, 162
  impedance diagram, 111
  invention of, 5
  measurements, 278
  orifice resistance, 88
  overview of extended example, 19, 111,
    144, 162

Orifice pulse-tube refrigerator (*cont.*)
  pulse tube, 208
  Rayleigh streaming, 208
  regenerator, 127
  thermal-relaxation resistance, 94
  total power, 144

P
Penetration depth, 7, 29
  animation, 86, 90, 95, 114
Phase, 60
Phasor diagram
  and acoustic power, 118
  for continuity equation, 79, 93
  defined, 61
  general thermoacoustic, 103
  of half-wavelength resonator, 84
  for momentum equation, 79, 87
Phasors, 61
Porous media, 174, 238
Power, *see* specific type (Acoustic power, Total
    power, Heat, ...)
Prandtl number, 8, 155
Pressure transducers, 260
Pressure vessel, 248
Prime mover, 50
Pulse tube
  acoustic power in, 136, 144
  exergy in, 159, 162
  Gedeon streaming in, 201
  impedance of, 107, 112
  jet-driven streaming in, 209, 245, 285
  joining conditions for, 190
  length of, 245
  purpose of, 19
  Rayleigh streaming in, 198, 204, 245, 285
  surface finish of, 245
  tapered, 205
  temperature profile in, 190
  total power in, 136, 144

Q
Quality factor, 63, 146, 257
Quasi-steady hypothesis, 181, 184, 241

R
Radial standing wave, 250
RC and *RLC* loads, 264, 276, 279, 287
Refrigerator, 50, *see also* specific type
Regenerator, 126, 174
  acoustic power in, 125, 136, 144

  animations, 17, 20, 125, 126, 136
  examples, 238
  exergy in, 159, 162
  Gedeon streaming in, 200
  hydraulic radius of, 174
  impedance of, 107
  microscopically etched, 240
  parallel plate, 240
  purpose of, 233
  stacked-screen, 101, 174, 238
  streaming, internal, 211, 286
  temperature profile in, 141
  total power in, 134, 136, 144
Resistance, thermal, *see* Thermal-relaxation
    resistance
Resistance, viscous, *see* Viscous resistance
Resonator, 78, 114, 246
  equality of acoustic and total powers, 136
  example, 107
  half-wavelength, 84
  harmonic suppression, 249
  Helmholtz, 81
  pressure vessel safety, 248
  shape, optimal, 246
Reynolds number
  complex, 175
  peak, 179
  in porous media, 175
  in steady flow, 178, 222
Rott's acoustic approximation
  for continuity equation, 67, 78, 92
  defined, 65, 77, 97
  for entropy and enthalpy, 68
  for equation of state, 68, 92
  for exergy and Gouy–Stodola theorem,
    158
  extensions beyond, 171
  harmonics added to, 220
  for heat-transfer equation, 68, 91
  for momentum equation, 68, 79, 86
  with streaming superimposed, 214
  for total power, 134
Roughness, *see* Surface roughness

S
Safety, 248
Scale models, 225
Screens, 174
  as flow straighteners, 245
  constructing regenerators from, 238
  effective thermal conduction, 177
  hydraulic radius of, 174
  as stacks, 238

Second law of thermodynamics, 38, 44, 51,
    156, 196
Shock waves, 222
Sign conventions, 80, 128, 184
Similitude, 222
Simple harmonic oscillator, 59, 62
Sodium, 232
Source/sink, volume-flow-rate
   and acoustic power, 121
   boundary-layer limit, 106
   defined, 106
   small-channel limit, 106
Spatial average, 85, 87, 98
Specific acoustic impedance, 115
Specific heat, 8, 47
Speed of sound, 47
Stack
   acoustic power in, 127, 143
   animations, 9, 14
   examples, 13, 233
   generation of acoustic power, 9
   impedance of, 107
   purpose of, 9, 233
   streaming within, 211, 273
   temperature profile in, 141
   total power in, 143
Standing wave, radial, 250
Standing-wave engine
   DELTAEC file, 294
   acoustic power, 129, 143
   animation, 9, 128
   and critical temperature gradient, 96, 128
   exergy, 163
   heat exchanger, 242, 244
   impedance diagram, 107
   measurements, 271
   minor losses, 187
   overview, 9, 107
   overview of extended example, 11, 163
   resonator, 83, 93, 182, 187, 246, 249
   stack, 107, 129, 234
   thermal-relaxation resistance, 93, 123
   total power, 143
Standing-wave refrigerator
   DELTAEC file, 296
   animation, 14, 129
   and critical temperature gradient, 96, 129
   impedance diagram, 111
   measurements, 273
   overview of extended example, 16, 111,
     217
   resonator, 83, 90, 249
   resonator resistance, 122
   stack, 236

Steady flow, *see* Streaming
Stirling cycle, 3
Stirling engine, 1, 3, *see also* Thermoacoustic-
    Stirling heat engine
Stirling refrigerator
   animation, 19
Streaming, 197
   added to Rott's acoustic approximation,
    214
   deliberate, 16, 212, 273
   Gedeon, 200, 263
   jet-driven, 209, 245, 285
   Rayleigh, 204, 245, 263, 285, 286
   through regenerator, 198
   regenerator internal, 211, 233, 240, 286
Stress tensor, viscous, 42
Surface roughness, 178, 245, 282
Symbols and notation, xv, 60
   complex, 65
   Rott's acoustic approximation, extensions
    beyond, 199
   time derivative, 168

T
Tables
   analogy, electric–acoustic, 80
   gas properties, 49
   lost work and entropy generation, 154
   lost work, standing-wave engine, 164
   relative sizes of impedances, 107
Temperature gradient
   and acoustic power, 124
   critical, *see* Critical temperature gradient
   and total power, 137
Thermal buffer tube
   acoustic power in, 136
   exergy in, 159
   jet-driven streaming in, 209, 245
   joining conditions for, 190
   length of, 245
   Rayleigh streaming in, 198, 204, 245
   surface finish of, 245
   total power in, 136
Thermal conductivity, 8, 100
   and choice of gas, 232
   effective, of screens, 177
Thermal expansion coefficient, 46
Thermal reservoir, defined, 50
Thermal-relaxation effects
   and acoustic power dissipation,
    122
   animation, 91
   in boundary-layer approximation, 91

Thermal-relaxation effects (*cont.*)
   deliberate imperfect contact, 11, 15
   lost work and entropy generation, 154
   resistance, defined, 104
   resistance, use of, 92, 121, 123, 155, 161
Thermal-relaxation resistance
   defined, 92
   examples, 93, 94, 123, 124
   thermoacoustic, 104
Thermoacoustic engines and refrigerators, 26
"Thermoacoustic heat flux", 135
Thermoacoustic-Stirling heat engine
   DELTAEC file, 302
   acoustic power, 125
   animation, 20, 110, 125
   efficiency, 277
   Gedeon streaming, 204
   heat exchanger, 243
   impedance diagram, 108
   interface, 247
   invention of, 5
   joining condition, 197
   measurements, 274
   minor losses, 188
   overview, 108
   overview of extended example, 23
   regenerator, 108, 125, 177, 238
   resonator, 188, 249
Thermoacoustics, 3, 26
Thermocouples, 262
Thermoviscous functions, 98
Total power
   and exergy, 158
   Rott's acoustic approximation for, 134
   with streaming superimposed, 216
Transducers
   power, electroacoustic, 250, 266
   pressure, 260
   temperature, 262, 263
Traveling-wave engine, *see* Thermoacoustic-
   Stirling heat engine, Stirling
   engine

Traveling-wave refrigerator, *see* Orifice
   pulse-tube refrigerator, Stirling
   refrigerator
Turbulence, 177, 183
   Blasius correlation for, 228
   and Iguchi hypothesis, 181
   and minor loss, 184

V
Valve flow resistance, 186
Viscosity, 8, 97
Viscous effects
   and acoustic power dissipation, 122
   animation, 86
   in boundary-layer approximation, 86
   lost work and entropy generation, 154
   resistance, defined, 102
   in tortuous porous media, 176
Viscous resistance, 87, 121, 155
   examples, 90, 107, 122
   thermoacoustic, 102
Viscous stress tensor, 42
Volume flow rate
   defined, 77
   and joining conditions, 194
   measurement of, 264, 266, 288
Volume velocity, 77
Volume-flow-rate source/sink, *see* Source/ sink

W
Wave propagation, 77
   animation, 72
   lossless, 71
   Rott's equation for, 101
Wave vector, 72
Wavelength, 7
Waves, 71, 77, 119
Website, 293
Womersley number, 179
Work, 32, 50

Printed in the United States
By Bookmasters